Embassy's Coastal Cruising Guide

Atlantic Coast

The finest marine facilities from Maine to Florida

Edited by
Donna Caruso

Embassy Marine Publishing
A division of Embassy Imprint, Inc.

Embassy's Coastal Cruising Guide

Embassy's Coastal Cruising Guide: The Atlantic Coast

Publisher & Editor-in-Chief
Mark C. Borton

Managing Editor
Donna Caruso

Assistant Editor
Steven J. Nicastro

Navigational Editors
Ian S. Quarrier
Robert L. Higgins

Operations Manager
Irene L. Roy

Manufacturing & Fulfillment
Janine M. Smith
Lisa M. Nigosanti

Art Director
Stephen T. Corsello

Graphic Designers
Laura L. Bloom
Sarah L. Jeffreys
Robert R. Van Keirsbilck
Carla J. Vachowski
Susan M. Spillman

Regional Sales Representatives

Call 1-800-999-1075

Eric Ramotowski ....Rhode Island
Dibbon Joy ....Connecticut
Ellen Schneider ....Stratford, CT to New York City
George Werner ....Long Island, New York
Craig Dozier ....Virginia
Nancy Reindollar ....North Carolina
Nancy Schilling ....South Carolina & Georgia
Ched Gladding...........Southeast Florida
Val Ely ....West Coast

© *Embassy's Coastal Cruising Guide: The Atlantic Coast* is the first in a series. Other volumes of *Embassy's Coastal Cruising Guide* are in development. Please call us Toll-Free at 1-800-999-1075 to order books, to place an advertisement, or for additional information.

Embassy Marine Publishing has made every effort to insure the accuracy of the information in this book. However, no publication can replace good judgement, experience, and first-hand knowledge, and the prudent mariner will not rely on any single aid to navigation. Embassy Marine Publishing hereby gives notice that all charts in this book or sold with it are for illustration purposes only and are not to be used for navigation. Embassy Marine Publishing therefore makes no warranty or guarantee for any use of these charts for navigation or any other use including their merchantability.

Any references to certain courses, anchorages, or navigational information in this book are intended merely as illustrations. The prudent mariner is well advised to consult the U. S. Coast Guard, the Defense Mapping Agency, and the National Oceanic and Atmospheric Administration for current navigational information. All mariners should have up-to-date navigational charts on board and subscribe to the Local Notice to Mariners published weekly and available free from your local Coast Guard station. The use of any and all such navigational references is at the mariner's own risk.

Similarly, inclusion in this book of such things as marine facilities and services, restaurants, places of interest, attractions, or sporting events is for informational purposes only, and no guarantee is provided as to their accuracy or current status, nor are they at all recommended. Neither Embassy Marine Publishing, nor any of its officers, employees, or agents shall in any way be responsible for any consequential, incidental, or exemplary loss or damage resulting from the use or misuse of any material in this book for navigation or any other purpose.

ISBN 0-930527-29-1 Printed in the United States of America, 1993

142 Ferry Road, Old Saybrook, CT 06475 U.S.A., (203) 395-0188

Embassy's Coastal Cruising Guide

Atlantic Coast

The finest marine facilities from Maine to Florida

TABLE OF CONTENTS

Atlantic Coast

PARTICIPATING MARINAS

How to Use Embassy's Coastal Cruising Guide

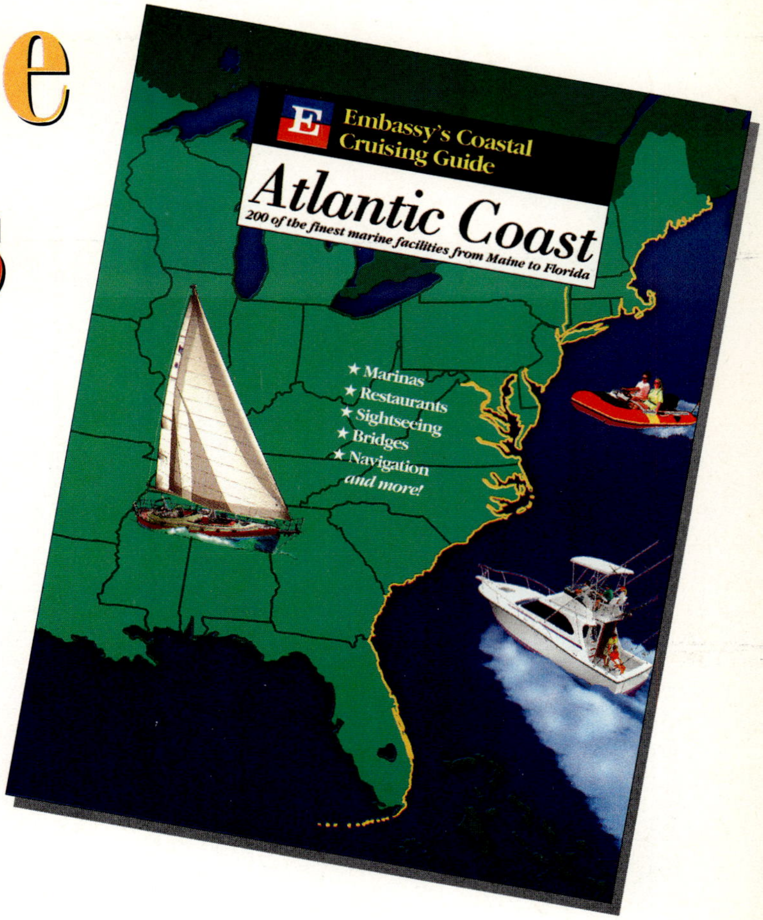

Nine years ago, as we headed down the Connecticut River from its headwaters on the Canadian border, we had no idea that the *Embassy Guide* series would lead us through 16 states, and over 5,000 miles of rivers, bays, and waterways along the Atlantic and Gulf coasts. With well over 100,000 hours of experience behind us, we have learned quite a bit about boating and how to build the best boating guides and chartbooks.

Embassy's Coastal Cruising Guide: The Atlantic Coast is the only book that covers the entire Eastern Seaboard from Maine to Florida and serves as an overview and introduction to the wonderful and varied cruising opportunities to be found in these waters. Many of these areas are detailed in our four regional volumes of *Embassy's Complete Boating Guide & Chartbook* now in publication, and two more now in production. Those familiar with our regional volumes will find *The Atlantic Coast* reassuringly familiar in style and organization. There are lots of charts and maps to orient yourself, everything is color-coded, information is where you need it, and as always, the book is filled with fun and interesting reading.

However, *Embassy's Coastal Cruising Guide* is not the same as *Embassy's Complete Boating Guide & Chartbook.* Were *The Atlantic Coast* to be as detailed as our regional volumes, it would be a foot thick and weigh 50 pounds. Rather, *The Atlantic Coast* is designed to give you basic navigational information, sightseeing highlights, and detailed information on some of the finest marinas on the coast from Eastport, Maine to Key West, Florida. Use this book while you are cruising through an area—then switch to one of our super-detailed regional volumes when you want to go gunkholing.

Only 150 of the finest marinas on the East Coast of the United States are included in this book. One of our representatives has been to every marina listed, and we can assure you that every one is a reputable company and has first-class facilities. Each of the participating marinas in this book has contributed to help cover the cost of producing this book. As such, it would be appreciated if you would mention to the marina operators that you read about them in *The Atlantic Coast.* In future editions, we will be expanding the text and including more facilities.

Our intent is to bring you the finest boating guides and charts in the world, no matter where in the world your boating takes you. For more information, or to make comments or suggestions, please call us at 1-203-395-0188 or write to us to, "Tell us what you think." Thank you and Happy Boating!

Mark C. Borton

Mark C. Borton
President and Founder

How to Use *Embassy's Coastal Cruising Guide*

Introduction (pages 8 to 57): Boating isn't just going from marina to marina. It is learning about and having fun in the areas you are visiting. To help you appreciate and understand what you see, we asked the experts to give you a perspective on the worlds of geology, fishing, birding, scuba diving, chartering and ecology. It is good reading and a great reference for identifying whatever it was that you just passed.

Reference Section (pages 58 to 85): Here you will find useful information such as a bridge index, weather info, hurricane tracking chart, VHF radio use, first aid, and basic boat handling.

State Chapters (pages 86 to 269): Each of the 14 states bordering the Atlantic on the East Coast of the United States is described in this section. Each State Chapter is color-coded for easy reference and begins with a full-page sketch chart of the area so that you can get your bearings. All the participating marinas are located on the State Chart and detailed in the Marine Facilities Listings on the adjacent page.

The chapters and text are organized from North to South, and include both navigational and sightseeing information in sequence. Participating marinas are featured with information as to transient dockage rates, restaurants and other amenities, provisioning and repair services available. A full-scale, full-color section of a NOAA chart is also included pinpointing the location of the marina. At the end of each chapter are all the important telephone and VHF numbers for the Coast Guard, radio telephone operators, etc., as well as announcements from local marine service companies. For more on the State Chapters, see page 10.

Product & Service Directory and **Index (pages 270 to 277):** Use this section of the *Guide* like the yellow pages of the phone book to find specific products or services. Here you will find diesel engine parts, sail lofts, insurance agents, surveyors and all manner of other things marine. You will find companies listed by type of business in the Directory, and by name in the Index.

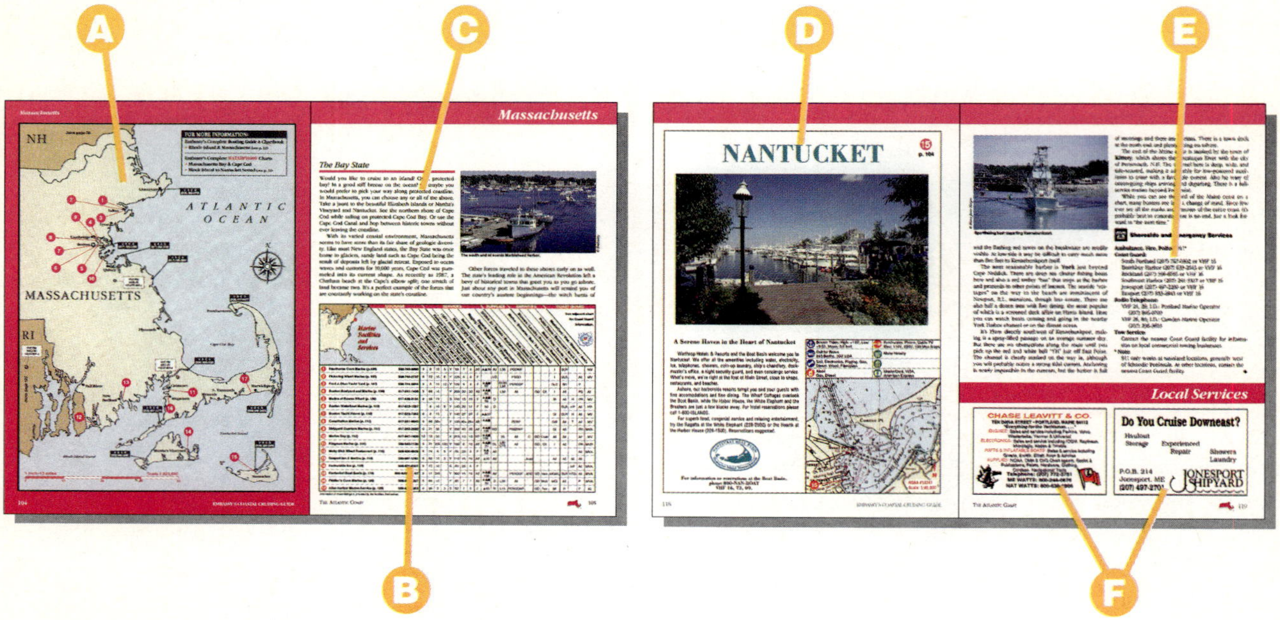

A State Charts give you the big picture so you can orient yourself and see where the marinas are located and the distances between them. Marina Key Numbers are cross-referenced with the Marine Facility Listing and the Marina Features. Coast Guard Stations and NOAA weather broadcast stations are also located.

B Marine Facilities Tables provide detailed information about the participating marinas and the services they offer. Marina Key Numbers are cross-referenced with the State Charts and the Marina Features. A page number reference is also provided showing on what page more information and the Marina Feature can be found. Most of the categories are self-evident, but it should be noted that marina operators sometimes report their approach and dockside depths optimistically.

C State Chapter Text is written from North to South, and includes both navigational and sightseeing information in sequence. This information will be most useful if read in advance of your trip with a NOAA chart or chartbook in hand so that you can see what is being discussed. Major waterways and destinations are set in boldfaced type for easy scanning. Telephone numbers are always included so you can get up-to-the-minute information on local attractions and events.

D Marina Features detail what is available from the participating marinas and in the surrounding area. The full name, address, telephone number and VHF channel are always listed so that you can contact the facility. The Marina Key Number is usually located in the upper right corner and is cross-referenced with the Marine Facilities Tables and the State Charts. The page number listed beneath the Marina Key Number directs you to the State Charts and Marine Facilities Listings for additional information. A full-scale, full-color section of the appropriate National Oceanic and Atmospheric Administration (NOAA) chart is provided to pinpoint the location of the marinas and allow you to navigate to their docks. A summary of the marinas services is also provided. Again, most of the categories are self-evident, except for the tidal information which is referenced to the *Tide Tables for the East Coast of North and South America* as published annually by NOAA.

E Shoreside and Emergency Services are listed at the end of each State Chapter. Here are all the important telephone and VHF numbers for the Coast Guard, radio telephone operators, marine towing companies, etc..

F Local Services section includes announcements from local marine service companies—important resources for every boater—which are not located at a marina and therefore not included in the Marine Facilities Table.

I'll never forget my first offshore passage. It was late in the year, hurricane season well past. We motored down the East River with New York City looming over us at twilight. By the time we poked out into the Atlantic, it was dark; we headed offshore under sail.

We hadn't even begun our watch system when the wind began to pick up—first 15, then 20, soon 30, finally gusting to 40. Waves were crashing into the cockpit, and it was cold. One of the crew started getting seasick. I was scared. Was this normal? Would it get worse? Would our 42-footer hold up?

Needless to say, the passage was not the cruise I had envisioned. And by the time we reached Norfolk, Va., the words "Intracoastal Waterway" sounded as safe and secure as a warm blanket during a thunderstorm. Bridges—Groundings—Strong Currents—No sailing. I didn't care. At that particular time, you couldn't have convinced me that anything was worse than the open ocean. Of course, I've since found that the Atlantic can be as friendly and fun as a good friend. And, sort of like a friend, if you really pay attention, you know when it's best to stay away. But I still hold a certain fondness for the Intracoastal Waterway for rescuing me on that trip.

Although I didn't realize it at the time, I had met the Intracoastal Waterway long before I arrived in Norfolk. We had left from Long Island Sound (Intracoastal Waterway) and sailed down the East River (ICW). Had we left from Maine, we would have sailed down the coast through the Cape Cod Canal (ICW again). And had we tucked in sooner, we could have entered one of the New Jersey inlets or taken the Chesapeake and Delaware Canal to Chesapeake Bay (all ICW). But, as is the case with many people, that was not what "Intracoastal Waterway" meant to me. Virginia meant the start of The Waterway; I thought the end was somewhere in Florida, and I kind of wondered if there was more to it. Apparently, I wasn't alone.

"No one really knows where the true start and end is," says Arlene Dietz, Director of the Army Corps of Engineers Navigation Data Center. "It was created over time. The fact is that the Atlantic Intracoastal Waterway is really the name of a certain project."

The Intracoastal Waterway is actually a series of projects encompassing many different bodies of water. They are sounds, rivers, swamps, canals, and channels—natural and manmade. Although it includes all navigable areas, some of them are decidely for shallow-draft boats, such as the protected waters off New Jersey and the inside route through the Florida Keys.

Just at the bottom of Chesapeake Bay starts the Atlantic Intracoastal Waterway. This stretch includes the

SOME *INSIDE* INFORMATION

Donna Caruso

Albemarle and Chesapeake Canal Route and the alternate Dismal Swamp Route, which is rarely used by commercial traffic. This section is what many, if not most, people talk about when they speak of the start of the Intracoastal Waterway, or as some delicately put it—The Ditch. If you start at Norfolk and add the miles all the way to Key West, you'll tally more than 1,243 statute miles. Marked at five-mile intervals on charts, these statute miles are referenced by establishments along the ICW.

Although the Intracoastal Waterway is somewhat of a jigsaw puzzle in present day, it did start out as a cohesive vision. In 1808, Albert Gallatin, then U.S. Secretary of the Treasury, submitted a comprehensive plan to bind the new nation together with a government-sponsored system of roads and canals. He proposed the construction of an inland waterway along the Atlantic Coast from Boston, Mass., to St. Marys, Ga. The principal work was to be the construction of four canals, which he estimated would cost $3 million. He thought that his whole program could be completed in ten years.

That unified program never became reality. Instead, the waterway Gallantin envisioned came into being through a series of local projects, such as the Cape Cod Canal; the Chesapeake and Delaware Canal; the Albemarle and Chesapeake Canal; the Dismal Swamp Canal, and the dredging of channels in existing rivers. The natural waters of the Intracoastal include Long Island Sound, Chesapeake Bay, a chain of six sounds in North Carolina, the Indian River in Florida, and a long list of other waterways that were either navigable as they were or dredged to be so.

As for industry, our predecessors found that, in fact, the offshore waters often continued to suit those involved in long-distance commerce. It turned out that the intracoastal slices were used more for local industry and by light-draft vessels and tugs unable to navigate long stretches of open ocean.

Many recreational boaters soon began to take advantage of the Intracostal Waterway as well. The Army Corps of Engineers has counted more than 10,700 recreational boats passing through the Great Bridge Guard Lock south of Norfolk annually. That doesn't include a couple more thousand recreational boats using the nearby Dismal Swamp Canal route. Nor does it account for vessels coming and going on the waterway at other points.

But the true beauty of the Intracoastal Waterway is its uniqueness. Although an intracoastal route does cut through Florida and snake along the Gulf Coast, there is nothing comparable on our West Coast. The ICW is largely protected by the unique barrier islands that wind their way down the East Coast. These islands, although far from static, take the brunt of the Atlantic Ocean's onslaughts leaving boaters with a buffered down version of weather. They can decrease winds up to 30 percent from the open ocean.

Were it not for the natural lay of the eastern coastline, the Intracoastal might not very well be there at all. So if you're complaining about inconveniences of the waterway (read, depths) remember that some coastal projects have been delayed to protect the natural coastline and ward off erosion.

The Intracoastal does have its downfalls, though. You have to wait for bridge openings; you have to be aware of commercial traffic and some strong tidal currents; you have to watch that you don't slide absentmindedly onto the bottom in certain areas. As usual when you take to the water, planning ahead forestalls most of these problems. Your trip down the Waterway should be alert but it doesn't have to be apprehensive.

If you're going to travel down the Intracoastal, you'll have some homework to do. First and foremost, always have detailed charts for all sections of the Waterway you're planning to cruise. Secondly, read them. You'll need to be especially aware of bridges and depths.

The minimum overhead clearance of fixed bridges over the Intracoastal is 56 feet, which is at Julia Tuttle Causeway in Miami; all other fixed bridges have a 65 foot clearance. For other bridges, you'll often have to request openings or check ahead for scheduled openings during restricted hours. (See pages 83-85 for ICW bridge schedules.)

Many of the bridges monitor channels 16 and 13 so a VHF radio is usually your best bet for bridge communications. Radio contact should include calling the bridge by its proper name, then giving your vessel's type, name, position, and direction. You may also request an opening by horn with one prolonged blast of four to six seconds and one short blast of one second, or you can signal visually with a white flag or light raised and lowered vertically.

VHF is also invaluable for determining the actions of commercial craft on the waterway. Channel 13 is used for navigational communications on commercial vessels; you may want to monitor the channel to determine the intentions of large craft.

If you take the Albemarle and Chesapeake Canal route, you'll encounter one lock, the Great Bridge Lock. If you take the Dismal Swamp Canal Route, you'll

encounter two: the Deep Creek Lock in Virginia and South Mills Lock in North Carolina. You should signal the lockmaster between three-quarters and one-quarter of a mile from the lock with two long and two short blasts of the horn. You'll know the lock is available when you get a green signal, usually a light or flag. You should use bow, stern, and spring lines and fenders when locking.

Depths are another part of the Intracoastal equation. Officially, the project by way of the Albemarle and Chesapeake Canal from Norfolk, Va., to Fort Pierce, Fla., has project depths of 12 feet, and from Fort Pierce to Miami depths of 10 feet. Unofficially, it's a good idea to have a sounding pole marked with your boat's draft and oars for your dinghy in case you have to search for deep-enough water. Any local knowledge you can pick up along the way will be helpful. Also, try to navigate mid-channel if other traffic is not present. Here are some specifics to keep in mind: The Great Dismal Swamp Canal is maintained at only six feet due to reduced use by commercial traffic; the Alligator River-Pungo River Canal may have low depths and sides of the canal are foul with debris and submerged stumps—try to navigate midchannel; shortly after entering South Carolina, the land cut between Miles 347 to 365.5 should be navigated with caution. Rock ledges abut both sides of the channel at narrow portions (one section is called the Rock Pile).

Aside from knowing what's up ahead, you should also think about where you'll need to be at a given time for provisioning, fueling up, or for any other reason. For example, if someone is leaving the boat, make sure you drop them off at a town near an airport or one with bus service. (I realized this the hard way when I paid $70 for a taxi ride.)

Mail can be delivered to General Delivery if you plan ahead and notify the post office that you want your mail held. Although bank machines can be found at many stops, you won't always find them in the less populated areas where you will still need good old cash.

Just as important, once you've planned appropriately, enjoy the waterway. The experiences to be had are as varied as the type of craft you'll encounter on your trip. In Virginia, there's Navy-proud Norfolk. As you cross the line into North Carolina there's Oracoke with its pirate legends and rich nautical heritage. It was here that Blackbeard the Pirate met his death off Teach's Hole, a channel named after him. And there's Beaufort, long hailed as the "Gateway to the Caribbean," where you can wander the docks and look at the varied home ports on transoms or watch the ceremonial folding of the flag every evening (not to mention the sea stories you'll hear by chatting with other boaters).

In South Carolina, there's Georgetown with its unique rice culture and the wonderful plantation houses you'll pass as you cruise the Intracoastal. Stop at Charleston and visit the famous Boone Hall Plantation. In Georgia, beautiful Savannah will show you some true southern hospitality. (Marina facilities are best in the suburbs.) And, of course, Florida holds everything from historic St. Augustine to the art deco of Miami Beach, not to mention the unique Keys that stretch into the horizon. And that's just one taste of the experience. In between it all is the unique natural setting—eerie swamplike runs, cypress stumps, the osprey and egrets, saltmarshes, wide sounds, and long barrier beaches. In North Carolina, you'll see the swamplands known as the Dismals and Pocosins, and the treeless prairie land with thick grass and wild flowers called the Savannas. As you move into South Carolina, the wide sounds will close down to more narrow passageways still bordered by the barrier islands. As you head through Georgia, you'll see tidal rivers and marshes covered with dense grasses, such as the beautiful Marshes of Glynn near Brunswick.

The Intracoastal Waterway will always hold a special spot in my log. On that first trip, we did find ourselves momentarily high and dry not long after we began our journey south from Norfolk. But it was sunny, the water was lake-calm, and I knew I was tucked in for the night.✦

Photographs © Mary Jane Hayes

Another Way to Take to the Sea

Art Kelsey
Sail Westbrook, Inc.

Jody Lexow
Jody Lexow Yacht Charters

Photo courtesy of New Jersey Division of Travel and Tourism

> There is nothing—absolutely nothing—half so much worth doing as simply messing about in boats . . . or with boats . . . In or out of 'em, it doesn't matter."
>
> —Kenneth Grahame, *The Wind in the Willows*

Regardless of prime rates, job-market statistics, gasoline prices, roofs that need repair, foul-weather reports, and all sorts of other stormy seas, there is still a great company of stalwarts who agree with Kenneth Grahame. These sturdy souls will always find a way to take to the sea.

Many will cruise in vessels they own. Others will find a vessel to rent or charter. Be they large or small yachts, power or sail, captained and crewed or bareboat, in locations all over the world, chartering is alive and well.

For the charterer there are many options. Likewise, there are several questions that are asked, in one form or another, during the planning for that ideal vacation and challenge of the sea aboard a chartered vessel.

What Kind of Yacht Should We Charter?

This is usually addressed in terms of the purpose of your charter, which can range from a wedding, cocktail party, or dinner dance to a business meeting, family vacation, or fishing trip.

The next consideration is the number of people involved and the length of the charter. Accommodations for eating, sleeping, and personal needs should be assessed carefully vis-à-vis the tastes and compatibility of your charter party. A party of six can have a great time for a three-day weekend aboard a 30-footer. An entire week aboard the same 30-footer might be a little taxing. Remember, on vessels of less than 50 feet there are rarely any secrets after the first couple of days.

Like any party or vacation, charters can range from the simple to the sumptuous, from haute-cuisine catered to bring your own picnic. Prices range accordingly. With this in mind, you would be wise to consider the amount of money you wish to spend before looking at boats as this makes the "shopping" process easier and more productive. You should also request a detailed cost estimate from the charter broker or company outlining all potential expenses, including the charter fee, dockage, service, catering, fuel, gallery fees, and repositioning of the boat.

An important consideration when looking at charter vacations is whether you want a bareboat charter or crewed charter. A bareboat charter means the group renting the yacht does the sailing. A competent boating background, as established by resumé or other means

prior to the actual rental, is absolutely necessary, both for reasons of passenger safety and potential repair expense. A skipper is designated by the group, and it is helpful for others on board to have boating experience, or plan to learn fast. This is an activity-oriented vacation, but part of the fun is finding a private spot to drop anchor and laze on the deck in the sun.

If you decide on a bareboat charter, you may want to pay special attention to the size of the vessel being considered. Experienced boaters will swear that larger vessels are easier to operate than smaller craft. These same seasoned sailors will also readily agree that the larger the craft, the greater the skill required. Chartering a large vessel in the interest of obtaining privacy without adequate skill can result in some seriously strained muscles and psyches, not to mention bank accounts (marine damage is never cheap to repair).

A bareboat experience, while still an adventure, can sometimes be taxing if sailing the boat becomes more work than play. In contrast, a crewed charter comes with a captain and/or crew, and often includes a cook. Particularly on larger yachts or in unfamiliar areas, a good crew can be essential. Even seasoned sailors enjoy a crewed charter. Most crews encourage guests to help with the sailing, and are happy to pass on their skill to the novice or compare stories with a fellow seaman.

If you are considering a crewed charter, you may be concerned about compatibility with the crew. This is where a trusted yacht charter broker can help. It is their business to know all the crews and boats they recommend. Crews are trained to serve, but not to intrude. Families with young children can really benefit from a crew willing to be maritime babysitters who can entertain the kids while mom and dad relax, not to mention the ease of having someone else cook.

Attention should also be directed next to some specifics about the particular vessel. This is particularly important for bareboat charterers, but also should be noted by those chartering with a crew. After all, whether you prefer to hop quickly from port to port or languish in a light breeze will determine the tempo of your vacation. If your choice is a powerboat, propulsion is certainly important. You will want to know whether it is a single or twin screw, diesel or gasoline, the fuel consumption, cruising range, and speed. In addition, electrical back-up systems should be assessed.

If your choice is an auxiliary sailing yacht, the same considerations are equally pertinent, with two exceptions. Auxiliary sailing vessels are powered by single engines. Their cruising speed is dictated by their hull design and ranges between 5 and 8 knots regardless of the size or type of engine. Power becomes most critical when maneuvering in close docking quarters or if caught in a strong tide with little or no wind. Sail inventory, running rigging such as reefing and furling systems, and number and size of winches, are obviously more important.

Whether you are in a powerboat or a sailing vessel, it makes little difference to Neptune. He sends the winds and rains on the stink-potter and the rag-bagger alike. Simply be certain that the vessel is appropriately equipped for the worst conditions imaginable. Then sail only when it seems sensible.

Basic navigational instruments and aids are absolutely essential. The exact kinds and quantities vary. Compasses, charts, parallel rules, dividers, pencils, and at least one good tide and current reference work should be minimum inventory on any vessel, charter or otherwise. Beyond these essentials are a VHF marine radio, radar, LORAN, depth sounder, knot meter, and log. The list can become quite lengthy and some items are more useful than others, depending on the skill, preference, and experience of the user. However, there is no substitute for knowing the basics of navigation. There is nothing worse on the sea than not knowing whence you have come, whither you are bound, or where in the world you are at present.

Safety and emergency equipment are both equally important. Safety harnesses, life preservers, fire extinguishers, flares, and basic spare parts are all items that can make a big difference in a crisis. If the charter-vessel inventory does not include these items, a carry-on kit is a worthy investment.

Where is the Yacht Berthed?

It is preferable that your charter be berthed in proximity to waters you'd like to sail. In the North, Maine offers a unique brand of beauty. But the rugged coastline and high incidence of fog make good piloting and navigational skills a must for bareboat charters. If you're unsure of your skills, go for a crewed charter. The best charter times are from June through the end of September.

The southern coast of New England all the way down to New York also offers some of the best boating in the world—ranging from the islands off Massachusetts to the Hudson River in New York. There is a variety of wind and sea conditions from early May well into October. There are dozens of ports and equally dozens of places to see and things to do on land, or ample opportunities to set a long-distance course and drive hard from dawn to dusk just for the joy of working with the sea.

Farther south, Chesapeake Bay offers a chartering season slightly longer than the more northern climes; charters are generally available from April through the end of October. More than 200 miles long, the Chesapeake offers sailing, fishing, the ports of Baltimore and Annapolis as well as isolated anchorages and small scenic towns. There are many rivers and estuaries you can explore on either the Virginia or Maryland shore of the Chesapeake.

One of the less well-known areas to charter on the East Coast is in North Carolina on the Pamlico and Albermarle Sounds. This area has the same season as the Chesapeake, but if you're sailing, winds are best at the start and tail end of the season. Steeped with pirate history and a unique southern atmosphere, you'll be pleasantly surprised should you decide to charter here.

Florida chartering is most appealing when the temperature is dropping in the North—November through April. There are many delightful areas to charter in the state, but most would agree, the chartering jewel is the Florida Keys. If you're bareboat chartering, shallow-draft boats will allow you to easily navigate the intracoastal route through the Keys. The Keys offer the best of scuba diving, snorkeling, and fishing (not to mention relaxing and watching the sunset).

And there is always the choice of chartering somewhere faraway or exotic. The Caribbean offers great year-round cruising. For the Mediterranean, Greece, and Turkey, the season is from April to October. For those with an interest in something off-the-beaten-track, cruising is also available in Australia, Alaska, Tahiti, and Thailand, among many other exciting places.

Overall, you should select your point of departure with attention to the location and the kind of cruising you and your party will enjoy the most. Remember that boats go more slowly than cars. An extra hour on the highway may buy you a whole day on the water, making it a lot more feasible for you to tie up or drop the hook in some of those ports you've always dreamed about approaching from the sea.

Who's Offering the Boat for Charter?

Whether it be a private owner, a charter broker, or a full-charter company with whom you do business, choose someone who knows and cares about the vessel you are chartering. A fairly complete inventory and checklist of equipment and a set of clearly written operating instructions for the vessel are indications of responsible charter management and can make your vacation a lot more pleasant. Likewise, a walk-through of all operating systems followed by a short shakedown cruise with someone who knows the vessel is invaluable. It will steal a couple hours at the beginning of your vacation but is well worth the effort in the long run.

Where Do We Find Charter Opportunities?

A charter broker is a good place to start. Brokers can advise you of the many options available. Also, many of the boating periodicals carry advertisements of charter opportunities regularly. Many even contain charter directories. An hour on the phone asking the appropriate questions will yield a host of possibilities. Then it's up to you to choose where you want to spend some time "messing about" in a boat. ✦

© Mary Jane Hanes

Provisioning for the Palate

Jan Robinson
Author, Chef,
and Captain

I've Been There!

The boat is in the water. You're in the galley. The crew is hungry. "Where's the can opener?"

Every time I stock the galley for a trip, I always have that anxiety attack. Did I forget something? Will we have enough?

Food for thought

Provisioning is one of the most important parts of planning a trip. The key is to be able to provide well- rounded, enticing meals without spending a whole lot of time in the galley. To do this, you'll need to do some planning ahead of time. Here are a couple of basic questions you'll want to address.

How many days do you want to provision for?
How many meals are you going to serve?
What kinds of food do you like?
Will you be able to reprovision along the way?

Start with the number of days you want to provision for—three, seven, twenty, or whatever. If you're heading to populated destinations, you may be able to replenish perishables and forgotten items en-route. Keep in mind, however, that it's a whole lot easier to stock all your staples before you head off rather than haul them a mile or so from the closest grocery along the way.

As a guide you'll probably eat breakfast, lunch, dinner, and one other lighter meal per day. Obviously tastes and appetites vary, but here is one common scenerio:

Breakfast is easy—take oatmeal, grapenuts, boxes of cereal, long-life milk, pancake mix, syrup, fruit juice, and fruit. For protein take one egg and one sausage patty or link per person, per day.

Lunch is a little more difficult to figure. It depends on how active the crew is. Plan a main dish such as a pasta salad, some cheese, crackers, a good five-grain bread, and something to drink.

Dinner: for seven days, you'll need seven main courses, seven vegetables, seven salads, and seven desserts, and beverages multiplied by the number of crew.

Plan before provisioning

Before you head off to the store, make a list. Plan healthy meals and consider easy backups, such as sandwichs, for those unexpected foul weather days. The following is a list of suggested staples.

peanut butter	baking powder	tea
jelly	cocoa	tomato sauce
grits	mustard	vinegar
pasta	relish	mayonnaise
honey	ketchup	butter or margarine
oil	raisins	unpopped popcorn
flour	spices	pancake mix & syrup
eggs	herbs	creamer
sugar	yeast	long-life milk
baking soda	coffee	beverages

To eliminate the need for a long list and many containers of spices, herbs and other staples, use mixes that you have put together yourself before leaving home. Also, remember that the water from your tank may not be as palatable as your favorite spring water, so you might consider bringing along some of the good stuff.

Remember that perishable foods will spoil faster on a boat, and some foods faster than others. Therefore plan to use meats like ground beef first, then chicken and pork, then beef. After that, plan on using canned meats, fully cooked smoked meats, pasta, rice and freshly caught fish. If you really want simplicity in the galley, pre-cooking pasta and other dishes can save a great deal of time and mess.

Creative stowage

Stowing your goods in an orderly fashion is one of the most challenging aspects of provisioning. For long cruises you'll need every inch of your ice box or refrigerator for those items that really do need to be refrigerated. So use other cool, dark places for items such as eggs, fruits, and vegetables. Potatoes, fruits, and vegetables can be stored in hanging baskets if you have the room. You can get dry milk or beans as substitutes, or long-life milk that can be dry-stored.

Also put some forethought into the order that you will eat your meals and store food accordingly. There is nothing worse than digging all the way to the bottom of the freezer for the evening's meal.

Tools of the trade

Since most galleys are somewhat limited in the amount of storage space available, you will want to keep cooking utensils and other basic supplies to a minimum. Here are a couple of basic pots and pans you'll want: a ten-inch skillet, a large saucepan, and a six-quart dutch oven. Try to use pots and pans that have the same size lids, as they can be difficult to store. If you don't have an oven, a pressure cooker can be used for meats, potatoes, vegetables, and many other foods. The meals come out tasty and retain vitamins.

The following is a list of kitchen utensils and supplies that you may find useful:

- vegetable peeler
- wooden spoons
- can opener
- church key
- kitchen scissors
- measuring cup
- measuring spoons
- disposable baking pans
- non-abrasive cleaner
- strainer
- mixing bowls
- grater
- spatula
- apron
- pot holder
- toothpicks
- salt & pepper
- aluminum foil
- charcoal lighter
- matches
- candles
- bar soap
- dish detergent
- disinfectant
- facial tissues
- paper towels
- ziplock bags
- garbage bags
- insect sprays
- sponges
- toilet tissue

Use non-disposable tableware, glasses, and mugs (forget about using any paper plates or disposable plastics as they take up too much room in the trash bin). For utensils use an inexpensive flatware. Keep in mind that presentation is important when you serve a meal, so take some bright, cheerful placemats, napkins, napkin rings, and a few flowers (even silk—they last longer.)

Eating and drinking is essential to making a trip a memorable experience. My goal is for you to enjoy the galley and still let you be on deck for the sailing, sunset, and the great cuisine.

Bon Appetit and Bon Voyage!

Atlantic Mackerel
Yellow Fin Tuna
Striped Bass
Winter Flounder
Atlantic Cod
Tautog
Black Sea Bass
Dolphin
Bluefish
White Marlin
Pollock
Swordfish

The thrill of a good fishing expedition awaits you anywhere along the coast, be it north or south, inland or offshore, on a powerboat or a sailing vessel. Of course, every type of fishing requires specialized tackle of some sort, and there is no simple way to be prepared for every contingency. The avid angler is most likely well-equipped, but even the the most casual of fishermen can handle most situations along the coast with only a few basic outfits.

One of the handiest outfits may already be in the hands of those who fish fresh waters—a one-handed spinning rig. The vast majority of your time on the boat is apt to be at a dock or mooring, and the opportunity to catch fish ranging from winter flounder in Maine to snook in Florida shouldn't be overlooked. Night is a particularly good time for such fishing as boat and dock lights draw bait fish—which, in turn, attract such predators as striped bass, weakfish, spotted seatrout, tarpon, snook, ladyfish, and lookdown. A spinning reel capable of holding at least 150 yards of 8- to 12-pound mono plus a matching 6 1/2- to 7-foot medium action spinning rod is ideal. The same rig will also serve at sea if you spot a floating board or weed line in blue waters of 70 degrees or more. Dolphin and triggerfish are attracted to such floating cover, and you can have great sport with them by casting bait or small lures.

For most boat fishing, a light conventional outfit is best. This should consist of a revolving spool reel, which can take at least 200 yards of 30-pound mono matched with a medium action 6- to 6 1/2-foot rod for bottom fishing and light trolling. This is a good choice for sailboats dragging a line astern in coastal waters for such species as bluefish, bonito, and little tunny. Almost any fish likely to hit a small lure can be handled with it, and there's sufficient power to drag the quarry in from a moving boat.

A boat spinning rig is another good choice, but not for trolling. Spinning reels have very little power and you'll have a hard time reeling in a fish while still moving forward. A boat spinning combo should consist of a medium spinning reel capable of handling about 200 yards of 20-pound mono matched to a 7- to 7 1/2-foot medium heavy two-handed spinning rod. This is what you'll want to have aboard when a school of blues or stripers is spotted in northern waters, or there's a chance to cast to large fish.

FISHING the Atlantic Coast

Catching dinner in Atlantic Coast waters.

With those three basic rigs, it's possible to handle 90 percent of the fishing you're likely to run into on a casual basis. Just multiply the outfits by the number of people aboard, and you'll still only be investing a few hundred dollars. If you should get into trolling for larger species, it's advisable to buy more specialized outfits geared to the size of the species you're seeking. But always consider how much you'll use the tackle before you buy. Unless it's used regularly, the same funds would be better spent on a charter boat a few times each year.

Now that you're geared up, psyche yourself up for some great and varied fishing. Here's what you can expect in Atlantic Coast waters.

Northern New England

Because cool waters prevail from Cape Cod north to **Maine**, most fisheries vary considerably from those to the south. Maine, in particular, has a relatively short sportfishing season with the height of the action from June through September. Those species readily available only south of Cape Cod from fall through spring can be caught in quantity during mid-summer off the Maine coast.

Mackerel, winter flounder, and harbor pollock (the young of the species, generally weighing 1 to 3 pounds) are often caught from docks and in bays, while banks not far from shore provide good shots at cod, pollock, and haddock. Offshore ledges in the Gulf of Maine, such as Jeffrey's, are noted for the largest specimens of the latter species plus some wolffish and an occasional halibut.

Giant tuna start hitting in July and continue into September. There's lots of competition for these giants, which are worth thousands of dollars when sold to Japanese buyers, but some are caught by trolling and chunking. Also catching on with the big game fraternity is shark fishing. Blue sharks are very abundant during the summer, along with a few large makos, and there's no need to go far offshore. Bottom structure 5 to 25 miles off is usually attractive enough for them.

Striped bass are abundant in the river systems, such as the Saco and Kennebec. Most are schoolies (small stripers), but specialists catch lots of jumbos on live eels and other large baits at the mouths of the rivers, primarily at night. Jumbo bluefish have been a feature of inshore fishing in Maine during the last several years, though they were completely absent from those cool waters for decades in the past.

New Hampshire has similar fishing along its very short coast. The nearby Isle of Shoals is a traditional hot spot for bottom fishing, and the IGFA world record cod of 98 3/4 pounds was caught there in 1969. The Piscataqua River provides excellent striper fishing.

Northern **Massachusetts** shares many of the same waters and adds yet more in Massachusetts Bay. Stellwagen Bank, one of the finest areas along the coast for giant tuna, cod, pollock, and wolffish, is close to Gloucester and other northern ports. Bottom fishing is best from April through June; giant tuna become the main target after that. Gloucester has been the major port for the giant tuna fleet, which chases those valuable fish throughout the Gulf of Maine from July to October if the seasonal quota hasn't been caught. Many giant tuna boats also sail out of such ports as Newburyport, Ipswich, Green Harbor, and Provincetown. Trips to offshore ledges, such as Jeffrey's, Cash's, and Georges Banks, are the best bet for large cod, pollock, and haddock during the summer.

Massachusetts has long been noted for striped bass fishing. There's excellent sport along the entire coast, but particularly off Cape Cod. From Cape Cod Canal to Provincetown, there are numerous shore and boat opportunities for stripers of all sizes from May to October. Bluefish have been just as important to inshore sportfishermen since they reappeared in the 1960s.

Southern New England

The southern side of Cape Cod is also devoted to stripers and blues. Pleasant Bay at Chatham is famed for big bass in shallow, protected waters. The offshore islands of Nantucket, Martha's Vineyard, and Cuttyhunk attract legions of striper and bluefish addicts for both surf and boat fishing. Nantucket's many offshore shoals provide great trolling opportunities, while famed Sow & Pigs Reef at Cuttyhunk has probably produced more giant stripers than any other single area (those not familiar with the area's underwater rocks are advised to charter boats at Cuttyhunk or New Bedford for that very specialized night trolling or eeling). The Martha's Vineyard Striped Bass & Bluefish Tournament is an annual fall feature that brings many anglers to the beautiful resort island.

Bottom fishing is also excellent off Cape Cod and southeastern Massachusetts for such species as winter and summer flounder, scup, sea bass, and tautog in inshore waters and for cod, pollock, and an occasional halibut offshore of Chatham. Giant and medium tuna are frequently found in the same areas late in the summer, and white marlin are an offshore attraction for Nantucket and Martha's Vineyard sportfishermen—along with school bluefin tuna, yellowfin, bigeye, and albacore tuna, and even an odd blue marlin. Swordfish were a regular catch in years past, but are a rarity these days.

The **Rhode Island** coast offers similar fishing in a rocky setting, which is ideal for stripers and tautog. Bottom fishing is often excellent for scup and summer flounder (fluke), but winter flounder have been so overfished by commercial boats that fishing for them has been banned during the prime spring season. Nearby Block Island provides much of Rhode Island's best fishing. The rip at the north end is a prime trolling area for stripers, and there are hot areas for bluefish, cod, pollock, scup, fluke, and many other species off the south end. A series of deepwater wrecks off Nantucket, including the **Andrea Doria**, are often fished by Point Judith party boats that catch large cod, pollock, and white hake from the 250 foot depths.

Giant tuna fishing at the mouth of Narragansett Bay and just off the coast at Rosie's Ledge is just a memory now, but they are still encountered at times in ocean waters off Montauk and Block Island as are school and medium bluefins. Sharking is a very popular sport here, too, particularly in June and July. Many boats also run far offshore to Block Canyon for marlin and tunas.

Offshore fishing in **Connecticut** is identical for boats from such ports as New London, Groton, and Niantic. The Race, that turbulent area where Block Island Sound and Long Island Sound meet off Fisher's Island, is a famous area for bluefish and stripers. The rocky shoreline of Long Island Sound is ideal for stripers of all sizes plus blues and tautog (blackfish). Don't ignore the far western end of the Sound, as the volume of school stripers is actually greatest

© Delaware Development Office

Surf fishing in Delaware waters.

there due to its proximity to the Hudson River, which has had many record spawns in recent years.

New York Bight

The right angle formed by **New York**'s Long Island and **New Jersey** creates a natural fish trap for species migrating along the coast, creating some of the best fishing in the world in one of its most densely populated areas. You'll find many cold-water species, such as cod, pollock, white hake, and silver hake (whiting) and semi-tropical inshore species such as triggerfish, Spanish and king mackerel, blue runners, and cobia. Underwater canyons also provide quantities of yellowfin, bigeye and albacore tuna, white marlin, dolphin, and a few wahoo. Blue marlin are far from common, even in the canyons, but some of the largest along the Atlantic Coast have been caught here—with both the New York and New Jersey state records being over 1,000 pounds.

Sport sharking started on Long Island and has become a very important offshore activity throughout the area in June and July. There are lots of big-money tournaments in both states, with makos being the featured species, though blues and threshers often take the early contests if waters are below 64 degrees, and huge tigers are normally summer winners. Hammerheads, duskies, and sandbars (browns) are other common catches, and the rare great white shark is as common in New York Bight as anywhere.

From July through October, big game attention shifts to canyons requiring runs offshore of 80 miles or more. Block Canyon is most popular with eastern Long Island boats and Hudson Canyon attracts those sailing out of western Long Island and northern New Jersey. Northern New Jersey boats also work the smaller canyons south of the Hudson such as Toms, Carteret, and Lindenkohl. Southern New Jersey sportfishermen target Spencer, Wilmington, Baltimore, and Poor Mans canyons. Trolling is most effective early in the summer, but overnight chunking for yellowfin tends to start in mid-August and is often best during September and October. Large numbers of 30- to 70-pound albacore are often trolled in the Hudson Canyon from late August into mid-September. There are many tournaments, but all pale before the $500,000 Tournament at Cape May in late August which actually pays out over a million dollars.

The Mud Hole, a deep water trench extending from just offshore of New York Harbor out toward Hudson Canyon, is a focal point of bluefin tuna fishing from late summer to early fall. The Oil Wreck area and Monster Ledge are the usual hot spots, and large fleets gather there for both school tuna and giants. Chunking at anchor along the steep dropoffs is the most popular method. Be sure to check on quotas and seasons, which are changed by the National Marine Fisheries Service every year.

The most popular inshore species in this area from May to October are bluefish, striped bass, and fluke (summer flounder), but there's also lots of attention paid to weakfish, bonito, little tunny (false albacore), winter flounder, porgies (scup), blackfish (tautog), sea bass, ling (red hake), whiting (silver hake), cod, and mackerel. Bluefishing continues into late fall, when the largest choppers of the season are often encountered on the surface under clouds of birds.

Montauk, at the eastern tip of Long Island, may well have the finest all-around fishery on the coast. Boaters there have a choice of productive local waters, including the rips off Montauk Point and Block Island and then out to many offshore areas that produce huge quantities of blue sharks, makos, tunas, and marlin.

The Hudson River shouldn't be overlooked, as it boasts a large population of striped bass. Striped bass fishing is best in the spring and fall, but both bass and blues can be caught around the mouth of the East River at downtown Manhattan and off the Statue of Liberty even in the summer.

Raritan Bay has been providing the best fluke fishing along the coast over the last few years. That bay between

Staten Island and New Jersey has much better water quality than during recent decades, and also boasts of fine striped bass and bluefish angling. Large sportfishing fleets sail out of the northern New Jersey inlets—Shark River, Manasquan, and Barnegat. There's somewhat less pressure in the central part of the state down to Atlantic City, Margate, and Ocean City, but another heavy concentration can be found at the southern end in Wildwood and Cape May.

In Delaware Bay, you'll find weakfish, including areas such as Fortescue to the west. And the bay is the normal northern end of the range for large black drum.

Delmarva Peninsula

Fishing changes gradually as you head south and encounter the Delmarva Peninsula. Parts of **Delaware**, **Maryland**, and **Virginia** all sit on this peninsula bordered by Chesapeake Bay to the west and the Atlantic Ocean to the east. You'll find southern species such as amberjack rarely encountered north of Ocean City, Md., and both red drum and tarpon, uncommon north of the Virginia barrier islands. Cobia are a common catch in lower Chesapeake Bay and at the buoys marking the shipping channels coming out of that bay.

Weakfish (grey trout) are very important for inshore fishermen throughout the region, but the closely related spotted (speckled) seatrout is also a common catch in Virginia. Though tautog are thought of as a cold water species, some of the largest (including the IGFA world record 24-pounder) have been Virginia fish. The mix of fish both in the ocean and in Chesapeake Bay is similar to South Jersey, but spot and croakers also become important bottom fish.

The Chesapeake and its many river systems produce most of the migratory striped bass population from North Carolina to Maine, and fishing for that species is very important, though severely limited as to seasons, bag limits, and sizes.

Offshore fishing is good for school and medium bluefins on inshore lumps such as the Hot Dog, Sausages, and Jack Spot. The canyons offshore aren't far. From Ocean City, Md., it's just 56 miles to the Baltimore, 53 miles to Poor Mans, and 57 miles to Washington Canyon. White marlin fishing is featured, and there are more blue marlin caught in these canyons than farther north. Yellowfin and bigeye tuna, plus dolphin and wahoo, are also common in the canyons.

Sharking is also very important here, and Delmarva gets a shot at makos a week or two before they arrive in New Jersey. The Delaware state record mako is a very impressive 942-pounder.

Cape Henry to the Outer Banks

Within a fairly short distance, fishing conditions change greatly south of Chesapeake Bay. Virginia fishermen enjoy great angling at the Chesapeake Bay Bridge-Tunnel, which acts as an artificial reef across the mouth of one of the largest and most productive bays in the world. Weakfish (grey trout), tautog, black drum, bluefish, and striped bass are among the prime species.

Virginia Beach waters hold cold-water Atlantic mackerel in March, and get a variety of semi-tropical species by summer. Wrecks and buoys located not far from Virginia Beach, plus the Chesapeake Light Tower and the Navy Tower well to the southeast, provide great opportunities for amberjack, cobia, sea bass, tautog, spadefish, and Spanish mackerel. Some of the largest tautog and sea bass ever recorded have come from the Triangle Wrecks and other wrecks in this area, including a 21-pound, 6-ounce tautog caught last June. The two 9 1/2 pound sea bass which are tied for the IGFA world record were both caught out of Virginia Beach. Kingfish (king mackerel) are an important trolling species in the summer and fall.

Offshore angling for marlin, tunas, dolphin, and wahoo is often excellent at the Cigar, the Weather Buoy, and in Norfolk Canyon. School bluefin tuna provide hot trolling action closer to shore in June. Many Virginia anglers are also into sharking, and some of the largest tiger sharks have been caught in local waters.

In **North Carolina**, the Gulf Stream is only about 20 miles from Cape Hatteras, though that short trip to oceanic fishing in warm, blue waters can be rugged in windy weather, which is common along the Outer Banks from

© Al Ristori

Three fishermen wrestle with a marlin.

fall through spring. Some of the finest and most convenient bluewater trolling in the world is out of Oregon and Hatteras Inlets. This area is probably the blue marlin capital of the East Coast, with large numbers of that great game being trolled in the Gulf Stream from May to September. White marlin fishing usually turns hot around Labor Day, and this is the northerly limit for a reasonable shot at a sailfish. Dolphin are probably the most popular species sought by fishermen sailing from Outer Banks ports, and they are caught in vast numbers along the weed lines that accumulate along the edge of the Gulf Stream. Also abundant and highly-valued for food and sport are yellowfin tuna (which normally average 20 to 60 pounds) and wahoo. Bigeye tuna up to 200 pounds or more are also hooked at times.

Few anglers fish for sharks off the Outer Banks, but those fish are as abundant there as anywhere, including makos inshore of the Gulf Stream during mid-winter. King mackerel trolling is featured during the fall, and tuna are available all winter—though most are caught commercially since few sportsmen are willing to brave the elements at that time.

The Outer Banks are just as noted for inshore fishing opportunities. Surf casting for red drum, bluefish, Spanish mackerel, spotted seatrout, and pompano is excellent at times, and a few specialists also catch large sharks from the beach at Cape Hatteras. Neighboring Ocracoke Island provides fine surf casting without the Hatteras crowds. Boaters can find excellent fishing in the inlets and close offshore for red drum and bluefish. The shallow waters of Pamlico Sound produce lots of large red drum and cobia as well as seatrout, flounder, spot, croaker, sheepshead, and many other small food and game fish. Tarpon specialists also have a shot at those fish during the summer.

Southern North Carolina to Northern Florida

The section of coast encompassing **Southern North Carolina**, **South Carolina**, and **Georgia** is much farther from the Gulf Stream, but there's still lots of effort directed toward the marlins and tunas from ports such as Morehead City and Charleston. The broad continental shelf also provides lots of opportunity for king and Spanish mackerel, bluefish, weakfish, sea bass, sheepshead, and other species. Fishing at the edge of the shelf is particularly productive for groupers, snappers, triggerfish, amberjack, porgies, and a host of other semi-tropical bottom fish plus a few cobia, dolphin, and other surface species. The vast network of bays, sounds, and rivers is also worth checking out for spotted seatrout, red drum, cobia, tarpon, and sheepshead. Striped bass below Hatteras are not migratory and tend to stay in their native river systems, though they may drop down to the mouths of inlets during late fall.

Florida's coast north of Cape Canaveral is quite different than the rest of the state's coast because the Gulf Stream swings well offshore and water temperatures drop much lower in winter. Most offshore sport occurs from spring through fall, and the area benefits from the migrations of such species as kingfish, Spanish mackerel, cobia, amberjack, wahoo, blackfin tuna, barracuda, sailfish, and tarpon. Blue and white marlin tend to be caught farther offshore than other species. Sharking is also popular from northern ports. Groupers, snappers, and other offshore bottom species provide good fishing year-round.

Red and black drum, snook, jack crevalle, spotted seatrout, and many other species are found inshore during the winter. The north-flowing St. John's River provides the southernmost populations of striped bass and shad on the coast. Except during severe cold spells, there's good light tackle sport throughout the winter in such central coast rivers as the Sebastian, Banana, and Indian.

Southeast Florida and the Keys

Offshore fishing changes greatly below Cape Canaveral as the all-important Gulf Stream flows close to shore. Rather than being the off-season, winter is prime time from Fort Pierce southward for sailfish and king mackerel. Unfortunately, that fishing tends to be best when northerly fronts move through and six-foot seas prevail. Sailfishing tends to be best from Fort Pierce to Stuart, though it's certainly more convenient off Jupiter, Palm

Beach, Pompano Beach, and Fort Lauderdale—where the Gulf Stream is only a few miles offshore. Dolphin become more common as the waters warm during spring, and are a regular catch during mild winters. Wahoo and barracuda are part of that Gulf Stream fare. Blue marlin tend to arrive just a bit farther offshore in late spring and summer, and are sought by skippers trolling live "bonito" (actually little tunny).

High spots closer to shore are good for African pompano, barracuda, and large kingfish. The latter are often trolled off Fort Pierce only a mile or two offshore in relatively shallow waters. Tarpon move up along the coast during the spring, and provide some memorable days in the fall when they attack schools of mullet off the beaches. Jumbo bluefish pass offshore on their way north in March and April, and some years chase bait right into the beaches—where they even bite swimmers who get in the way. Cobia are spotted swimming with rays as they migrate in both directions, and reef fishing is productive all year for snappers, groupers, and a host of colorful tropical species. Wrecks produce lots of amberjack, though those fish have been depleted due to the recent intensive commercial fishery.

Tarpon are warm-water fish, and not common during the winter except at the hot water outlet of the Fort Lauderdale power plant. However, snook are year-round (when the seasons are open) in southeast Florida inshore waters. Pompano are a prime target in the surf and along the Intracoastal Waterway.

The **Florida Keys** rank with the world's great fishing areas. Bonefish and permit become common in the lower portion of Biscayne Bay, just below Miami Beach. Fishermen enjoy great sport with them as well as barracuda, blacktip, and lemon sharks, and even an occasional large mutton snapper. Large tarpon move in during the spring northerly migration, but the smaller fish can be found in some areas year-round.

Sails and kings are also important to boaters along the Atlantic side, and there's a great run of dolphin in the spring. Underwater humps off Islamorada and Marathon produce lots of blackfin tuna and amberjack. Sharking has become popular there in recent years, and some huge makos are caught—particularly in March.

Reef fishing is much better here than farther north along the coast, with yellowtail snappers averaging far more than the 12-inch minimum. Black groupers, mutton snappers, barracuda, amberjack, and "smoker" kingfish are also common catches on the reef.

Blue marlin are most common on the 200-fathom line off Key West, with October being the usual prime month for blue marlin trolling. Key West has a harbor tarpon fishery which produces great numbers of silver kings (many weighing well over 100 pounds) from February to July. Guides usually obtain shrimp boat "trash" (small fish and crabs) and chum those tarpon at anchor in protected areas just minutes from downtown Key West. A yellowfin tuna fishery involving chumming with live bait also has been developed off Key West in the months of December and January.

At the end of the line, in the Dry Tortugas, you'll find the best grouper and snapper fishing. There are no facilities here, but you can anchor in the lee at Fort Jefferson.

Before You Cast

Be sure to check licensing regulations in area you plan to fish. Licensing of saltwater fishing is becoming very common from Maryland to Florida—and it is only a matter of time before licenses will become a fact of life in northern states. Check conservation regulations as well; they have become complicated in many states—especially in Florida.

Though set seasons for sportfishing only apply in a few cases (such as snook in Florida, summer flounder along the coast, and striped bass in Maryland and Virginia), there are closures when quotas are met for such species as king mackerel and bluefin tuna—and minimum sizes and bag limits for many fish.

The local tackle shop is one of the best sources for information on regulations or general fishing tips in the area. You can ask questions about what's running and pick up local tips. And finally, you may find some regional fishing publications that will guide you to the best the area has to offer. ✦

Robert Mack
Associate Editor, *Scuba Diving*

DIVING ALONG THE EAST COAST

Crustaceans or crinoids, sport fish or sponges, historic wrecks or hysterically colorful reefs—your undersea dreams can be fulfilled by any of many spots along the East Coast. A wealth of wonderful opportunities for everyone from novice to expert lie in the waters just off the coast. The waters may not always be warm or calm, and the visibility changes with region and season, but the diving can be spectacular.

Only those with significant diving and boating experience should attempt braving most of these waters without an experienced local dive guide. Currents can be treacherous and the undersea world is in a constant state of transformation. Whether you plan to use your boat or dive from shore, you can usually hire someone—often for a very reasonable price—who'll steer you toward the best of what the area has to offer. No matter how many years of experience you have under your weight belt, your dive day will still be enriched by locating a personal guide or going along with a group organized by a local dive shop.

Remember, too, that diving is governed by both state and local laws. In many areas, these laws are enforced, sometimes to the tune of pocket-emptying fines or even jail time. Find out what the laws are where you're planning to dive and obey them—everyone will be much better off.

© Bill Lovin

The gun mount of a U-352 attracts marine life and divers.

Down Under in the Northeast

For a long time, divers who enjoyed New England's rich waters felt like they had to defend themselves. Let's face facts: the water is cold—never far above 65°F, and in winter dropping below freezing in some places. Visibility rarely extends to 40 feet and is more often closer to 15 feet. But these days more and more bubbleblowers are lured by these cold waters and their somber beauty.

Because the water's so cold, it holds a lot of oxygen. That breeds plankton, which colors the water its typical green and reduces its clarity. But this rich plankton soup serves as foundation for a food chain that runs from anemones and assorted bottom fish to sharks, whales, and seals. If you can't force yourself to don a full wetsuit or a dry suit to try diving in New England, you'll never realize the thrill of watching a speedy striped bass zip across your line of vision or understand the inspiring beauty that can be found on a wall of fragile near-transparent anemones waving their soft pink hearts aloft in the current.

© Bill Lovin

The waters off the East Coast hold a wealth of diving opportunities.

And you won't get to savor the flavor of the lobster you just saw strolling around that morning.

A nearly unlimited number of coves, bays, and inlets mark the rugged coast of **Maine**, with a couple of thousand islands lying just offshore. The state's 2,500 miles of shoreline offer exciting and underexplored dive sites teeming with sea life. Some of the strangest marine creatures on the planet thrive here: the lowly lumpfish, the grotesque goosefish, and the scarifying sculpin are just a few of the prehistoric-looking creatures that thrive in Maine's cold water. In addition to the abundance of bizarre undersea dwellers, other things set this state apart: the Gulf Stream swinging away from the East Coast at Cape Cod causes water temperatures within the Gulf of Maine to drop dramatically, and the four-foot tides south of the Cape double on the north side increasing as you go up the Maine coast. Most Maine diving takes place off the southern coast and in Casco Bay simply because there are more dive shops in this area, but dive sites abound all the way to the Canadian border for the intrepid scuba hound.

Massachusetts ranks as one of the top 10 states in diving popularity. The rocky north shore and Cape Ann offer exquisite diving into a glacially carved submarine world of canyons and ridges full of seastars and soft corals, an innumerable variety of anemones, and schools of pollock and cod. Cape Cod has snared a good percentage of the 2,000 shipwrecks found off the state's coast; the cape extends so far out into the ocean that diving at its tip offers you up-close and personal views of whales and ocean sunfish. In Massachusetts Bay, you'll drop into a vastly underexplored wilderness where you can catch your dinner with the Boston skyline visible in the distance; off the south shore kelp-covered rocky reefs and famous wrecks dominate the scene. And there's plenty of the same marvels to be found in Martha's Vineyard.

Rhode Island has more to offer the visiting diver per acre than any other state in the union. If you're a fan of lobster you can catch the limit for a nominal fee while you're vacationing in the state. Spearfishermen can seek the prized striped bass, which travel swiftly through Rhode Island's waters; wreck divers can visit everything from wooden schooners to World War II-era U-boats. Narragansett Bay is a treasure trove for avid bottle collectors; shore divers find easily accessible sites ranging from 10 to 150 feet in depth. But for many the favorite activity is capturing the tropicals borne north by the Gulf Stream. Since they're doomed by the approach of winter, divers altruistically take home sealife for their saltwater aquariums. Clarity ranges from zero to 80 feet, best at offshore sites in late summer when the Gulf Stream swings close to shore. This is strictly full wetsuit or dry suit diving. Water temperatures seldom rise above 60°F even in the hottest months of summer, and deeper dives have sharp thermoclines, stratums in which temperature declines at least one degree centigrade with each meter increase in depth.

Connecticut has added a great deal to its popularity with out-of-state divers by granting nonresident lobster permits. Since the state fronts on Long Island Sound, most dives are shallow and water clarity can be low. Pollution control measures have significantly cleaned up the waters in recent years, though, and Connecticut divers stay busy thanks to rocky reefs and abundant marine life. Swift tidal currents and heavy boat traffic require cautious diving at many sites.

Visitors to **New York** can take their pick of the rocky and shallow Long Island Sound or the deeper wreck dives off the sandy Atlantic coast of Long Island. Water temperatures range from a summer surface high of about 72°F to a winter low right near the freezing mark; as many as three thermoclines drop the temperature on deeper dives, so wear a protective thermal suit. Divers find the Rye Breakwater, just east of White Plains, a good spot for observing the plentiful marine life of Long Island Sound. A sand-and-mud bottom about 20 feet down is interrupted by the breakwater, where crabs sidle over the rocks, eels slither out from crevices, and blackfish poke about looking for invertebrates to crunch. At night you'll find lobster roaming the rocks. Further east on Long Island, the dives range from wrecks of paddlewheelers and World War I battle cruisers to drift dives through fish-filled inlets. Crane Neck Point is probably the best dive in the area. Located on the eastern tip of the bay, marine life is rich here, with kelp-covered boulders sheltering the oyster toadfish, who croak loudly during the June mating season; in shallow water look for pipefish, the skinny northern relative of the seahorse. Following the boulders out to 25 feet you can spot lots of large anemones, yellow sulfur sponges, pink-hearted hydroids, and perhaps a skate or dogfish. Diving at low tide will help you avoid the occasionally strong currents.

On the Atlantic shore of Long Island the Montauk jetties offer good opportunities for macrophotography; blue-claw crabs can be seen among the grasses of the sandy bottom, while eels and blackfish congregate around the jetty. For advanced divers, few spots on the island can beat Shinnecock Inlet. Although currents can be swift and boat traffic heavy, lobster appear at slack tide and striped bass begin returning to these shallower waters in late spring. As fall arrives, tropicals get swept up the Gulf Stream and striped bass make their last run for the year, some as large as 50 pounds.

Some of the most exciting wrecks on the East Coast are located just south of Long Island. The *Oregon*, a 518-foot luxury liner which sank after a collision in 1886, sits upright but broken at a depth of 120 feet just off Fire Island; the engine and boilers stick up almost 40 feet from the sandy bottom. The freighter *Iberia*, victim of an 1888 collision, has become a pile of heavy machinery, twisted steel, and cargo cartons that offers refuge to anemones, lobster, eelpout, and blackfish. Perhaps the most famous of the Long Island wrecks, the *U.S.S. San Diego* fell victim to a U-boat mine off Fire Island during World War I. The 504-foot vessel lies intact, but upside-down, at 110 feet; huge guns protrude from underneath the hull. Rusting pieces of steel plating have fallen, creating many entrances into the ship, but penetration is dangerous due to problems with disorientation and live ammo still aboard.

Mid-Atlantic Variety

It's difficult to draw too many generalizations about diving the Middle Atlantic region. The variety of dives you can make boggles the mind, ranging from the famous and spectacular wrecks off the coast of New Jersey to the artifact and oyster hunts within Chesapeake Bay.

So many ships have sunk off the coast of **New Jersey** over the centuries that some divers refer to the area as the New Jersey Triangle. Unlike the Bermuda Triangle, though, few of the vessels sank unexplainably; most were caught in terrible storms, collided with other vessels in crowded shipping lanes, or were U-boat victims. These wrecks offer many attractions, including artifact diving, lobstering, spearfishing, photography, or simply looking. Since the shifting sands change with every storm, the wrecks are seldom picked over; new sections and artifacts become exposed at regular intervals. The sandy bottom provides little anchorage for marine life so when a ship sinks, the natives flock to it; most of the New Jersey wrecks are covered with huge mussels, sea anemones, and other forms of encrustation. Fish ranging from bottom-dwelling flounders to cobia cruising overhead can be spotted within the vicinity of several wrecks, and you can't forget about the lobsters, able to extend their range southward by hiding within.

May through October is the most reliable time for diving, when water temperatures rise to a high of 70°F on the surface—although two or three thermoclines drop it down to the mid-50s on the bottom. Visibility averages 20 feet, but can approach 50 feet on a good day. Gloves and a good dive knife are necessities—and above all, remember that it takes special training and equipment to enter a wreck.

Almost all shore diving off New Jersey takes place around jetties, which lure marine life from the surrounding miles of sand bottom. Divers will find more than 100 jetties along the coast, each with its own specialty, shape, and restrictions—check locally. Clarity ranges up to 20 feet and depth rarely surpasses 15 feet. Divers should be especially cautious of both currents and boats around the jetties.

Even the tiny state of **Delaware** offers some great offshore diving. A boat is necessary at the Lewes Breakwater, unconnected to land by a jetty; there are two breakwater structures, each marked by a lighthouse. The outer one offers a superb mile-long run where tautog and flounder are common. Currents can reach 2 knots at high tide, visibility averages about 8 feet, and maximum depth is 40 feet. The Indian River Inlet farther south in the Delaware Seashore State Park is another possibility, but you need permission from the park to dive. Farther offshore are areas such as Fenwick Shoals, where a clean sandy 50-foot bottom draws schools of bluefish, kingfish, and sheephead.

The **Maryland** coast usually sees a great influx of divers on weekends during spring and summer months, many of whom come for the unparalleled experience of eating just-picked oysters. Others arrive to dive among stripers and tautog on the state's rock jetties, while still others dive for artifacts and fossils in Chesapeake Bay. Although water clarity is usually poor, any number of items can be found within groping distance on the bottom, since Chesapeake Bay has served for centuries as thoroughfare to Baltimore. The 30-mile stretch from Chesapeake Beach to Drum Point is known as Calvert Cliffs, one of the country's richest fossil beds; divers often find the fossilized nine-inch teeth of ancient 100-foot long sharks that used to roam these waters as well as alligator teeth, dolphin earbones and vertebrae, sand dollars, scallops, and whale bones. Water temperature on shallow dives can reach as high as 70°F during the warmer months but will drop as low as 50°F below the thermocline.

Virginia offers the clearest water in the Delmarva Peninsula; artifact diving with underwater metal detectors, oyster diving and visiting offshore wrecks—where the visibility can reach beyond 50 feet and summer water temperatures may rise close to 80°F—are some of the most popular diving activities. In the northern part of Chesapeake Bay, the Little Foxes Islands are a haven for underwater photographers, who find striking subjects like stingrays with four-foot wingspans. Much shore diving in southern Virginia takes place at Ocean View, where you can enter off the beach into water less than 10 feet deep and search for relics left from a turn-of-the-century hotel that burnt down. For better visibility and more exciting sealife, you need to go offshore—to places like the Chesapeake Bay Bridge Tunnel, where marine life thrives around the rock pilings supporting the massive structure. Flounder, tautog, bluefish, striped bass, and a wide assortment of seastars, whelks, mussels, oysters, and clams are common sights; black sea bass, sharks, and rays are also sometimes present. Currents can be quite heavy toward the boat passageways between the islands so be sure to check with local divers who know the safe spots away from the boat lanes. Visibility is best in fall and winter—up to 30 feet.

© Norman Despres

Wrecks and Reefs in the Southeast

With some of the best wrecks anywhere, **North Carolina** has earned its reputation among divers as the Graveyard of the Atlantic. But both South Carolina and Georgia also offer a selection of fascinating ocean dives. Luck with the weather and currents plays a big role with all dives. Summer water temperatures in the region can surpass 85°F so wetsuits are worn for protection rather than thermal insulation during high season.

From the early 1500s through the Civil War, the shifting sands of North Carolina's Outer Banks earned the fear and respect of every sailor who traveled the area. Clashing currents, roving shoals, natural magnetic irregularities, wars, and tempestuous storms have created one of the most hazardous shipping lanes on Earth. But it wasn't until World War II that an additional deadly element was added to this melting pot of sea disasters: submarine warfare. Of the 87 vessels lost along the North Carolina coast during the war, 29 were victims of Nazi submarine wolfpacks. And it's these relics of the greatest war ever fought that lure present-day divers. Of all the 2,000 ships entombed in the area, none seem to hold as much fascination as the victim ships of World War II and their submarine enemies.

The wrecks themselves can't take full credit for the state's outstanding diving—the Gulf Stream is also responsible. Swinging close to the coast, it improves the water clarity from a visibility of 10 to 30 feet inshore to more than 100 feet at offshore sites. Summer water temperatures can reach well into the 80s, and the Gulf Stream brings such a tropical assortment of marine life that most wrecks resemble an outpost of the Caribbean stuck on the sandy, barren seafloor. Beautiful corals, sponges, moray eels, angelfish, butterflyfish, spiny lobster, anemones, arrow crabs, grouper, amberjack, sharks, manta rays, starfish, sea turtles, and triggerfish are all frequent visitors between Cape Hatteras and Cape Fear.

The wrecks of Civil War blockade runners and modern steel vessels, as well as artificial reefs, comprise the best dive sites in **South Carolina**. At first glance the murky water off Charleston might not seem like the world's most appetizing dive site—but once you get a few miles off-

© Florida Department of Tourism

An underwater statue adds to the excitement of diving at John Pennekamp Underwater Park in the Florida Keys.

shore, the picture changes dramatically. The silt and topsoil flushing out of the state's many rivers causes zero visibility right offshore, but the dirt settles out after a few miles. Clarity can reach 60 feet or more at many offshore sites. For most sport divers, intentionally created reefs are big draws. A chain of these reefs stretches along the coast from 3 to 13 miles offshore, all lying at depths between 40 and 75 feet. Just 10 miles out of Charleston, the Caper's Island Reef is built around the carcasses of a landing craft, a trawler, and several other old vessels. About 20 miles south at the Kiawah Island Reef, the wreckage of a tugboat provides the most excitement, as the intact galley and cabin allow exploration. Thousands of auto tires in heaps around the wrecks expand the live-bottom area. Pairs of triggerfish weave around the wrecks, moving their big, flat bodies with delicate ripples of their upper and lower fins. Small tropicals like blennies and hawkfish dart into the sanctuaries of the coral-covered tires as you approach, and flounder often hide or rest in the sandy bottom. In the spring, big fish like cobia school around the immense wreck of the *Betsy Ross*, a 442-foot Liberty Ship intentionally sunk 16 miles off the coast of Hilton Head. The upper deck is at 75 feet, with the vessel touching bottom at 110 feet.

Although many sport divers have yet to discover it, one of our country's richest National Marine Sanctuaries lies 16 miles off **Georgia's** Sapelo Island. Gray's Reef encompasses 17 square miles of limestone ledges lying at depths between 70 and 110 feet. Rock reefs are heavily encrusted with corals, sea whips and fans, basket sponges, anemones, and tubeworms. Blue angelfish, triggerfish, and schools of spadefish and snapper swim along the reef top, while grouper and cardinalfish lurk in the caves and recesses that undercut the ledges. Amberjack, cobia, and barracuda patrol the upper water. Gray's Reef is also one of the world's largest breeding grounds for the endangered loggerhead turtle, frequently seen along with jewfish, rays, and enormous spiny lobsters. Visibility varies greatly, attaining a maximum of about 50 feet. Georgia also has a large system of artificial reefs; diving is recommended on those farther than 12 miles offshore, since strong tidal currents and estuary conditions like zero visibility mark the nearshore shelf.

Fascinating Florida

Florida has more coastline than any other state besides Alaska, the only living coral reef in North America, and the greatest collection of fossils accessible to divers of any state in the union. The list of Florida's attractions has made for books in themselves, but one statistic tells it all: 57 percent of all divers traveling within the continental United States make Florida their destination.

While many divers overlook the northeast coast of Florida in their haste to reach more acclaimed southern spots, you can find some terrific summer dive sites here. Visibility improves from its winter low of 15 feet to an average of 30 feet, and the water warms up to 80°F. Plan on a long boat ride, since most spots are well offshore, but you'll be rewarded with lobster and abundant game fish for your trouble. With Liberty Ships, World War II bombers, a 19th century gunrunner and both natural and artificial reefs offshore, the Daytona Beach area promises some super dives. Summer is the dive season, when visibility reaches about 25 feet and the water temperature approaches 80°F. As in the rest of northern Florida, novices should plan a trip with experienced local divers in order to hit the hotspots. The Fort Pierce area—known as the Treasure Coast because of priceless finds from the 1715 Spanish Silver and Gold Fleet discovered just offshore—is also gaining fame as a sport diving center. Three major reef lines lie within 300 yards of shore, making the area the main place in Florida for easily accessible beach diving. The deeper reefs begin just 12 miles out from the Fort Pierce inlet and harbor an abundance of larger marine life as well as a diverse array of tropicals, including blue chromis and damselfish. Summer clarity ranges from 10 to 50 feet, and water temperatures reach into the 80s.

Without question, the Palm Beach area offers the wildest diving on Florida's Atlantic coast; the mighty current of the Gulf Stream comes closer to shore here than anywhere else, bringing with it a rich assortment of marine life. Unlike more northern locales, Palm Beach divers go out year-round. Weather and seas are more predictably calmer in the summertime. But winter, although prone to higher seas and colder water, is considered by many the best dive season since prevailing northeast winds slow down the current and big pelagics move in—count on seeing leopard rays, huge schools of amberjack, and a plethora of sea turtles. Add to the excitement that most diving requires a boat ride of just 15 minutes or so, and that water clarity almost always surpasses 50 feet and often exceeds 80 feet. Do be aware, though, that the current requires special procedures. Drift diving—letting the current sweep you along with a live boat keeping track and picking you up downcurrent—is usually practiced, but maverick and counter currents and winds spring up. Be prepared for anything.

Fort Lauderdale is another fast-growing diving center, a diver's smorgasbord with hundreds of dive sites along three distinct reef lines as well as a good variety of wrecks. Miami has long been known for its shallow areas of extensive coral and visually stimulating deeper reefs. In recent years, thanks to a local artificial reef project, it's also become a wreck diver's mecca with many huge, intact vessels in deep water where visibility approaches 100 feet most times of the year.

But if popularity with divers is your criterion, the **Florida Keys** win hands-down the title of best diving in America—more than half a million divers and snorkelers visit the area every year. The only living coral reef off North America parallels the Keys four to seven miles offshore, creating a surreal world of elaborate coral structures, dazzling fish, and untold shipwrecks. One reason divers relish the Keys is the warm water—about 85°F in the summer and a minimum of 69°F in the winter. While good diving days take place year-round, late spring through autumn promises the most predictably clear weather. On calm days, visibility can reach 100 feet, although 60 feet is more the norm.

The Keys can be broken down into several different areas. Key Largo is home to John Pennekamp Underwater Park, founded in 1960 and still going strong with 40 species of coral and 650 fish species flourishing in the fringing reefs (four to seven miles offshore). The Upper Keys around Islamorada offer a number of reefs sporting purple seafans, bright orange sponges, and big brain corals in addition to large groups of schooling fish, including rare deepwater species like blackcap basslets and jackknife drums. The Middle Keys, centered around Marathon, include everything from acres of shallow coral gardens to spectacular Looe Key, probably our most breathtakingly beautiful National Marine Sanctuary. Key West has great reefs easily reached in half-day trips, untouched areas like the Marquesas, Cosgrove Shoal, and the Dry Tortugas. With dive sites ranging from shallow reefs to deep dropoffs to wrecks, there's a little bit of everything at the southernmost tip of the continental United States.

Prepare and Enjoy

Diving anywhere along the East Coast can offer you a rich and varied marine environment. Be sure to do your research: contact local dive shops and Chambers of Commerce, which are usually most helpful in getting you acclimated and into their waters. Plan your dives in accordance with weather conditions; you can have a great dive on what seems like a dismal day, but it's less likely along the eastern seaboard. Above all, dive safely—know your equipment, practice your skills, and enter the undersea world with care, realizing that the same vistas you're widening your eyes at today have been there for years and will continue to exist for others to enjoy as long as everyone obeys local diving laws and does a lot more looking than touching. ✦

© Clyde H. Smith

Rivers in the Sky:
ATLANTIC COAST BIRDING

Paul Kerlinger, Ph.D.

As a "river of birds" flocked over the boats of Christopher Columbus and his crew, they realized landfall was imminent. And indeed it was; they were approaching the West Indies. But Columbus and crew may not have pondered that their diary description would become the first known European observation of bird migration in the New World.

What they described was undoubtedly the autumn migration of songbirds and shorebirds engaged in the long, overwater trip between North and South America, a phenomenon that has occurred for millennia. It is a description of the amazing migration process we see along the Atlantic Coast that involves birds flying as much as a quarter of a million miles in a lifetime, or tracking more than 200 miles a day at speeds of 30 miles an hour. Although travel patterns and distances vary, most of the birds of the Atlantic Coast are migratory, caught in the natural pull to take advantage of the most abundant food supplies of different seasons. And this "on-the-move" lifestyle gives an added dimension to observing birds in the region.

Along the Atlantic Coast, you'll find that there are vast expanses of ocean devoid of avian life, but there are also smaller patches in which birds and other sea life are abundant. These patches may include several types of sea birds, marine mammals, fish, and invertebrates. Finding these patches or flocks of birds at sea can be an unforgettable experience. The sight of a pod of dolphins or whales feeding just under the surface with seabirds flying directly above is one of the sights of a lifetime.

Keep in mind that birding on the water is a very different game from birding on land. On land, birding consists

© Embassy

Common Tern cradles chick in nest.

of searching among trees and shrubs, or on the ground. Birding aboard a boat involves searching a vast horizon. The best way to locate birds while sailing or under power is to slowly scan the horizon repeatedly. Dots located on the horizon soon take form. First there's movement, then wings, and, finally, plumage.

Armed with a good pair of binoculars and a decent field guide, you can expect to enjoy the dozens of species of birds that frequent the waters of the western Atlantic. Although you may enjoy the solitude of the water, you will never be truly alone. Not a day will pass without seeing a sea bird. Here's a good taste of what to watch for in Atlantic Coast skies.

Shearwaters, Storm-Petrels, and the South Polar Skua

Truly offshore, or pelagic, species come ashore only to lay their eggs and raise their young. They live mostly in the waters of the continental shelf, various undersea canyons, and the Gulf Stream. Among the most pelagic of species are a group of summer visitors to the Atlantic Coast, some having flown thousands of miles from their breeding colonies in the South Atlantic or Antarctic waters. This group includes shearwaters, storm-petrels, and the South Polar Skua.

Shearwaters are among the best fliers in the animal kingdom. Flying within a few inches of the waves, their wings sometimes cut the surface of the water—hence, their name. Most species are blackish or dark brown, with some white coloring. The amount of white coloring and its position is the best way of separating the different species. Another difference between the species is the way they fly due to their different sizes and shapes. Three main species are **Sooty Shearwaters**, **Cory's Shearwaters**, and **Greater Shearwaters**.

Sooty Shearwaters have no white underparts and are smaller and slightly thinner winged than Cory's or Greater Shearwaters. With their faster flight, they can be discerned with relative ease, even as far away as a half mile. The Sooty and Greater Shearwaters that visit our coastline engage in an epic migration. At the end of the austral summer after breeding on islands in the southern Atlantic off South America, they head northward toward the equator. When they encounter our East Coast, they head northeastward where they often feed and rest within a few miles of land. Eventually, these birds complete a circle by flying farther north, crossing the Atlantic to the coast of Europe, and then heading south again. This about-face in the North Atlantic leads them southward and back to their breeding sites the following year. Over and over again the cycle continues. An adult shearwater may fly for more than a quarter of a million miles in its lifetime.

Learning to discriminate among the shearwater species takes practice, but if you are lucky enough to find a feeding area with two or more species (often with humpback whales and Wilson's Storm-Petrels), you now have a fantastic opportunity to really learn the differences among these species.

The **Wilson's Storm-Petrel** is the most common of the summer sea birds. They can be identified by their small size and the way they fly. Walking on the water with their wings held above their bodies, these tiny sea birds feed on plankton and bits of animals that can be found on the water surface where whales or fish have been feeding. The Wilson's Storm-Petrel breeds in Antarctic waters and migrates north into the North Atlantic Ocean. Sometimes they can be seen from shore, but most often they inhabit food-rich areas on the continental shelf. There are tens of thousands of these birds in our waters each summer.

A closely related species, the **Leach's Storm-Petrel**, looks very similar. It, too, is black, but its tail is slightly forked, and instead of a broad white rump-patch, it has a very narrow white rump-patch. This species breeds from Maine northward and is rarely seen south of Massachusetts.

The **South Polar Skua** is a summer visitor to our coast. You can sometimes see this gray or brown, gull-sized bird stealing food from other sea birds, or near a flock of gulls following a boat. It has long, pointed wings, which have been adapted by most sea birds. If you compare sea birds to birds that live in forests, the differences are obvious. A sea bird simply would not be able to maneuver in a forest with its long, pointed wings. And forest birds with their short, rounded wings are not adapted for constant flight over the ocean.

Gulls, Terns, and Pelicans

Another type of visitor to our East Coast waters is more commonly seen. This type includes birds that breed on our East Coast and migrate south in winter, in some cases staying in the United States, in others going farther south to the tropics. They include the gulls and terns, and some **Brown Pelicans**.

The gulls and terns are our "bread and butter" birds and are almost always present along the coast. They are the most numerous of birds in our near-shore waters with species like the **Laughing Gull**, **Herring Gull**, **Forster's Tern**, and **Common Tern**. The smaller **Least Tern**, also known as the Little Tern, can be seen over the ocean, but is more likely to be seen in back bays, marshes, tidal creeks, and freshwater lakes and ponds. The **Arctic Tern** migrates over open ocean to the Southern Hemisphere, a total distance of more than 20,000 miles per year, and is less commonly seen than the other terns.

From New York to Florida, the Laughing Gull is the most common of our gulls during summer. They can be seen from the beach to more than 20 miles at sea; they are also the easiest to feed from the hand. With their black heads and the slate gray mantle (top of the wings and back), adult Laughing Gulls are easily recognized. Immature birds are less distinctly marked and are often dark over most of their bodies. Adults begin to molt—lose and regrow feathers—in July to late August or early September. At that point, the black head is nearly gone and takes on a dusky appearance. You should not feed these birds. Gulls are now a problem in many beach communities because they have deplorable manners.

The larger gulls, such as the **Herring**, **Ring-billed**, and **Great Black-backed**, distinguishable by their white heads, are seen along the coast in summer, but many do not breed then. These species often breed in more north-

Brown Pelicans now breed as far north as Maryland.

ern waters and migrate south to the Atlantic Coast. During autumn, winter, and early spring, these are the more commonly seen gulls on our coast as Laughing Gulls migrate southward.

Gulls and terns nest in colonies, sometimes numbering in the tens of thousands of pairs. The Laughing Gull colony that inhabits the marshes behind Stone Harbor, N.J., numbers about 12,000 pairs. The Ring-billed Gulls and Herring Gulls that are so common on the coast in winter nest in large colonies inland around the Great Lakes. Tern colonies, such as the one at Great Gull Island in eastern Long Island Sound, host tens of thousands of breeding Common Terns and the largest population of the endangered Roseate Tern in North America. Sailing by these colonies in early to mid-summer, it is easy to observe the comings and goings of adults carrying fish for their young. Terns sometimes violently "mob" gulls invading their colonies to eat eggs or chicks (and are known to attack humans!). These colonies depend on their isolation, so treat them like sanctuaries. In many areas it is illegal to enter a colony.

The largest of the birds frequenting our near-shore waters is the **Brown Pelican**. This species can wander several miles out to sea, but they are mostly seen within a mile of shore. After numbers of these birds dropped precipitously in the 1950's and 1960's, the species was listed as "endangered" by the United States Fish and Wildlife Service. DDT was named as the cause. After DDT was banned in 1973, pelicans rebounded and they now breed as far north as Maryland. A few hundred birds even set up "housekeeping" nests near Barnegat, N.J., in the early 1990's, but have failed to raise young.

Alcids, Loons, and Sea Ducks

Along the coast from Maine through the Canadian Maritimes, the "alcids" are a somewhat common breeding bird. The winter southward movement of alcids and other more northerly breeding birds into New England and Mid-Atlantic waters gives more birders a chance to see these interesting species. When these birds move southward and farther out to sea in winter, they show up in the waters from Massachusetts to Maryland. **Puffins** and **Black Guillemots** nest and spend their summer in the waters close to shore along the Maine and Nova Scotia coast. **Thick-billed**, **Common**, and **Razorbill Murres** breed slightly farther north, starting in Newfoundland, but winter farther south. All are readily distinguished as being alcids because of their relatively small size (crow size or smaller), black and white coloration, and their characteristic flight—flying just above the water with rapid, but not very graceful wingbeats due to their short wings. These shorter wings make them very different from almost all other pelagic birds.

Loons are some of the more bizarre birds found on the ocean during migraton. Although they breed on freshwater lakes in the northern United States and Canada, loons are one of the more interesting of "sea birds." **Red-throated** and **Common Loons** are found in our near-shore waters. Red-throated Loons are more commonly seen of the two, especially in late autumn through spring. One of the largest known aggregations of Red-throated Loons can be seen each year during spring migration at the mouth of Delaware Bay between Cape May, N.J., and Cape Henlopen, Del. At this time, several thousand loons "stage" before continuing their migration to the Canadian tundra, where they breed. Birds begin to aggregate in the rips off Cape May and Cape Henlopen in February and peak in numbers in March. They feed in the rips by diving for a minute or more at a time. Lines of hundreds of these birds can be seen working a single rip, all facing into the current. A few Common Loons are also usually present. They are larger and more "bull-headed" than Red-throated Loons, and can be differentiated in flight by their feet sticking out behind them.

The migration of loons occurs during both autumn and spring, although it is more noticeable during autumn. At almost any place along the Atlantic Coast from Maine to Virginia, loon migration occurs within three miles of shore, so it is mostly a coastal migration. Places where the land juts out into the ocean serve to concentrate the numbers of these migrants, as it does for many other migrants. Montauk Point at the eastern end of Long Island is an excellent place to see these birds. Other hotspots include the southern New Jersey shore, the Maryland shore, and the outer tip of Cape Cod.

Other bird species that hug the coast during their southbound migrations include sea ducks such as **scoters**, **Oldsquaw**, **eiders**, and **Red-breasted Mergansers**. In addition, an observer within a mile or two of shore can also see thousands of **Brant**, **Canada**, and **Snow Geese**, cormorants (mostly **Double-crested Cormorants**), gulls of various species, an occasional jaeger, and even a few alcids. These migrations become concentrated in some of the same places as loon migration.

Scoter migration is fun to watch and these ducks are easy to identify. Males are black; females are brown. They fly in flocks of ten to more than 500, usually within 3 to 10 feet of the water. Their flocks undulate upward and downward to conform to the surface of the waves. One reason why so many loons, scoters, pelicans, cormorants, and others migrate at such low altitudes is to take aerodynamic advantage of the air near the water's surface. The aerodynamic advantage of "ground effect" is related to reducing drag and increasing lift. The speed of flight for scoters, loons, and some other sea ducks is slightly greater than 30 miles per hour. Migration that starts at or just before dawn and lasts until noon results in flights of 150 to 175 miles per day, assuming the birds have no tail-wind. With a tail-wind, these birds can fly more than 200 miles in a morning's migration. If you watch these migrants, you will also notice that on days when they fly into headwinds, these birds fly at lower altitudes than with tailwinds. They do this to maximize the distance they travel (or minimize the energy they use) during a given period of time. Some of the waterfowl that we see during winter along our coasts migrate at night, so we rarely see large numbers flying south during daylight.

© Clay C. Sutton

Snow Geese hug the coast during migration.

Land Birds Take to the Sea

So far we've looked at water birds—those that can land on the water and float without running the risk of drowning or fly almost endlessly. There are also hundreds of landbird species that migrate over water. Although millions of land birds make a portion of their migratory flights over the ocean, they are not always spotted because these migratory flights usually occur at very high altitudes and much of the flight is at night. Shore birds and most songbirds that leave the coast of Nova Scotia or Massachusetts during September or early October do so just after dusk and they climb to very high altitudes. Radar studies from Puerto Rico, the West Indies, and Bermuda show that these birds fly at 3,000 to more than 10,000 feet above the ocean. This makes them virtually invisible. During thunderstorms or adverse winds, these birds sometimes descend, land on boats, or perish in the sea. It is at these times that they are most visible.

How can a **Blackpoll Warbler** or a **Semipalmated Sandpiper** weighing less than an ounce fly over 2,000 miles without landing? These birds weigh so little that, could they be mailed, it would cost a single, 29-cent postage stamp anywhere in the United States! It is amazing that these small songbirds regularly cross large bodies of water. In spring and autumn millions cross the Gulf of Mexico and parts of the western Atlantic Ocean. This is probably the migration seen by Columbus and his crew.

There are even a few hawks that can be seen over the Atlantic, including **Ospreys**, **Peregrine Falcons**, **Merlins**, **Bald Eagles**, and **Northern Harriers**. Peregrine Falcons and Merlins (a smaller falcon) are often observed chasing and catching warblers, cuckoos, sparrows, or other small birds that are tired. These predators eat their prey while still on the wing or they land on a boat to finish their meal. Most of these species will rest on boats, buoys, and offshore oil rigs.

Winter Birding for the Hardy

After autumn migration, the types of birds that you'll encounter along the Atlantic coast are very different than during migration or during the summer. The thought of venturing 20+ miles out to sea in more northern states in winter may sound a bit bizarre; this type of birding attracts a hardy bunch. However, winter birding in the coastal waters of the Canadian Maritimes and the United States offers some wonderful opportunities. For example, puffins and puffin-like birds abound along the New England and mid-Atlantic Coast. Puffins can be seen in winter as far south as Cape May, N.J., although they do not occur at these more southerly latitudes in large numbers. They are more common in the waters off eastern Long Island to Cape Cod (including Cape Cod Bay) and Maine. These are the same alcids that breed from Maine northward.

Another alcid that breeds even farther north and comes to our waters during winter is the tiny **Dovekie**.

The size of a starling, this little black bird flaps very rapidly and is easy to miss. Like the other alcids, they are seen most frequently when they take flight as a boat approaches.

Winter birds also include a few rarities with names like skua, jaeger, fulmar, kittiwake, and phalarope. Some types of gulls are also seen farther south than usual, including **Little**, **Black-headed**, **Glaucous**, and **Iceland Gulls**. They are most common farther north along the New England and Canadian coasts. The more common **Bonaparte's Gulls** are quite small and are more commonly seen in our waters. These birds can be found around fishing or shellfishing boats.

When you see a flock of several hundred or even a thousand gulls around a fishing boat, you are looking at an opportunity to see some of the more rarely sighted birds. Although most of the birds will be Herring or Great Black-backed Gulls, others will occasionally be mixed in. Such birds may include jaeger and skua, the predators and "bullies" of the seabird community. The **Great Skua**, called "Bonxie" in Europe, is a large bird with a wingspan of about five feet. It is a predatory bird and will kill other birds. It is recognizable by its large size and its dark coloration with white wing patches. **Parasitic** and **Pomarine Jaegers** are also dark, but they are smaller and less predatory. They are obvious in a crowd of gulls because they will usually be darker and they will be chasing the gulls, attempting to make them give up their food.

You may even want to attract flocks of gulls to your boat as a way to sight some of the rare birds. As you probably know, gulls actively search for food around boats; stale bread can draw them in from several miles. You'll get a lot of gulls, but also some other types of birds on occasion. Although this takes a little patience, finding a skua or a fulmar off New York, Maryland, or New Jersey is truly special. A small amount of vegetable oil can also attract birds; it is most successful for finding shearwaters in summer.

Perhaps the rarest of sea birds off our coast are the **albatrosses**. These monstrous sea birds, with wingspans exceeding seven feet, probably visit American waters every year, but few people see them. I know of two sightings of albatrosses (Yellow-nosed and Black-browed) in New Jersey waters. One was seen by a fisherman (and an excellent birder) who was standing on a jetty. The other was seen by a museum curator who was watching birds from the Cape May-Lewes, Delaware ferry. I have heard fishermen talk about seeing huge birds skimming low over the water and I wonder how many of these birds are albatrosses. The possibility of seeing an albatross or another rarity is what inspires many birders to make trip after trip to sea.

Birding from a Boat

To bird from your boat, a pair of seven- or eight-powered binoculars with a wide field of view is essential. Although ten-powered binoculars afford more power for viewing distant birds, it is difficult, if not impossible, to hold them steady. Also, these binoculars are often heavier. Selecting the right pair is a personal process; there are no binoculars that are ideal for all viewers. Try different pairs before you purchase a pair. One final point about binoculars—if you buy a cheap pair, you will probably have to buy another pair before long. This means that you really did not save on the cheap binoculars. Furthermore, you will not get much pleasure from a cheap pair. Remember, sea birds are beautiful and you want to be able to actually see them.

The other investment that you should make is a field guide. Field guides are books that have illustrations, a small amount of text, and maps that help birders identify what they see, tell them what they can expect to see during their trip offshore, and introduce them to the enormous variability among birds. Field guides become like an old

Osprey adults bring food to their young.

© Clyde H. Smith

Sea birds fish at sundown.

friend or a comfortable pair of shoes. As a starter, any of the standard field guides to the birds of eastern North America or all of North America will do very nicely. The Roger Tory Peterson Field Guide to Eastern Birds is the old standby. It is a classic that will serve you for many years. The National Geographic Society field guide to the birds of North America is also an excellent choice. Either of these will get you started and both include many land birds that can be expected over the ocean. For the intrepid boater who wishes to become an expert and who may have thoughts about birds on other oceans, *Seabirds: An Identification Guide*, by Peter Harrison (Houghton Mifflin), is the book. This investment will serve you anywhere in the world. It is a thick book with hundreds of species of sea birds. At first glance you will be overcome by the incredible diversity of birds that can be found on the ocean, but in time you'll gain an appreciation for such diversity.

That's all the additional equipment you'll need, which makes birding a relatively inexpensive pursuit. But don't let that fool you—you will get hours of pleasure and excitement by watching individual birds or migrating birds. And, if you keep watching the sky, you may just see a "river of birds" making the exact same migration as those spotted by Columbus. ✦

Paul Kerlinger is Director of the New Jersey Audubon Society's Cape May Bird Observatory.

Ice and Islands:

E. Robert Thieler

The Diverse GEOLOGY of the Atlantic Coast

Scientific American, 1888

A 19th century depiction of moving the Coney Island Hotel in New York due to barrier island erosion.

In nature, few things stand alone. And so it is that the Ice-Age glaciers echoed their song down the Atlantic Coast of what is now the United States. While they scoured the northeastern coastline depositing land masses in their wake, the glaciers' advances and retreats sent sea levels rising and falling all along the coast. In the farthestmost reaches of present-day Florida, the changing sea levels helped tropical coral reefs and mangroves build islands. And for almost 1,300 miles of Atlantic coastline, the result was a necklace-like chain of barrier islands.

From outer space, it is the barrier islands that stand out, appearing thin and delicate along the coastline. But their appearance belies a strength and flexibility held by few other geologic features on the face of the Earth. While they stretch out like fortresses against the ocean from Long Island to Miami, their everchanging nature is what has helped them survive. And that survival began due to the ice action in the North.

New England's Icy Past

For two to three million years, repeated glaciations molded the New England coast into what we see today. Deposits of glacial debris, called terminal and recessional moraines and outwash plains, dominate the southern part of the New England coast.

The great Wisconsin glacier—named for the state in which its deposits were first recognized—was the last great sculptor of the New England coast. The glacier covered the area north of Long Island some 20,000 years ago. At the time, Maine, New Hampshire, Massachusetts, Rhode Island, and Connecticut were buried beneath 1,500 to more than 3,000 feet of ice. The large ice sheet scoured much of New England completely clean of any overlying sediment, leaving rounded hills of erosion-resistant igneous rocks such as granite. It carved deep valleys where relatively soft metamorphic and sedimentary rocks were present. The power of this ice can be seen clearly in Maine, with its headlands and islands, such as Mount Desert Island in Acadia National Park, and its long, narrow estuaries from Portland to Penobscot Bay.

During the glacier's peak, its weight depressed the continent beneath it. Sea-level was nearly 350 feet lower than today, and the shoreline was located some 100 miles south of Long Island, on what is now the continental shelf. The glacier began its retreat as the Earth began warming. 18,000 years ago, at the beginning of the Holocene epoch (the period of time leading up to today) the glacier had reached the present-day Connecticut coastline. When it began to retreat, the ice sheet left a long pile of debris known as a terminal moraine at its southern extremity. This moraine forms Long Island, the longest island in the country, as well as Block Island, Martha's Vineyard, and Nantucket.

© Joseph T. Kelley

Headlands in Maine etched by glacial activity.

As the glacier continued to retreat northward, it left a trail of recessional moraines, formed as the glacier paused in its retreat, and outwash plains, debris from meltwater streams. Combinations of these moraines and outwash deposits presently form the mainland coastline of Rhode Island and Cape Cod. In fact, the same recessional moraine that runs across southern Rhode Island can be traced east to the Elizabeth Islands off southwestern Cape Cod, through Woods Hole and the Cape Cod Canal, all the way to the Cape Cod National Seashore.

By about 13,000 years ago, the melting ice reached the Maine coast. The ocean flooded the still-lowered coastal areas, extending inland as much as 35 to 40 miles from the present shoreline. The last of the ice had melted out of the Maine mountains by about 12,000 years ago. With the great burden of the ice removed, the land actually rebounded vertically and caused a rapid retreat of the sea. Recent geologic research suggests that the shoreline during this period was located east of Cape Ann, Mass., at a present water depth of more than 150 feet.

As the land stopped its upward rebound, continued worldwide warming and subsequent melting of glacial ice caused the modern drowning of the New England coast. Today, you can see where glacial sands and gravels were worked into islands at Old Orchard Beach, Maine, and Plum Island, Mass. The muddy deposits left by the brief intrusion of the sea into what is now inland Maine have also been reworked by erosion, producing Maine's extensive mud tidal flats.

The Florida Keys: A Tropical Counterpoint

The Florida Keys, too, owe their origin to the great ice sheets of the high latitudes, but in a much different way. As the ice sheets advanced and sea-level fell, much of Florida was exposed. Then it was flooded with warm, shallow seas as sea-level rose during interglacial periods. During the periods of high sea-level, coral reefs flourished laying the foundation for what are now the coastal lowlands and keys of southern Florida. The Keys themselves sit on the high points of old barrier reef systems. There are few natural beaches on these islands; usually the shoreline is bare limestone or lined with mangroves.

The word "key" comes from the Spanish "cayo," which means a small, low-lying island. The Keys consist of 97 low-lying limestone islands extending in an arc around the Florida Peninsula from Miami to Key West. The islands, as well as many of the submerged banks between them, are composed of two limestone formations—the Key Largo Limestone and the Miami Oolite—that formed during periods of high sea-level.

The Key Largo formation is old coral reef with quartz and carbonate sand at its core. A variety of fossil coral heads, as well as the preserved shells of many different clams and snails can be found in the limestone. The Miami Oolite has a much different origin. It is composed primarily of particles called ooids, which are sand-sized, sphere-shaped grains of calcium carbonate. Similar ooid shoals are presently quite common in the Bahama Islands.

The natural environment of the Keys is breathtakingly beautiful and home to many unique and fragile interrelated ecosystems. On the seaward side of the Keys are the barrier reefs. These reefs are the Keys' primary defense against waves. The spectacular branching moosehorn and staghorn corals common here are well adapted to the harsh environment of the barrier reef. Landward of the barrier reefs are the patch reefs, which are small, some-

© Florida Division of Tourism

The Florida Keys sit on the high points of old barrier reef systems.

what circular features found in shallow water. These two types of reefs make up the only large living reef systems in the continental U.S. Behind the Keys lies shallow and muddy Florida Bay, which is an important nursery area for shrimp.

The Barrier Island "Necklace"

Barrier islands comprise nearly three-quarters of our East Coast. They are home to some of the oldest resorts in the United States, including Coney Island, N.Y. and Atlantic City and Cape May, N.J. The Intracoastal Waterway snakes its way among many of these islands and the estuaries and rivers they shelter from the sea.

Barrier islands are long bodies of sand that form offshore islands, separating the open ocean from the mainland. Barrier island chains (strings of three or more barrier islands) front about 2 percent of the world's shores. The longest barrier island chains in the world are found in North America, where they extend like a fine necklace draped from the south shore of Long Island to Mexico's Yucatan Peninsula.

Although delicate and even fragile in appearance, these islands are amazingly durable and flexible. They absorb ocean waves that would otherwise hit the mainland; they protect the wetlands so vital to the health of our coastal waters and offshore marine fisheries. But if the waves are too large or if sea level rises, the islands actually move landward, more or less intact. These attributes make barrier islands one of the most dynamic and some would argue, intelligent!—geologic features on Earth.

A typical barrier island is a maximum of 30 feet thick and its above-sea-level portion is made up entirely of sand. Inlets, some of which are infamous for their navigational difficulties, are the channels that separate adjacent islands. Through inlets, tidal waters from the ocean meet river waters from the continent. Inlets open and close frequently on time scales ranging from decades to millennia. New inlets form during big storms as water returns rapidly to sea from lagoons or sounds.

It is generally assumed that a given chain of barrier islands will always have roughly the same total cross section of inlets. This is because the volume of tidal and river water that must flow through the inlet is more or less constant. The result? When one inlet forms another inlet usually closes.

Some inlets also migrate along the shore. When Oregon Inlet opened on the Outer Banks of North Carolina in 1843, for example, a lighthouse was built on its south side. The inlet began to migrate southward at about 100 feet per year and the lighthouse fell in the water. A second small light met the same fate. Finally, a substantial lighthouse, Bodie Island Light, was constructed on the north side. It didn't fall in, but within a few decades it became useless for navigation because it was too far away from the inlet. Today, the light is nearly two miles north of Oregon Inlet.

On the ocean side of each barrier island is the shoreface, a steeply dipping, concave surface that extends from the beach to the continental shelf. The shoreface typically extends to water depths of 30 to 40 feet, at which point the sea floor becomes much more gently sloping as it merges with the shelf. The shoreface is an integral part of barrier islands. Fair-weather waves push sand landward that eventually becomes dunes. During storms, large volumes of sand move in a seaward direction across this surface. If the storm is big enough, sand will move well beyond the shoreface and out to the continental shelf. Such sand is often permanently lost from the beach and island.

The beach, which is really the uppermost part of the shoreface, is like a giant sand pump. Every grain of sand on the above-sea-level portions of a barrier island once came across the open-ocean beach. Sand comes across either by storm overwash or by wind. In the case of wind-blown sand, the sand may go back and forth to the sea depending on the direction of the wind.

Barrier islands are a product of rising sea level. Thus, once again enter the ice ages as the ocean transgressed and regressed across the Atlantic Coastal Plain. Many of the narrow, subtle sand ridges on the lower Coastal Plain, particularly in Virginia and the Carolinas, are former barrier islands. They were stranded ashore when the sea level receded during a past glacial advance. Many coastal communities such as Myrtle Beach, S.C., and Morehead City, Harkers Island, and Beaufort, N.C., occupy these former islands, taking advantage of the safety afforded by the high elevation.

To understand the evolution of the modern barrier islands, we must, as we did in New England, step back in time some 18,000 years when sea level began to rise to its present position. Along the eastern seaboard, from Long Island to Miami, the shoreline was at or near the edge of the continental shelf. Rivers such as the Delaware and Susquehanna drained across what is now the continental shelf, dumping their sand loads near the shelf edge.

When the river valleys were flooded by the rising sea, estuaries were formed and the formerly straight and uncomplicated shoreline became highly irregular. The ridges between the former river valleys then became headlands projecting out into the ocean. Waves chewed into the headlands; currents distributed the eroded sands upcoast and downcoast forming long sand spits that extended across the mouths of the estuaries. At the same time, the wind was piling up large sand dunes forming ridges parallel to shore.

Two events, one relatively frequent and one gradual, eventually caused the sand ridge-covered spits to become islands. Storms, which occur almost daily on a geologic time scale, eventually breached the spits from land, leaving them to fend for themselves. The breaching of Nauset Spit at Chatham, Mass., in 1987 is an excellent modern illustration of this process. Simultaneously, the gradual sea-level rise, driven primarily by the melting ice in the high latitudes, flooded the lowlands behind the sand dune ridges, isolating them as islands. This is happening today behind the Currituck Peninsula near the Virginia-North Carolina border.

Courtesy Duke University

A view from space of North Carolina barrier islands.

Once the spits became islands, a whole new set of evolutionary processes took over. The main sediment supply was no longer the adjacent eroding headlands. Instead, sand pushed across the shoreface by waves became the main sediment source. Rivers no longer supplied sand to the shoreline since their mouths were miles inland at the heads of estuaries. As sea level rose, the newly formed islands adjusted to their new source of sand and began to migrate in a landward direction.

Barrier islands respond in predictable and sensible ways to all kinds of natural events such as storms, rising sea levels, retreating shorelines, migration of inlets and many other events. Waves, tides, wind, sand supply, sea-level change, underlying geology, and vegetation all play a role in the barrier island scenario.

Waves carry sand from the continental shelf or adjacent islands to the barrier islands. Storm waves then wash over the island and bring huge volumes of sand into the interior of the island.

© John D. Pearce

Looking down a barrier beach in Florida.

Over time, wind is capable of carrying large volumes of sand from the beach into the dunes. Prevailing directions of wind are important in determining the overall volume of sand. This is demonstrated by the difference in width and sand volume of two adjacent islands in the Cape Lookout National Seashore; Shackleford Bank and Core Bank, N.C. The dominant winds in this area are from the north during winter and the south in summer. Shackleford Bank is a wide, high, sand-rich island oriented east-west. Thus, for significant periods of time each year, sand is blown into the vegetated dunes. Core Bank, a narrow low-lying island with few dunes, is oriented north-south. On most days of the year, sand blows either up or down the island but not across. As a consequence, little sand is blown from the beach to the island interior and few dunes form.

Tidal amplitudes (half the tidal range) also affect barrier islands. Some studies suggest that tides control the spacing of inlets, which is another way of saying that tides control the length of islands. With large tidal amplitudes, a large volume of water must flow in and out of inlets. The inlets are closely spaced; the islands are short. The Georgia islands, where normal tides are seven feet, are good examples of short islands. With small tidal amplitudes, less water must flow in and out of the lagoons; the inlets are widely spaced and the islands are long. The North Carolina Outer Banks and the Florida shoreline north of Cape Canaveral have long islands and the tide range is less than three feet.

Sea-level change plays a fundamental role in island migration. As already mentioned, barrier islands are a product of a rising sea level. No sea-level rise means no valleys flooded to form estuaries, and no islands built across the mouths of estuaries.

The underlying geologic framework of barrier islands controls the location of some islands. Islands underlain by mud, for example, are prone to more rapid retreat than are islands underlain by sand. Some Delaware islands have incorporated the core of old ice-age barrier islands formed more than 100,000 years ago. On the Outer Banks of North Carolina, stretches of gravel beaches almost always correspond to the offshore location of old deposits of river gravel laid down when the sea level was lower.

Vegetation also plays a major role in barrier island evolution. For example, onshore winds blow lots of sand into the islands but when the winds reverse and blow offshore, not nearly as much sand is carried towards the sea. Dune vegetation holds the sand in place. Sea Oats, the southern U.S. dune plant, tends to form in clusters. As a result, there are gaps in the dunes used by storm overwash to bring sand onto the island and build up its elevation. American Beach Grass, found in the northern islands, propagates in such a way that it causes continuous dunes to be built up without gaps, as on Plum Island, Mass. The lack of dune gaps prevents most storm overwash and the islands do not build up elevation as efficiently as southern islands.

Since essentially all barrier islands worldwide are eroding, we could justifiably conclude that they are all disappearing. If that was the case, we have indeed been fortunate to exist just at that moment in geologic time when barrier islands exist. But the theory of barrier island migration spares us from this somewhat myopic view.

Island migration wasn't even recognized until a couple of decades ago. During the 1930s and 1940s, shoreline erosion became evident on the North Carolina Outer Banks and a massive dune building effort was begun to save the barrier islands. A continuous dune was constructed almost from the Virginia border to well beyond Cape Hatteras. Then, during several studies done in the 1960s by the National Park Service, it became apparent that the islands were not eroding away. Instead, they were actually migrating and shoreline erosion was part of that process! The Park Service announced in 1971 that henceforth they would keep their hands off barrier islands and let nature continue its work.

The most efficient and rapid mechanism of island migration involves the simultaneous erosion of the open ocean side and overwash across the lagoon side (the landward building out of the lagoon shoreline). During times of rapid sea-level rise this must have been the primary way that barrier islands survived and migrated.

The mainland shoreline must also move back or the islands will no longer be islands. In fact, once in a while the migrating islands do catch up with and "smash"—in a geologic sense—into the mainland. That is what happened at Myrtle Beach, S.C. where the mainland fronts the open ocean. Given enough time and enough sea-level rise, it's probable that the island will break away from the mainland a few thousand years from now. Along the northern New Jersey coast are a series of short headlands separated by strings of barriers. Before the construction of seawalls during the last 100 years, these headlands were eroding and thereby furnished sand to the adjacent islands. The Sea Islands of Georgia are really two islands in one, a modern one and an ice-age one formed many thousands of years ago. It just so happens that at this instant in geologic time the modern islands have just migrated into and joined up with the older islands.

The best illustration of the potential speed of barrier island migration is illustrated by the northern end of Assateague Island, Md. Until 1933, this island was one and the same island as neighboring Ocean City. In that year, a hurricane formed a new inlet which was immediately fixed in place by the installation of jetties by the U.S. Army Corps of Engineers. The jetties dramatically reduced the sand supply for Assateague Island. Sand destined for Assateague instead built out the beach along southern Ocean City. The new island to the south responded to its loss of sand supply by immediate and dramatically rapid landward migration. Now the open ocean surf zone in front of Assateague Island is landward of where the lagoon shoreline was in 1933. In just 30 years this island has moved several hundred yards, a complete island-width, landward! Within a couple of decades, migrating Assateague may begin to fill in the Intracoastal Waterway behind it. Farther south on the Delmarva Peninsula, several islands are migrating landward almost as rapidly as Assateague, but with little influence of humans involved.

A fundamental change in barrier island processes has occurred within the last few hundred years. Virtually all barrier islands in the world are eroding on their ocean side. On islands with open water (as opposed to salt marsh or mangroves) behind them, erosion is also occurring on the lagoon side. Observations of undeveloped barrier islands everywhere show that the islands are slimming down, even those islands that were building seaward just a millennium ago.

Two causes have been suggested for the narrowing of barrier islands. One is that humans, by damming rivers, jettying inlets, seawalling beaches, and dredging channels and harbors, have brought on a sand crisis. The shortage of sand is causing worldwide erosion. Another view is that the island thinning is due to a recent increase in sea level. If this is the case, the narrowing distances between ocean and lagoon will bring on rapid and frequent cross-island overwash and allow rapid island migration in the future. Thus, this reasoning holds that the islands are effectively preparing for an upcoming acceleration in the rate of sea-level rise. Very likely, both of the suggested causes are behind the erosive state of barrier islands.

While many areas along the U.S. East Coast are unaltered by people and provide unique views into the processes that continue to shape the shoreline, there is a common theme that runs along the coast: we are very much in danger of loving it to death. To protect our buildings on barrier islands from the ravages of shoreline retreat, we build seawalls, groins, revetments and jetties. In a few decades, these structures (especially seawalls) cause the destruction of the beach.

When storms come and waves wash thick layers of sand onto the interior of the island, we clean off our streets and carry the sand back to the beach or to the nearest concrete plant. The island's preparation for sea-level rise is thus completely negated. Instead of allowing the islands to prepare for sea-level rise, we try to hold things neatly in place.

When we dredge sand from inlets and dump it far out to sea on the continental shelf, we take sand directly away from barrier islands and alter the sand supply. And the jetties next to inlets also increase the erosion rates. To stem the jetty-caused erosion, we build seawalls and groins and wipe out the beaches.

The legacy of our poor interaction with dynamic barrier islands is a long and never-ending story that gets worse each year. Unfortunately, it will very likely not end until a truly big storm hits a truly dense island population. The building boom of the last two or three decades has coincided with a lull in severe hurricane activity along the U.S. Atlantic Coast. But meteorologists predict that we are returning to a period of more frequent and more intense Atlantic hurricanes. The potential for a major disaster is huge. Such a disaster would undoubtedly slow down the mad rush to our barrier islands. Then, perhaps we will rethink the way in which we interact with these beautiful, dynamic, and important wonders. ✦

CONSERVATION *of* *a* COASTLINE

Roger D. Stone

Our Atlantic Coast remains a biological hotspot. Its wildlife ranges from turtles nesting on beaches to migrating shorebirds. The Chesapeake Bay area alone supports some 3,000 plant and animal species. Estuaries, such as Long Island Sound and North Carolina's Albemarle-Pamlico region, can produce far higher per-acre yields of protein than the richest farmlands. Saltmarshes act as nutrient-rich nurseries for plants and animals. And the coral reefs growing in the warm waters of the Florida Keys show off the diversity of life.

Yet no one can deny it—the rapid but disorderly growth of the U.S. economy has placed great stress on the ecosystems of the coastline. Sewage from cities and waterfront homes, toxic effluents from factories, and chemical-laden runoff from farms, golf courses, and yards are only a few of the forms of pollution draining from the land into the rivers, estuaries, and bays along the coast. And sediment from eroding soil silts up harbors and stifles aquatic plant life. The situation is serious, but it is far from hopeless.

History shows us the ample abundance of the region. Along many stretches of the Atlantic Coast, aboriginal Indians didn't need to farm. They lived quite well off the plants and animals on land or close offshore, all readily available for harvesting. From Ponce de Leon to Giovanni da Verrazano to Captain John Smith, early European explorers of the coastline were almost unanimously euphoric in their expressions of wonder about the glorious natural world they encountered.

Late in the seventeenth century, Jasper Danckaerts, a Dutch traveler to Brooklyn, NY, feasted on local oysters "not less than a foot long." William Wood, another seventeenth-century chronicler, enthused over 20-pound lobsters and 18-foot sturgeon, and said of Massachusetts oysters: "This fish without the shell is so big that it must admit of a division before you can well get it into your mouth." So rich was God's blessing on Long Island, wrote yet another tourist of the time, that settlers there could "live as happily as any people in the world."

Today, the entire coastline still holds much of its historic beauty. Between the two ends of the coastline lies a considerable diversity of climate, topography, vegetation, and wildlife. Some species are claiming ever larger shares of the total territory. The brown pelican and the glossy ibis, traditionally "southern" birds, have extended their ranges far northward. The great black-backed gull, once a rock-ribbed New Englander, now breeds ever farther to the south. But the coast can still be divided ecologically into sub-regions, each one with its own distinctive ecosystem characteristics.

Northern New England

The rockbound coast of northern New England, from Maine to Cape Cod, forms one region. In 1498, the Italian seaman John Cabot, in command of a British ship, reached the Maine coast in search of the mythical kingdom called "Norumbega." During the sixteenth century many sailors followed in Cabot's wake. But it was not until 1605, when Captain George Waymouth anchored in Georges Harbor, that a European navigator actually touched down along the Maine coast.

When Waymouth described ample nearby supplies of "great Cods and Hadocks that were well fed, fat, and sweete in taste," he became one of the first to record the quality of the Gulf of Maine's marine ecology. Strong currents from the Bay of Fundy, where the tide can rise and fall 20 feet or more, roil this shallow sea. Freshwater flowing in from Maine's many rivers adds to the turbulence. The result is an excellent habitat for plankton—small marine organisms that feed vast numbers of finfish, shellfish, and even a colony of right whales that summers near Brier Island on the Bay of Fundy's Canadian side.

Other marine mammals feast on the smaller fish. Waterfowl and seabirds—including puffins and razorbills at the southern extremity of their ranges—also fare well in what is frequently described as one of the world's richest marine environments.

Once past Portland heading west, you encounter progressively lower shores, sandy beaches, and evergreen giving way to deciduous woodlands. But all the way down to Cape Ann near Gloucester, Mass., the coast still emits strong echoes of Down East. Whale watching has become big business in Massachusetts Bay, where large congregations of humpbacks and finbacks are easily found during the height of the summer tourist season. And the water stays cold even as far south as the low, shallow north shore of Cape Cod where the Pilgrims stepped ashore in 1620.

Cape to Cape

Cape Cod marks the beginning of a second section of the coast that extends southward past Cape Hatteras to the South Carolina border. This shoreline is strung with barrier islands. In places where these outer islands are protected from human intrusions, they serve as noisy breeding grounds for gulls, terns, and less familiar species such as Wilson's Storm-Petrel and the Piping Plover. Dunes clad with distinctive vegetation rise from the mainland shore. Within this mid-Atlantic zone are no fewer than five important, if badly buffeted, estuarial systems—Long Island Sound, the Hudson, the Delaware, the Chesapeake, and North Carolina's Albemarle-Pamlico region.

For the boater moving southward from Massachussets, the warm shallow waters of Buzzards Bay offer a convenient passageway into this complex environment. Migrant striped bass and ravenous bluefish keep summer anglers well occupied in these waters.

Ahead lies Long Island, a low-lying, 128-mile-long sandbar. Although now home to some three million people, the island still harbors rare orchids and other treasures in the 100,000 acres remaining of undeveloped pine barrens. Sections of adjacent Long Island Sound sometimes suffer greatly for want of oxygen during the summer, victim of torrents of pollution flowing from Connecticut's farms, industries, and cities. The Sound's shellfisheries have dwindled and the succulent bay scallop has vanished from many areas. Somehow, though, the lobster population is staging a valiant comeback. And many finfish species continue to thrive in the Sound's well-traveled waters.

The New York Bight, with its great underwater canyon scoured by the Hudson of ancient times, also remains a rich fishery. The area is particularly notable for the schools of migratory bluefin tuna that reliably congregate here each year from July through September. Some weigh 1,000 pounds and fetch prices of more than $10,000 for the sashimi market in Tokyo. The Hudson, too, remains an important habitat for fish species, such as striped bass and shad which move from salt to fresh or brackish water to spawn. Benefitting from tight regulation, stripers have swarmed back into the river in recent years. Small numbers of caviar-producing sturgeon still inhabit these waters as well.

Delaware Bay, past Cape May, seems more a habitat for large tankers and freighters than wildlife. Yet each spring, there occurs on these shores one of nature's great miracles. A million or more ponderous horseshoe crabs, relics of the dinosaur era, emerge from the water to lay eggs near the high-tide line. The event coincides with the arrival of hordes of tired and hungry migrating shorebirds (principally sanderlings, red knots, and ruddy turnstones) that gain back their strength from these greenish eggs.

Until recently, few coastal ecosystems anywhere were more productive than Chesapeake Bay with seemingly inexhaustible supplies of oysters, clams, and finfish. A century ago, Chesapeake blue crabs could be bought, literally, for a dime a dozen. At the same time, the popularity of crabs has caused the Chesapeake to be heavily fished. Still today, this estuary produces enough of its last big cash crop for crabs to remain a staple and beloved (if increasingly costly) foodstuff in Baltimore and Washington.

From Assateague to Cape Charles, the barrier islands of Virginia's western shore are either public property or kept in near-virgin condition by The Nature Conservancy. If Captain Smith would recognize these places, he would be less at ease farther south. With the exception of the Hatteras and Cape Lookout National Seashores and a few other enclaves, the Carolina coasts have become heavily settled. Much money is spent in efforts to maintain the contours of this shoreline, to preserve the famous candy-striped Cape Hatteras lighthouse from falling into the sea, and to keep the inlets open for commercial marine traffic.

© Mary Jane Hayes

Nature and industry exist side by side on the East Coast.

Within the North Carolina inlets is the extensive Albemarle-Pamlico estuarine system. Waterfowl and shellfish still thrive in these quiet black waters and herons of many species stalk their marshes. Off the beaten track these days because of changes in the coastal inlet pattern, these sounds remain cleaner than the Chesapeake. At sea, to the south and east of storm-battered Cape Hatteras, are the fearsome Diamond Shoals. There, more than 600 ships have been driven ashore and, with the northward-flowing Gulf Stream again running close by, fishermen harvest resident populations of such tropical species as the tasty grouper.

The Lower Coast

In South Carolina and Georgia, the coast turns marshy and muddy. Palmettos grow in sandy soils on offshore islands; black skimmers shave the wavetops to feed inshore; in narrow channels through the marshes alligators flop into the water at the sound of an approaching boat. Gnats and mosquitoes abound. Developed islands, such as Hilton Head and St. Simons with its grand old Sea Island resort, alternate with little changed islands such as Daufuskie. Near Charleston, SC, is the virtually beachless Folly Beach which, like Ocean City, Md., must either spend millions to replenish its sands or risk losing its flocks of summer frolickers.

Florida, a 58,560-square-mile limestone sponge, is an ecosystem of unique importance, badly wounded by overdevelopment but still a place of exquisite beauty and wonder. Herds of ponderous sea turtles, manatees, as well as an extremely colorful assortment of various birds and marine life share the state's East Coast with ever more numerous swarms of human residents and tourists as well as with the deep blue water of the fast-moving Gulf Stream. Farther inland, the 200-mile "river of grass" flows gently into the swampy Everglades. This highly distinctive national park, home to great flocks of roseate spoonbills, white ibis, and snowy egrets, has undergone extreme stress from the adjacent sugar-cane industry as well as from the rapidly-growing city of Miami nearby.

All the way from Quoddy Head to Key West, the coastline has been despoiled in a variety of ways. When tankers run aground and oil spills, or when stormy seas damage or destroy shorefront apartment buildings, the degradation becomes glaringly visible. More often, it remains hidden under tranquil blue water, its diverse causes stemming as much from the changing chemistry of the atmosphere and runoff from faraway farms and factories as from the hubbub of nearby urban and resort development. No brief account can provide a full description of how these interconnected forces affect the Atlantic's coastal waters, yet they can be effectively summarized as follows. Chief among the causes and consequences are:

Resource Abuse and Land Misuse. Harvesting of marine species beyond tolerable limits has become dishearteningly commonplace. The pattern is true of species ranging from the once-plentiful haddock and pollock in the North Atlantic to queen conch and spiny lobster, which have all but vanished from the Florida Keys. Habitat destruction at the water's edge has been as damaging as overharvesting to the well-being of marine populations. Construction of cities, towns, resorts, and residential communities often involves the sacrifice of wetlands, mangroves, or other areas of particular importance to marine and bird life.

Unwise Beach Development. Along many parts of the coast—New Jersey, Maryland in the Ocean Beach area, Virginia Beach and the Carolinas, not to speak of Florida—houses or condominiums are often built on sites that are dangerously exposed to storm damage and erosion, a phenomenon already afflicting 95 percent of the U.S. shoreline as the sea level continues its rise. Where natural erosion is exacerbated by overdevelopment, often efforts to "nourish" beaches with new sand, or protect them through the construction of jetties or groins are counterproductive. Bulwarks constructed to protect coastal lands from the rising seas also prevent saltmarshes from naturally "migrating" inland as they become inundated by the rising sea.

In many instances, notably along North Carolina's Outer Banks, those who buy the new beach condominiums face unrecognized but possibly fatal threats. Road facilities to evacuate people from areas with storm flooding danger are often inadequate. Occupants of many coastal high-rise buildings are also exposed to fire hazards because of the access difficulty, especially when storms flood roads.

Pollution. Physical damage caused by coastal development represents one aspect of the problem. In the notorious example of Boston Harbor, one consequence of industrial development is toxic pollution. Some of it comes from factories disposing of excess chemical contaminants such as polychlorinated biphenyls (PCBs) or polyaromatic hydrocarbons (PAHs), or toxic heavy metals (e.g. zinc, cadmium) by dumping them into adjacent waters. These substances settle to the bottom of bays and harbors. Stubbornly resisting biodegradation, they persist in heavy concentrations and poison many forms of marine life. Tumors and lesions found on bottom-feeding fish and crabs provide stark evidence of the damage. In dredging operations, toxic substances are sucked up from the seafloor and redeposited where they may do additional harm.

With industrial usage of toxics under tightening control everywhere along the coast, an increasing share of its toxic pollution comes from what are called nonpoint sources. Runoff from golf courses, farms, and even lawns contains residues from toxic pesticides and insecticides. Stormwater flowing from city streets, freeways, construction sites, parking areas, and paved driveways carries oil, grease, lead, and other toxic metals. Such runoff also contains human and animal wastes that contribute bacteria, viruses, and other pathogens to adjacent waters. Fish and shellfish contamination can lead to hepatitis, gastroenteritis, and other human illnesses. According to one recent Environmental Protection Agency (EPA) report, half an inch of rain in Providence, R.I., dumps so many toxics and pathogens into Narragansett Bay that parts of it are closed to shellfishing for a week afterwards. Airborne toxic chemicals from as far away as Egypt have been found in Chesapeake Bay. Everywhere, the air is increasingly a pathway for coastal pollution.

Another pollution problem results from the flow of nutrients commonly used as fertilizers—nitrogen, phosphorus, potassium—into Atlantic coastal waters. They appear as part of the runoff from farms, contained in raw or only partially treated human or animal waste or, once again, from atmospheric sources. In normal quantities these nutrients are vital to the growth of aquatic plants that play important roles in estuarial ecosystems. In excess, they increase the rate of growth of algae and other very small and primitive plants or animals, sometimes resulting in explosive "blooms" or "tides" of them. In very thick concentrations, they color the water red, brown, or yellow depending on the species.

As they decay, algae soak up oxygen, already scarce because of the amount used in the biochemical oxidization of the nutrients themselves. Oxygen depletion in the water may kill fish and other species. Mass die-offs can occur under extreme circumstances. When surface algea blooms, it clouds and darkens the water; ecologically important seagrass beds in many regions have consequently died for want of sufficient sunlight.

Certain species of dinoflagellates, a class of organism that may proliferate when nutrient supplies grow, manufacture a neurotoxin that can enter the food chain and stun or even kill people who eat infected shellfish.

The good news is that there is plenty of citizen action and political pressure to stabilize the situation along the coastline. No place along this shoreline suffers from the lack of careful scrutiny and recommendations for environmentally sound development plans. Among the principal techniques being employed by citizens, public agencies, or partnerships are:

Scientific Research. This the basis for most wise action. Along the coast, an integrated network of institutions studies and records local as well as global change, and feeds findings into the decision-making process. The College of the Atlantic in Bar Harbor, Maine, and the New England Aquarium in Boston specialize in whale studies.

The Woods Hole Research Center and the Woods Hole Oceanographic Institution are among those studying the pressures on marine life in Buzzards Bay and its rivers and harbors. The Marine Sciences Research Center at the State University of New York (Stony Brook) has done extensive work on Long Island Sound. And so it goes all the way down to Florida. The workload is often shared between distinguished academic centers, such as the University of Maryland's Laboratory for Coastal Research in College Park and the Duke Marine Lab in Beaufort, N.C., and federal efforts such as NOAA's wide-ranging studies and the EPA's estuaries program.

Saving Individual Species. With the science in hand, sharply focused private groups can take on the challenge of trying to prevent the disappearance of threatened forms of life. Notable among these efforts are those to keep Florida's 1,200 remaining manatees from being chewed to death by boat propellers; the Florida-based Caribbean Conservation Corporation's determined program to safeguard beaches where sea turtles nest; the Baltimore-based Committee to Preserve Assateague's valiant fight to protect the Piping Plover nesting areas on the heavily-used island; and the Maine Audubon Society's successes in repopulating islands along the Maine coast with puffin and razorbill breeding colonies.

At many points along the coast people have worked hard to help ospreys, brown pelicans, and other affected birds continue to recover from the ravages of the insecticide DDT. Often single species programs such as these help protect whole ecosystems.

Saving Critical Habitats. Back in the 1980s, Lena Ritter of the North Carolina Coastal Federation waged and won a stirring fight to keep a little island called Permuda, which sits astride the Intercoastal Waterway, from being consigned to hotel development. The victory she and fellow residents won, which saved adjacent shellfishing grounds and the livelihood for many local people, is but one of scores of similar efforts from one end of the coast to the other.

In certain instances, as in that of the Critical Area program established in Maryland for the sake of Chesapeake habitats, state governments and agencies have pitched in as well. At the John Pennekamp State Park in Florida, recently installed guest moorings enable boaters to enjoy the reefs without killing coral by dragging their anchors through it. The Interior Secretary fashioned a compromise agreement to shore up the besieged Everglades National Park; it won praise both from environmentalists and from sugar growers in the region. This is but one example of new federal determination to practice "ecosystem management" with regard to coastal landholdings. Another is the performance of the reborn U.S. Army Corps of Engineers, once a dredge-we-must agency with sweeping powers, now ever more an ally of environmentalists.

Countering Helter-Skelter Development. On Cape Cod in the late 1980s, voters approved a resolution calling for a moratorium on development until planning could be completed. Few major hotel or condominium projects on North Carolina's Outer Banks go unchallenged by spirited and politically adept local groups. Florida has a useful law obliging local municipalities to present environmentally sound development plans before they become eligible to receive state funds for key public works. The federal Coastal Zone Management Act, though badly underfunded, is a creative piece of legislation providing for federal-state-local partnerships to conduct ecologically sensitive planning in coastal areas.

Federal and state fishery management programs, often hotly contested by the fishermen themselves, seek to restore populations of depleted species. The boating and marina industries, which for too long avoided the problem of yacht pollution, have launched new efforts to make it easier for individual boaters to comply with waste-management regulations. Regulatory agencies have gotten better at applying sensible controls to what comes out of smokestacks and drainpipes.

For all the positive forces at work, there are also strongly negative ones. Precisely because of its continuing appeal, more and more people want to crowd against the Atlantic shore. Beachfront villages become towns. Towns like Maryland's Ocean City achieve six-figure populations, and then sprawl stretching outward from big cities gobbling up vast stretches of coastline. Land developers, sometimes aided and abetted by dishonest local politicians, prefer to chew into virgin land rather than recycling parcels that have already been used. In recent cases such as *Lucas vs. South Carolina Coastal Commission*, subject of a recent and much-debated Supreme Court ruling, individual property rights have won out over broader community concern for environmental quality.

Preserving the remaining ecological values of the Atlantic coast and restoring it where possible to the levels of abundance at which the early European explorers marveled—these will be major challenges for the years and decades ahead. ✦

Roger Stone is vice-chairman of ECO, a business magazine about the environment, and author of numerous books, including the The Voyage of the Sanderling.

INTERNATIONAL FLAGS AND PENNANTS

ALPHABET FLAGS

Flag	Code	Meaning
Alfa	•—	Diver Down; Keep Clear
Bravo	—•••	Dangerous Cargo
Charlie	—•—•	Yes
Delta	—••	Keep Clear
Echo	•	Altering Course to Starboard
Foxtrot	••—•	Disabled
Golf	——•	Want a Pilot
Hotel	••••	Pilot on Board
India	••	Altering Course to Port
Juliett	•———	On Fire; Keep Clear
Kilo	—•—	Desire to Communicate
Lima	•—••	Stop Instantly
Mike	——	I Am Stopped
November	—•	No
Oscar	———	Man Overboard
Papa	•——•	About to Sail
Quebec	——•—	Request Pratique
Romeo	•—•	
Sierra	•••	Engines Going Astern
Tango	—	Keep Clear of Me
Uniform	••—	Standing into Danger
Victor	•••—	Require Assistance
Whiskey	•——	Require Medical Assistance
Xray	—••—	Stop Your Intention
Yankee	—•——	Am Dragging Anchor
Zulu	——••	Require a Tug

REPEATERS

1st Repeat

2nd Repeat

3rd Repeat

CODE — Code and Answering Pennant (Decimal Point)

NUMERAL PENNANTS

1	•————
2	••———
3	•••——
4	••••—
5	•••••
6	—••••
7	——•••
8	———••
9	————•
0	—————

Weathering the Atlantic Coast

Atlantic Coast weather can challenge your navigation skills. Local conditions are as varied as the states from Maine to Florida, but certain weather conditions are common to the coastal area. They range from "Nor'easters" to hurricanes, fog to rain, and calms to windy cold fronts.

Fog and Rain

Visibility is generally fairly good along the coast, but it can be hampered by **fog**, **rain**, and **haze**. Fog is usually the most restrictive to navigation, often reducing visibility to less than a quarter of a mile.

Fog forms when warm, moist air moves across colder water **(advection fog)**, when very cold air moves over warmer water **(steam fog)**, or when moist air is cooled to near its dew point by radiation **(radiation fog)** or rainfall **(precipitation fog)**.

The entire Atlantic Coast is susceptible to radiation-type fog often forming shortly after sunset on cool, calm, clear nights. This fog generally does not extend any great distance seaward, but may seriously restrict harbor activities. It usually burns off during the morning hours.

Waters north of Cape Cod, and especially Maine, are known for coastal fog, a type of advection fog. In summer, decreasing water temperatures combined with warm air from the south or southwest can create large patches of fog that last for days. Advection fog also occurs farther south due to the warm airs of the Gulf Stream blowing over the cooler near-coastal waters.

Fog often occurs over land and water in opposition. So don't use poor visibility at inland or sheltered harbors as a guide to conditions at sea or in approaches. The areas at the heads of bays and rivers may be clear while the fog is quite thick outside.

In addition to fog, **rain** can reduce visibility over both land and water along the coast while haze and smoke can restrict visibility over land.

Cold Fronts

These often sharp boundaries between relatively warm air and cold air usually approach the Atlantic Coast from the west and head north. Ahead of the front, winds are usually squally and often blow out of the south through southwest. Atmospheric pressure falls moderately; seas become choppy; showers, and sometimes thunderstorms, appear.

During the spring and summer, a line of thunderstorms, known as a squall line, may develop. These instability lines can form 50 to 300 miles ahead of a fast-moving front and may contain tornadoes or waterspouts.

Once the front passes, winds shift rapidly to the west and northwest. Strong gusts and squalls continue. Clearing usually occurs a short distance behind the front as the cold air moves in. Cold fronts can move through quite rapidly. Their frequency decreases with latitude.

Nor'easters

These winter-type storms are most intense from fall through spring. Called extratropical cyclones, they generate hurricane-force winds and vary in size from 100 miles to nearly 1,000 miles in diameter.

Many such systems develop in the Gulf of Mexico, move across central or northern Florida, then up the East Coast. While these storms are usually well forecasted, they can develop explosively, particularly off Cape Hatteras.

The most dependable early indicator of such a storm is falling atmospheric pressure. A definite weather change is likely if you observe pressure falls exceeding 2 mb every 3 hours; a drop of 5 mb in 3 hours indicates a strong change; 10 mb in 3 hours warns of an impending extreme event.

December through February is the heart of the storm season when, even in southeastern waters, an average of 4 to 6 storms per month arise. In more northern states, especially Maine, as many as 10 of these storms can hit the area in a single month.

Tropical Cyclones: Depressions, Storms, and Hurricanes

The most disastrous storms of the North Atlantic are the notorious tropical cyclones—better known as hurricanes in their fully developed stage. Tropical cyclones include tropical depressions, tropical storms, and hurricanes.

A hurricane starts as a depression and escalates to a tropical storm before it reaches its fully developed hurricane stage. In small hurricanes, the diameter of the area of destructive winds may not exceed 25 miles. In some of the greatest storms, the diameter may be as large as 500 miles. At the center is a comparative calm known as the "eye of the storm," which may be as little as 7 miles in diameter but is rarely more than 30 miles.

An average of nine tropical cyclones form each year, reaching at least tropical storm density. About five of these reach hurricane strength, and about two reach the United States. The entire Caribbean area, the Gulf of Mexico and its the coastal regions, and the Atlantic Coast are subject to these storms during the hurricane season, from June through October (peaking in September). Infrequently, hurricanes have developed in May and November, as well as during other months.

A hurricane is to be avoided at all costs. Fortunately, tropical depressions are well tracked and reported on by the National Hurricane Center in Miami, Fla., warnings are issued using the following terminology:

Tropical Depression: A tropical cyclone in which the maximum sustained surface wind is 33 knots or less.

Tropical Storm: A tropical cyclone in which the maximum sustained surface wind ranges from 34 knots to 63 knots.

Tropical Storm Warning: A warning for tropical storm conditions are expected in a specified coastal area within 24 hours or less.

Hurricane: A tropical cyclone in which the maximum sustained surface wind is 64 knots or more.

Hurricane Watch: An announcement that hurricane conditions pose a possible threat to a specified coastal area within the next 36 hours.

Hurricane Warning: A warning that sustained winds of 64 knots or higher are expected in a specified coastal area within 24 hours or less.

What to Do in a Hurricane

Here are some general guidelines to follow when a hurricane is forecast.

If you are at or close to shore:

- Monitor the weather reports closely. You can track the hurricane's course using the coordinates given by the National Hurricane Center to determine if it will hit your area.
- If possible, move your boat into a hurricane hole, or well back into a winding river or canal.
- Keep your boat as far away from other boats as possible.
- Run additional dock or mooring lines. Either use bow and stern anchors and lines to keep the boat well away from docks, or keep it close to the dock and rely on fenders to protect the boat.
- Use plenty of chafing gear in chocks. Use two or more anchors.
- Seal windows, door, and hatches with duct tape to keep boat watertight.
- Shut sea cocks. Plug unvalved through-hulls.
- Remove anything on deck that can be blown away.
- Remove any loose items from the cabin.
- DO NOT stay on the boat.

There are three schools of thought concerning the handling of a boat while at sea during a hurricane. They are:

- Active: Take an evasive course and get out of the storm.
- Passive: Shut down and wallow.
- Defensive: Just maintain steerageway, keeping the wind and seas either off the bow or the quarter.

Weather Broadcasts

Your best access to up-to-date weather information is through the broadcasts by NOAA's National Weather Service. Coastal marine warnings and forecasts are issued every six hours over VHF-FM radio. On the East Coast, broadcast times are at 4:30 a.m., 10:30 a.m., 4:30 p.m., and 10:30 p.m. (add one hour during daylight savings time). These forecasts include winds, sea heights, and weather conditions for the next 36 hours, as well as small craft advisories, gale warnings, storm warnings, and tropical storm or hurricane warnings.

Continuous taped weather transmissions are also broadcast 24 hours a day. They include weather and radar summaries, wind observations, visibility, sea and lake conditions, detailed local forecasts, and other information. The transmissions repeat taped messages every four to six minutes; they are updated every two to three hours. These routine transmissions are interrupted for severe weather warnings.

Weather broadcasts for the Atlantic Coast are received primarily on the following VHF channels: WX-1 (162.559 MHz); WX-2 (162.400 MHz); WX-3 (162.475 MHz). You will usually be able to receive the broadcasts within 40 miles of the transmitting antenna.

You can receive Marine Services Charts (MSC) for your area that include locations of weather radio transmitters, Coast Guard stations, and marine observation sites as well as locations, times, and frequencies of marine broadcasts from the National Ocean Service, Distribution Branch (N/CG33), Riverside, MD 20737; (301) 436-6990.

Marine Warnings and Displays

Warning flags and symbols are an immediate signal for caution. Use these as an adjunct to other information, not as a primary source. The warnings, flags in daytime and lights at night, are displayed at prominent points along the coastline, such as Coast Guard stations, lighthouses, and docks. They are as follows:

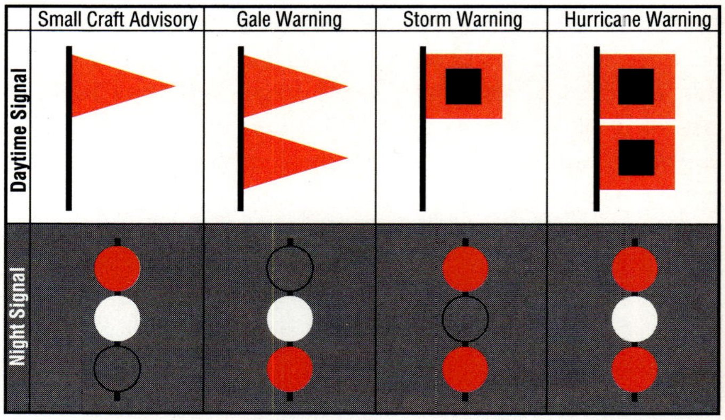

Weather Maps

You can also get weather information from maps in newspapers or on television, or charts transmitted by the National Weather Service by radiofacsimile. To receive the NWS charts by radiofacsimile, you can tune into the proper frequency for your area during the correct transmission time. For the frequencies and schedules for different areas, order the NOAA publication "Selected Worldwide Marine Weather Broadcasts" from the Government Printing Office by calling (202) 783-3238.

Embassy's Hurricane Tracking Chart

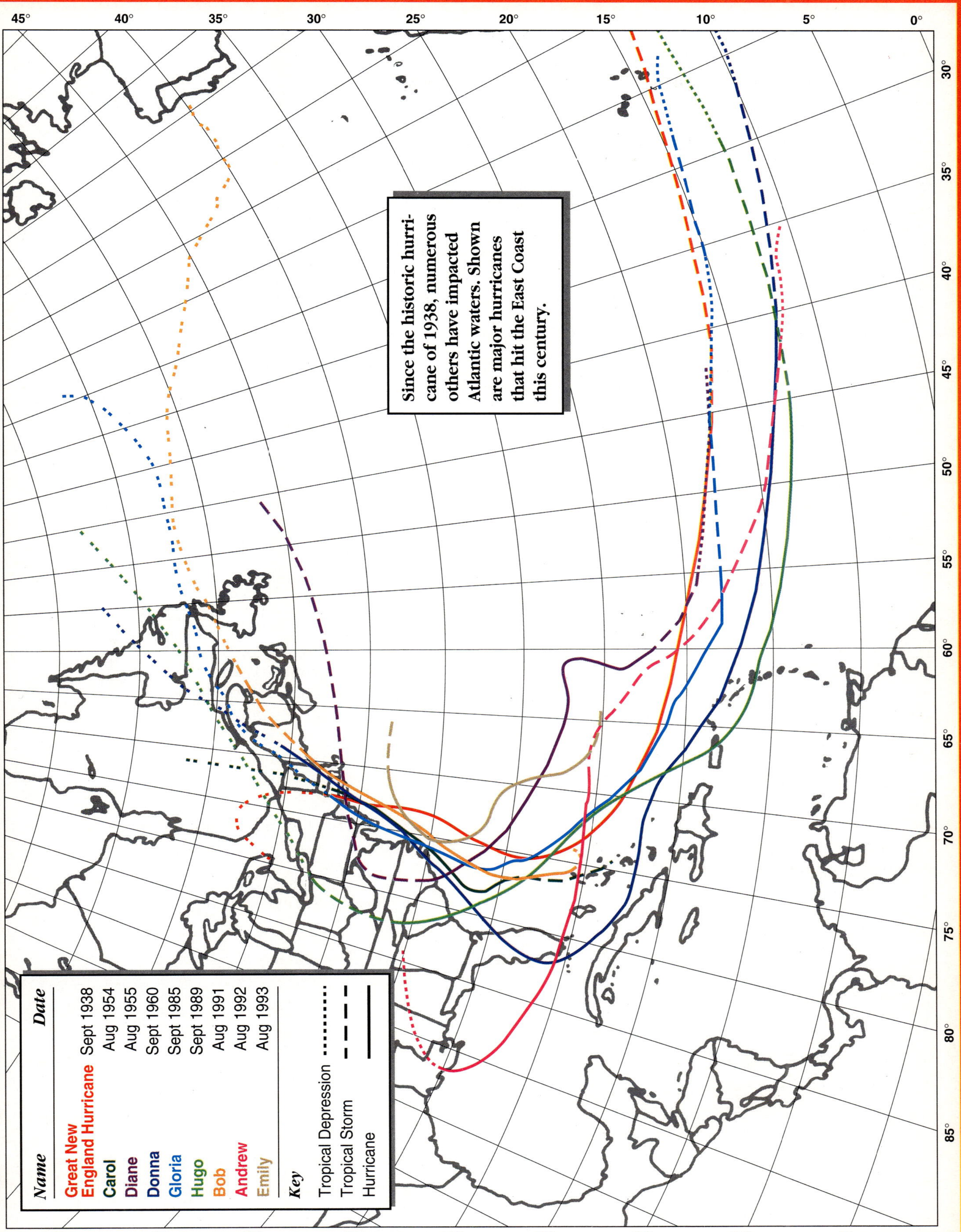

Experienced skippers aren't the only people who climb aboard boats. Many of you are passengers, just along for the ride. Someone else runs the boat and takes the responsibility. But what if the skipper were to fall overboard, or get sick or injured while you're "just along for the ride?" Suddenly, YOU are in command. Suddenly, YOU need to know the basics of how to run the boat, how to use emergency equipment, and what to do in case of trouble. The time to learn is NOW, before you take command.

Learn the Basics of Boating

Before you start, locate the lifejackets or personal flotation devices (PFDs), fire extinguisher, signal devices and other safety gear. Practice throwing PFDs or a line, anchoring the boat, and boarding from the water. Ask questions about the boat and how it operates. The more you know, the safer you'll be if something does happen.

How to Start Manual or Pull-Start Engines

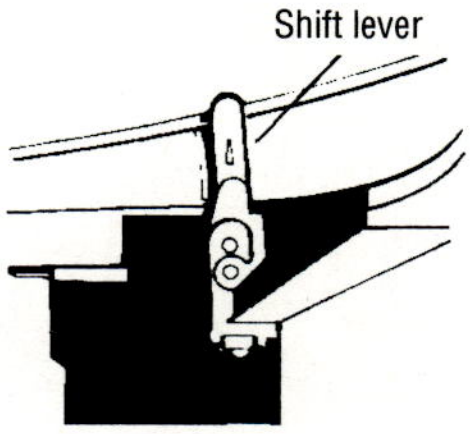

1. Put the **shift lever** in the **neutral position**—usually the middle or straight up.

2. If the **motor is cold**, pull the choke knob all the way out. If the **motor is warm**, keep the **choke in**, but use the choke if the motor fails to start after a few tries.

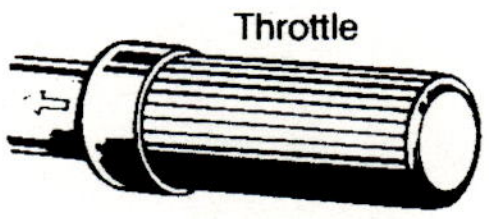

3. Set the **throttle control** to the **start position**.

4. Pull the **starter rope** slowly until the starter engages, then pull forcibly. Repeat if necessary.

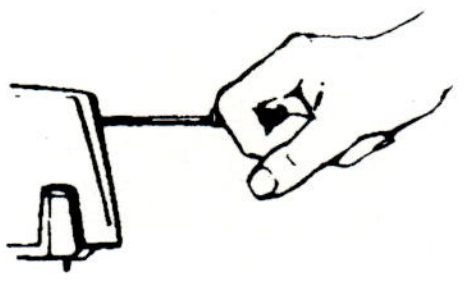

5. When the motor starts, push the **choke** in gradually until the motor is running smoothly.

CAUTION: *Never stand to pull the starter rope. If the motor has been left in gear, or the expected compression isn't there, you could lose your balance and fall overboard.*

How to Start an Electric-Start Engine

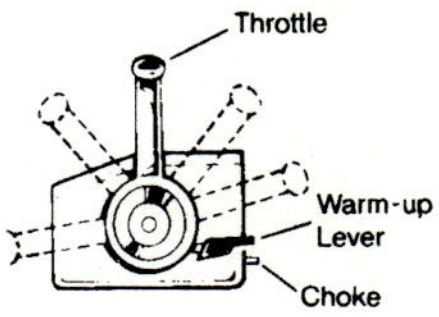

1. Squeeze the release lock on the **shift/throttle lever**, and move it to the **neutral position**, usually straight up.

2. If the **motor is cold**, move the **warm-up lever** to the **start position**.

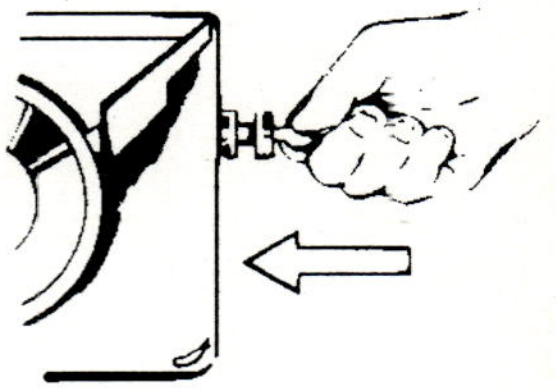

3. Turn the starter key to the **ON** position, and **push the key in** (or operate the choke switch) for **full choke**.

4. Hold the key in and continue to turn it to the start position. **Release** the key or the choke toggle as soon as the motor starts.

5. Gradually move the **warm-up lever** back to the **run position.**

6. If the **motor is warm**, use the same procedure, but **don't use the choke** unless the motor fails to start after a few tries.

How to Shift and Steer the Boat

Manual or Pull-Start

1. Look around to see which direction you want to go, and **move the tiller** in the **opposite direction**.

2. Turn the **throttle** to the **shift position**, move the **shift lever** to **forward** (or reverse), and **accelerate by turning the throttle handle** slowly to the fast position.

Electric Start

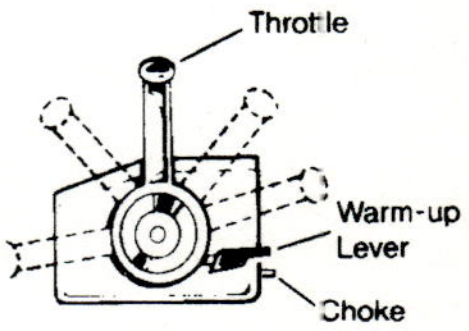

1. Squeeze the release lock and **push the lever forward**, and the **boat will move forward**. The farther forward you push the lever, the faster the boat will go. Pull the lever back, and the same thing happens—in reverse. Move very slowly in reverse, or you can flood the back of the boat.

2. Steering is the same as in an automobile. Turn the steering wheel (helm) in the direction that you want to go.

When You Come to a Stop

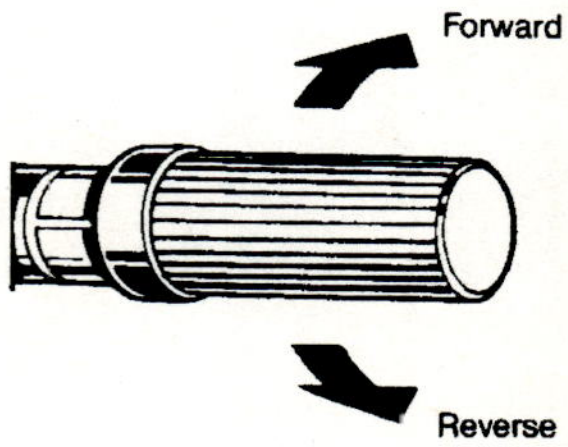

Don't aim your boat at a person in the water or at a pier or another boat. If you misjudge, you are likely to collide. Approach slowly and come alongside. Novices tend to approach too fast.

A powerboat settles quickly and slows rapidly after the power is shut down. However, while it will continue to move some after the motor has been stopped, you will lose steerage and won't be able to maneuver the boat very well. Therefore, you should aim the boat where you want to stop before you cut the motor or shift into neutral. And shift to neutral slightly earlier than you think you should. If you fall short, you can always shift back into forward gear for a low-speed boost to get you into position.

To stop a manual or pull-start motor, turn the twist grip to the stop position, and push the button that kills the engine. To stop an electric start motor, turn the key all the way to the left to shut it off—just like an automobile.

Man Overboard

1. Don't panic.
2. Get to the driver's seat and **shut off the motor.**
3. Locate the person in the water quickly and **keep your eye on him.**
4. Restart the motor and **approach the person slowly from downwind** (with the wind in your face) so you don't drift into the person after the motor stops.
5. Get to where you can **throw a lifejacket or anything that floats**, and **stop the motor**. If your throw misses, throw another lifejacket. Don't waste time trying to pick up the bad throw.

6. Tie a rope to the boat and **throw the other end to the person** in the water.
7. When the person grabs the rope, **pull him slowly to the boat.**

8. Try to **get a lifejacket on the person**, or at least tie the rope under his arms and fasten it to the boat so he can rest.

9. If you are not strong enough to lift the person into the boat, the best thing to do is sit tight and **call for help.** Lower (don't toss) the **anchor** to prevent drift.

How to Signal for Help

1. Raise and lower your outstretched arms slowly and repeatedly. If possible, hold a brightly color object in both hands.

2. Use a **mirror**. On a sunny day, the reflection can be seen for miles.

3. Signal SOS (... – – – ...) with a flashlight, horn or whistle.

4. Fire a **flare** or smoke bomb that releases a bright orange smoke.

5. Yell. Your voice will carry a very long way over the water.

6. Use **VHF radio channel 16** or **CB radio channel 9**. See **Electronic Communications Aboard** on page 72.

Used with permission of the National Safety Council and the U.S. Coast Guard.

Emergency Repairs Afloat

This chapter was prepared with assistance from the Coast Guard to aid you in repairing your boat should unexpected trouble set you adrift at sea. Most manufacturers provide specific trouble-shooting manuals and repair charts directly applicable to their products. These manuals and charts are generally excellent and should be used if you have them. If they are not available, these suggestions and hints may help you to get under way and back in port. You need not be a mechanic to make the repairs illustrated here. Nearly all causes of engine trouble are simple and fixable. Follow the hints one at a time, in order, and always reset adjustments before going on to the next item. Use your head and proceed calmly and methodically. The results of your work may save your boat and your life and will be gratifying when you get under way "on your own."

CAUTION: *The information is intended for emergency useonly. Permanent repairs should be made immediately upon your return to port.*

Before You Troubleshoot

Drop your anchor and stabilize your position to prevent you from drifting out to sea or onto rocks or shoals. Check to see if you have enough fuel. Never make more than one adjustment at a time when troubleshooting. Keep calm. Think, and proceed slowly and with caution.

The Electrical System

WARNING: *Before working on any electrical parts or doing anything that may cause a spark, get rid of any gas fumes. If the smell of gasoline persists, find the source and repair it before going on.*

1. If the starting motor will not operate, check the **battery** with a hydrometer or voltmeter to see if it is dead or has low voltage. If you have no other way, the battery can be checked by touching both terminals with a piece of metal such as a long screwdriver or the handles of a pair of pliers. A good fat spark indicates a live battery.

WARNING: *If you must check the battery by making a spark, make sure that your hands are dry and keep your face away from the battery. Drawing a heavy spark may cause a hydrogen burn and should not be done when other options are available.*

2. Check to see if the **battery terminals** are loose or corroded. Disconnect the terminals, clean both the terminals and the ends of the **battery cables** until they shine: reconnect, and tighten to insure good contact. Check both cables for good conduction; clean and repair as needed.

3. If your instruments (fuel gauge, temperature meter, etc.) operate when you turn the key in the ignition, then the **ignition switch** is all right. If the gauges do not work, check the terminals on the back of the switch (and the other ends of each wire) for tightness and good connection. Tighten, clean, or repair as needed.

4. Check the connections to the **solenoid** and to the **starter motor.** If the connections are good, connect the small and the large terminals on top of the solenoid, using a screwdriver with an insulated handle. If the starter motor turns, then the ignition switch is bad and must be replaced, but the engine can be started using the screwdriver as described above.

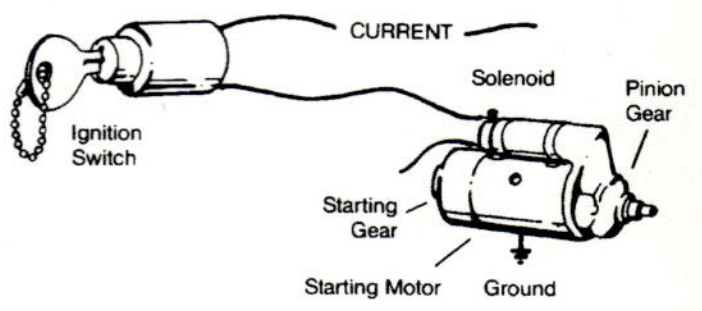

5. If the **starter motor** operates but will not engage, loosen the starter motor from the engine by removing the attaching bolts. Check the small **pinion gear** or **"bendix"** to see if it is stuck on the shaft. If it is stuck, free it up with a hammer and lubricate it with a few drops of oil from the engine dipstick.

6. If the starting motor operates but the engine will not start, look for corroded, dirty, loose, or damaged connections in the **primary electrical circuits**: the small wires from the junction box to the ignition switch, and the wires to the coil and to the distributor.

7. Look for loose or broken secondary circuit wires to the **spark plugs** and from the **coil** to the **distributor cap**. Wipe away any moisture from the wires or on the spark plugs that could prevent them from operating properly.

8. Check the **distributor** to see if the points are opening and closing as the engine is cranked. Stop the cranking with the points in their maximum open position. The opening should be about .02 inches, or about the thickness of a match book cover. See if the **contact button** is in place, free to move, and will contact with the **rotor**. *Don't forget to replace the* **rotor** *after you inspect the distributor.*

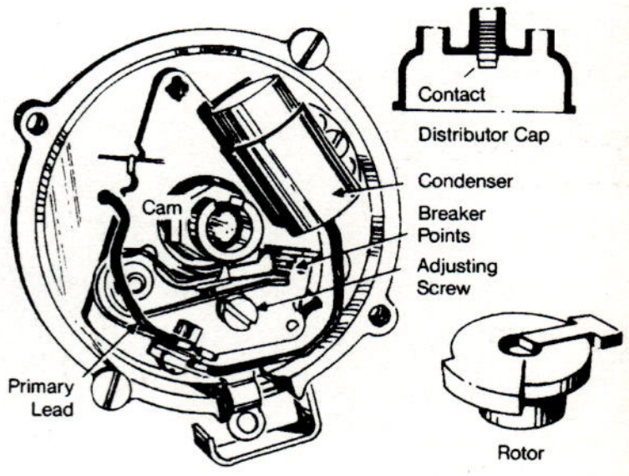

9. Using a pair of insulated pliers, try a **quick spark test** by holding a **spark plug** 1/4 inch from the **engine block** while cranking the engine with the ignition switch **ON**.

CAUTION: *To* ***prevent shock****, do not hold the spark plug wire with bare hands while cranking the engine. You can touch the spark plug wire to the engine block or use a "****sparky****."*

10. If the spark occurs, then the trouble may be with the **spark plugs** or the fuel system. If no spark occurs, then remove the secondary wire from the distributor cap to the coil. Repeat the 1/4 inch test, and if the spark occurs, the trouble lies between the center terminal on the **distributor cap**, the **rotor** contact, or the **spark plugs** and wires. If no spark occurs, then the trouble is in the **primary circuit**.

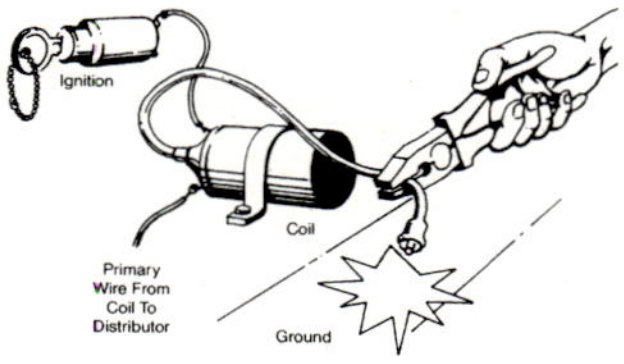

11. If you have an **overheated engine**, it may not start again until allowed to cool.

The Fuel System

WARNING: *Before making any checks or repairs during which fuel may be spilled, be sure you have switched off all electrical equipment, have a container ready to catch any spilled fuel, and have someone stand at the ready with a fire extinguisher.*

1. Always check your **fuel tank** before leaving port. Running out of fuel is the most common type of engine failure. A good rule to follow is never to cruise farther than one-third of your fuel supply. Save two-thirds for the return trip.

2. See if fuel is reaching the **fuel pump** by checking that the **fuel filter** or **sediment bowl** is filled with fuel. If no fuel is in the filter, and there is fuel in the tank, then the **shut-off valve** may have vibrated closed, or a **fuel line** may be clogged. Disconnect the inlet side of the pump and blow through the line. Look for dents in the fuel line and air leaks in the fuel pump gaskets or in the line connections. Make sure that the **tank vent** is open.

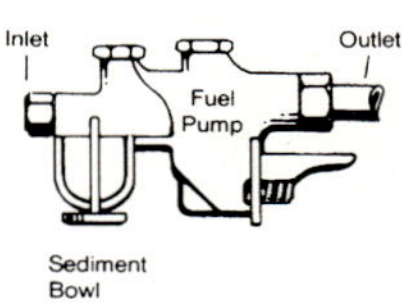

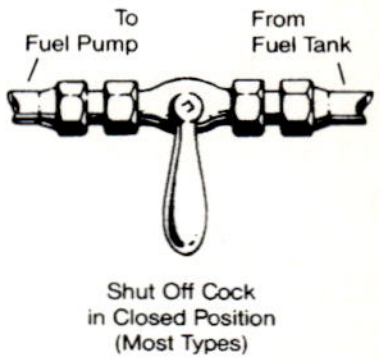

3. To see if fuel is reaching the **carburetor**, remove the **sediment bowl** from the **fuel pump** and check that the **screen** in the pump body is clean. Then disconnect the outlet from the fuel pump to the carburetor and see if the fuel flows out when the engine in cranked. Make sure that the **ignition** is **OFF** when making the test. Pulling out the coil wire will ensure that the ignition is off.

4. Make sure fuel is reaching the cylinders by removing a **spark plug** and checking that the tip is moist with fuel. If there is no fuel on the plug, the **idling jet** may be clogged. Refer to your owner's manual to determine which hole is the idling jet. **Never touch the main fuel jet.**

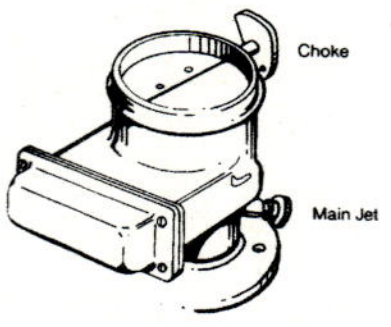

5. The **choke** must close properly, especially on dual **carburetors.** If the linkage permits one choke to remain open, both chokes will be out of service.

6. If the **spark plugs** are wet with fuel, this indicates that the **engine was flooded** by using the choke too long. The correct way to dry out the **cylinders** is to open the **throttle** wide. Put the **choke** in the open position (non-choking), and with the **ignition ON**, crank the engine several times. This draws nothing but air through the carburetor. **Always replace the flame arrestor before cranking the engine.**

7. Check that the **overspeed trip switch** hasn't automatically triggered air and fuel shut-offs because of excessive engine speed. If so, push the **reset button**.

Diesel Engines

Air flow problems are sometimes the cause of trouble with diesel engines. Some diesel engines are equipped with **automatic air dampeners** that close off the air supply in case of overheating or loss of oil pressure. Check the **air inlet cleaner** and **screen** to see that the passage to the **intake manifold** is clear.

A fuel problem is the other major cause of trouble with diesel engines and tends to fall into these basic categories:

1. The **fuel tank** is empty.
2. The **shut-off valve** in the fuel line may have vibrated itself closed.
3. Water in the fuel will prevent the engine from firing. Drain any water in the system by opening the cock at the bottom of the **fuel filter**. Prime the fuel system with the **prime pump** or by cranking the engine.
4. Remove blockages by cleaning or changing all **air** and **fuel filters**.
5. Check and tighten all **fuel lines** from the tank to the pump and seal any **air leaks**. Check the **gaskets** on the fuel filter and **strainer** housing or cap. Disconnect the **fuel return line** and allow fuel to flow until no air shows in the fuel.

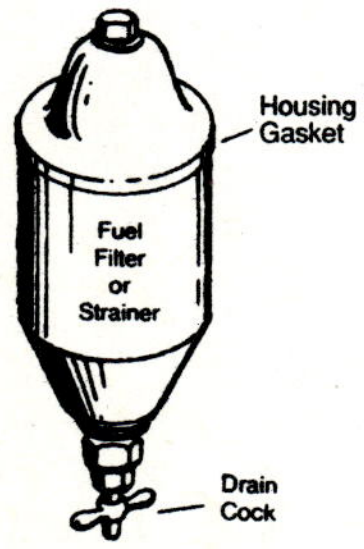

6. If no fuel is reaching the engine, check to see if there is an electrical **stop switch** or **solenoid** in the fuel line near the fuel pump that cuts off the fuel when shutting down the engine. A short-circuit may have closed the solenoid. Disconnect the wires and try to start the engine.
7. When changing **fuel tanks**, some **air** can enter the lines and cause the engine to stop. Use the **prime pump** to increase the fuel pressure, and try to restart.

Mechanical Repairs

Most emergency mechanical repairs can be made with what you have on board. With a little imagination, you can make almost anything needed to get under way. Here are a few suggestions:

1. Carry these basic tools and spare parts:

adjustable wrench	circuit tester
slip joint pliers	electrical wire
pipe wrench	spark plugs
vise grip pliers	fan belt
screwdrivers (all sizes)	points for distributor
box end wrenches	condenser
hammer	coil
spark plug wrench	fuel pump
flashlight, bulbs & batteries	fuel & air filters
electrical or duct tape	engine oil
knife	

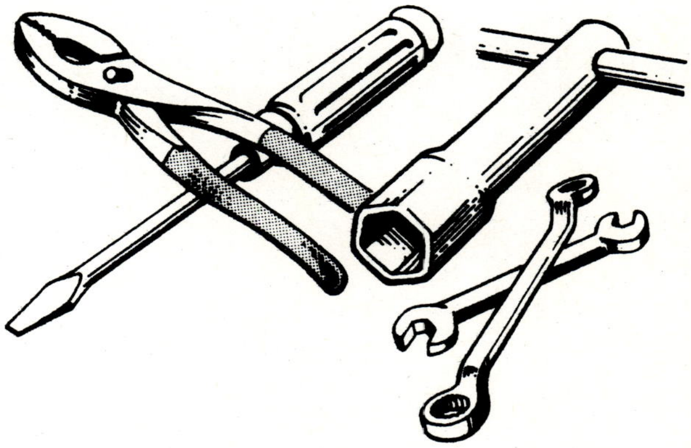

2. A broken **fan belt** can be replaced with a piece of line or **rope**. Tie the line around the **pulleys**, securing the ends with a square knot and tighten the moveable pulley.
3. Broken **pipes** and **hoses** can be repaired with first-aid **bandages**, **rags**, or **canvas** and secured with tape, rope, or your belt.
4. If you lose your **rudder**, make a **sea anchor** emergency steering device by using **buckets**, **deck covers**, or anything else that you can tie a

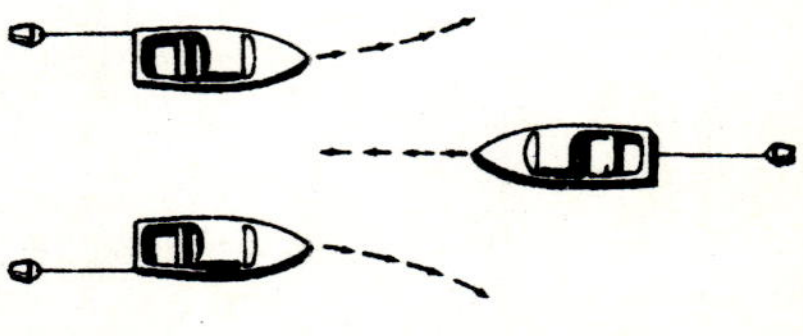

line to and drag **behind your boat**. **Shift** the line from **side to side** to control your direction, and **go slow**.

5. If you develop an **oil leak**, simply catch the oil in a bucket or pan and **pour it back into the engine**.

Sinking

If you find water in your boat, don't panic: there may be a simple explanation and remedy. Follow these steps:

1. Put on your lifejacket.

2. Get signaling and survival equipment ready.

3. If not readily apparent, find where the water is getting in. Most boats have openings in the hull that are necessary for the operation of the vessel, but they can malfunction. Some of these openings include:

- engine cooling water intake
- head (toilet) suction and discharge
- galley (kitchen) drain
- bait box water intake
- propeller shaft
- bilge pump or self-bailer.

4. If water is being pumped in by the operation of the engine, shut off the engine and make repairs.

5. Look for damage to the hull and stuff any holes with rags, blankets, pillows. Wedge them down with a deck cover or paddle. Tear off a railing if necessary. A damaged boat is better than no boat at all.

6. Get under way if possible and head for the nearest land or other boats.

Abandoning Ship

1. Stay with your boat as long as possible. Most boats will not sink completely but remain awash at water level. The boat will give you something to hold onto, and it is much bigger than you and will be much more likely to be seen by rescuers.

2. Collect your survival gear, and signal for help.

3. Keep everyone calm and together. Use a length of rope to connect everyone facing each other. Stay warm by huddling together and moving as little as possible.

4. Don't try to swim for shore unless you are absolutely sure of your abilities. Distances in the water are very deceiving, and cold water will sap your strength very quickly.

Signaling For Help

See **Embassy's Coastal Navigation Guide** on page 82, and **Electronic Communications Aboard** on page 72.

First Aid

The following is provided as a reference only. If the situation is serious, take the victim to a hospital immediately. Telephone numbers for emergency assistance are provided at the end of each chapter.

Breathing Difficulty

When breathing stops:

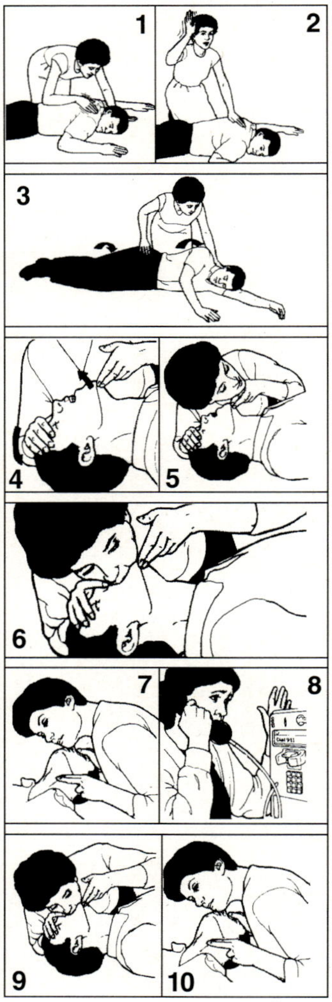

1. Check for unresponsiveness. Tap or gently shake victim. Shout, "Are you OK?"

2. Shout, "HELP!" Get the attention of people who can phone for help.

3. Position the victim on his back. Roll the victim toward you by pulling slowly and evenly from the victim's hip and shoulder.

4. Open the airway. Tilt the head back and lift the chin with your fingers under the bony part of the jaw.

5. Check for breathlessness. Look, listen, and feel for breathing for 3 to 5 seconds.

6. Give two full breaths. Keep head tilted back. Pinch the nose closed. Seal your lips tightly around the victim's mouth. Give two full breaths for 1 to 1-1/2 seconds each.

7. Check for pulse at the side of the neck. Keep head tilted back. Feel for carotid pulse for 5 to 10 seconds.

8. Phone Emergency Medical System for help. Send 2 people, if possible, to call an ambulance. Give location of emergency and condition of victim.

9. Begin rescue breathing. Keep head tilted back. Give 1 breath every 5 seconds. Look, listen, and feel for breathing between breaths.

10. Recheck pulse every minute. Feel for carotid pulse for 5 to 10 seconds. If victim has a pulse but is not breathing, continue rescue breathing.

Bleeding

Bleeding may be stopped by applying firm pressure directly over the wound to form clotting or by digital pressure on a pressure point in the affected arm or leg, or by a combination of these.

Poisoning

First Aid for a Conscious Victim

1. Dilute the poison by having the victim drink a glass of water or milk if the victim is conscious and not having convulsions. Discontinue dilution if it makes the victim nauseous.

2. Save the label or the container of the suspected poison for identification. If the victim vomits, save a sample of the vomited material for analysis.

3. Seek medical assistance by calling the poison control center or a physician. You should post the number of the poison control center for your area on your telephone. If you do not have the number, call 911 or dial 0 (operator).

First Aid for an Unconscious Victim

1. Maintain an open airway.

2. Call for an emergency squad as soon as possible.

3. Administer rescue breathing and CPR if required.

4. Save the label or container of the suspected poison. If the victim vomits, save a sample of the vomited material for analysis.

5. Do not give fluids to an unconscious person.

6. Do not induce vomiting in an unconscious person. If the victim is vomiting, position him on his side and turn his head so that the vomited material drains out of the mouth.

Animal Bites

Wash the wound thoroughly with soap and water. Rinse with clean running water. Apply sterile dressing. Always consult a physician at once. Confine animal to escape-proof quarters. Notify police.

Bruises

Apply an ice bag or cold pack. If the skin is broken, treat as a minor cut.

Burns and Scalds

For burns of limited extent: Apply cold water. Cover with sterile dressing. For extensive burns: Treat for shock. Remove loose clothing. Do not remove clothing that sticks to burned area. Consult a physician. Never apply butter, oil or any preparation to burn. If burn covers a considerable area or if fever or blisters develop, see a physician.

Cuts

Wash area with soap and water. Wash away from– not into– the cut. Apply direct pressure over cut with sterile dressing until bleeding stops. Apply an antibacterial ointment and sterile dressing.

Eyes

Foreign bodies: Remove only those foreign bodies lying on the surface of the eye. Lift off with the corner of a clean handkerchief or flush eye with water, using an eyedropper or bulb syringe. Do not rub eye. Never remove anything embedded in eyeball. Consult a physician.

Fainting
Keep person lying down with head slightly lowered. Loosen any tight clothing about neck. If person does not respond within a short time, summon a physician.

Fractures
Deformity of injured part usually means fracture. If fracture is suspected, do not attempt to move injured person. Call a physician at once. Treat for shock.

Choking
1. Ask, "Are you choking?"
If victim cannot breath, cough, or speak...
2. Give the Heimlich Maneuver:
- **A.** Stand behind the victim.
- **B.** Wrap your arms around victim's waist.
- **C.** Make a fist with one hand. Place your fist (thumbside) against the victim's stomach in the midline just above the navel and well below the rib margin.
- **D.** Grasp your fist with your other hand.
- **E.** Press into stomach with a quick upward thrust.

3. Repeat thrust if necessary.
4. Sweep the mouth if victim has become unconscious.
5. Attempt rescue breathing.
6. Give 6 to 10 abdominal thrusts. Repeat steps 4, 5, and 6 as necessary.

Frostbite and Chills
Handle gently to avoid injury. Bring person into a warm room and give them a warm drink (non-alcoholic). Immerse body part in lukewarm water but not hot water, or gently wrap in warm blankets. Do not rub or expose to fire, or put in hot water.

Heat Cramps
Symptoms: Cramps in muscles of abdomen and extremities. Treatment: Same as for heat exhaustion.

Heat Exhaustion
Symptoms: Cool, clammy skin with body temperature about or below normal. Treatment: Keep person lying down with head lowered. If conscious, give saltwater solution to drink (one teaspoon salt to one glass of water) in small amounts and at frequent intervals.

Heat Stroke
Symptoms: Hot, dry skin and extremely high body temperature. Treatment: Repeatedly sponge bare skin with cool water or rubbing alcohol or apply cold packs or place person in a tub of cold water (do not add ice) until body temperature is sufficiently lowered. Do not give stimulants. Consult a physician immediately.

Insect Bites
Remove stinger if present. Apply cold applications and soothing lotions, such as calamine. If person has a history of allergic reactions to insect bites, take to a physician immediately.

Nosebleeds
Place person in a chair with head erect. Loosen clothing at neck. Saturate towel with icewater and apply to bridge of nose, while holding nostrils together tightly. Keep changing cold towels at intervals of one minute. If blood continues to flow freely, call a physician at once.

Poison Ivy
Wash exposed area well with naptha (yellow) soap. Do not use brush or other rough material. Then use rubbing alcohol, if available. Apply calamine lotion. If area spreads, swells, or forms large blisters, see a physician.

Puncture Wounds
Encourage a little bleeding to wash out the wound by applying mild pressure at the wound. Then treat the same as a cut. Always see a physician. A tetanus injection is usually necessary.

Scrapes
Wash with soap and water. Blot dry and treat the same as cuts. If scrape is deep or dirty, see a physician.

Shock
Keep person lying down. Cover only enough to prevent body heat loss. Get medical help.

Splinters
Wash area with soap and water. Sterilize needle point by passing it through a flame and use it to tease out splinter. Apply antibacterial ointment and sterile dressing.

Sprains
Elevate injured part to minimize swelling, and apply ice bags or cold cloths immediately after injury. Cold applications of Epsom salts may be repeated every two hours. If swelling is pronounced, do not attempt to use injured part until seen by a physician.

Unconsciousness
Never attempt to give anything by mouth. Never attempt to induce vomiting. Place patient lying on side with head on arm. Loosen tight clothing; maintain body heat with blankets. Summon a physician at once. Be sure patient is breathing. If not, give artificial respiration.

Illustrations reprinted with permission of the American Red Cross.

Hypothermia and Cold Water Survival

What is Hypothermia and How Does it Kill?

Hypothermia is subnormal temperature within the central body. When a person is immersed in cold water, the skin and nearby tissues may cool very fast. However, it may take 10 to 15 minutes before the temperature of the heart and brain begins to drop. When the core temperature reaches 90˚F, unconsciousness may occur. When the core temperature drops to 85˚F, heart failure is the usual cause of death. However, a person in cold water may drown because he loses the use of his limbs, and his consciousness becomes clouded.

How Long Can One Survive in Cold Water?

Survival in cold water depends on many factors. The temperature of the water is only one of them. Others include body size, fat, and activity in the water. Large people cool more slowly than small people. Fat people cool more slowly than thin people. Children, because they are small, cool faster than adults.

If swimming or treading water, a person will cool about 35 percent faster than if remaining still. The so-called "drown-proofing technique" requires putting the head into the water and will cause a person to cool about 85 percent faster than if floating still with head out of the water.

A person of average size, wearing light clothing and a PFD (personal flotation device), may survive 2-1/2 to 3 hours in 50˚F water by remaining still. This survival can be increased considerably by getting as far out of the water as possible and covering the head. Getting into or onto a boat or anything that floats can be a real lifesaver.

Temperature

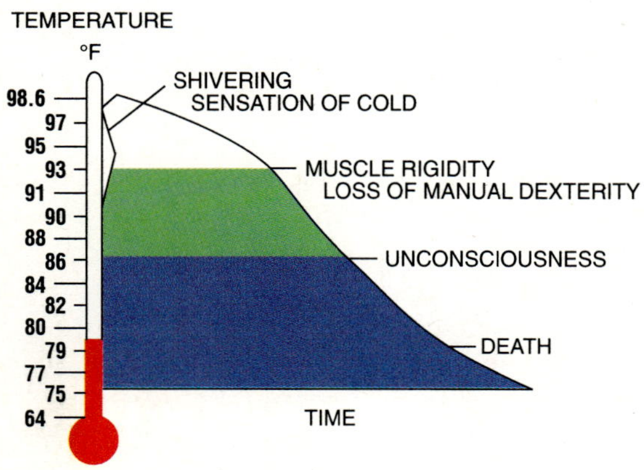

What Do I Do if an Accident Does Happen?

In case of accidental immersion in cold water, remember that water conducts heat many times faster than air. Most boats will float even when capsized or swamped. Therefore, get in or on the boat to get as far out of the water as possible. Wearing a PFD is a MUST. It will keep you afloat even if you are unconscious. Remain still, and if possible, assume the fetal position or the Heat Escape Lessening Position (H.E.L.P.).

H.E.L.P.

H.E.L.P. will increase your survival time. About 50 percent of the heat is lost from the head. It is therefore important to keep the head out of the water. Other areas of high heat loss are the neck, sides, and groin.

If there are several people in the water, huddling close, side-by-side, in a circle will also help preserve body heat.

Huddle

The following table shows predicted survival times for an average person in 50° F. water:

Situation	Predicted Survival Time (Hours)
WITHOUT PFD	
"Drown-proofing"	1.5
Treading Water	2.0
WITH PFD	
Swimming	2.0
Holding Still	2.7
H.E.L.P.	4.0
Huddle	4.0

Should I Swim for Shore?

This is a most difficult decision. It depends on many things. Some good swimmers have been able to swim 8/10 of a mile in 50°F water before being overcome by hypothermia. Others have not been able to swim 100 yards. Furthermore, distances on the water are very deceptive. Therefore, STAY WITH THE BOAT. Even a capsized boat is easier to see than a person in the water. This will make it easier for rescuers to spot you. DO NOT SWIM unless there is absolutely no chance of rescue and your are absolutely certain you can make it. If you do swim, use a PFD or some other flotation aid.

This Won't Save Your Life Unless You Are Wearing It.

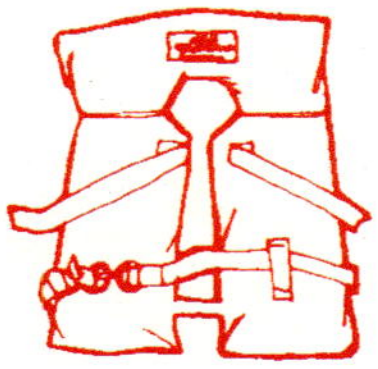

First Aid for Hypothermia Victims

Incorrect treatment of hypothermia victims may induce a condition known as "after-drop." This is caused by improper rewarming, allowing cold, stagnant blood from the extremities to return to the core of the body. When this cold blood returns to the core of the body, it may drop the core temperature below the level that will sustain life. For the same reason, Hypothermia victims must be handled gently and should not be allowed to walk.

1. Move the victim to shelter and warm as rapidly as possible.
2. Gently remove all wet clothing. The small amount of heat energy the victim has left must not be expended on warming and drying wet clothing.
3. Apply heat to the central core of the body (head, neck, sides, groin). Place the victim on a hard, flat surface. This will allow the administration of cardiopulmonary resuscitation (CPR), should the need arise.

A. Wrap warm moist towels— or other textiles— around the victim's head, neck, sides, and groin. As the packs cool, rewarm them by adding warm water (about 105°F). Check the temperature of the water with the elbow. It should feel warm but not burn.
B. Hot water bottles and heated blankets can also be used.
C. An effective field measure is for one or two of the rescuers to remove their own clothing, using their bodies to warm the victim's naked body. A sleeping bag or blanket should be used to conserve the body heat.
D. If the victim appears dead, CPR and rescue breathing should be administered. Never put an unconscious person in a bathtub. In cases of mild hypothermia, dry clothing and shelter may be all that is needed before the victim appears normal. However, all hypothermia victims should be seen by a doctor.

Do Not

1. Do not give the victim anything to drink, especially not alcohol.
2. Do not rub frozen body areas, especially not with snow.
3. Do not wrap a hypothermic person in a blanket without an auxiliary source of heat unless it is to protect him from further heat loss before treatment.

How Can I Avoid Hypothermia?

Since most boaters who die in water-related accidents had no intention of going in the water, the obvious answer is to avoid those behaviors that cause accidental immersion. Therefore, DO NOT:

1. Stand or move around in a small boat.
2. Overload your boat or distribute the load unevenly.
3. Decelerate suddenly, allowing the stern wake to overtake and swamp the boat by washing over the transom.

Some of the information in this chapter has been extracted by the U.S. Coast Guard from "Man in Cold Water" with permission of Dr. Martin Collis, University of Victoria, British Columbia, Canada.

Electronic Communications Aboard

Calling for help in an emergency is the single most important use of your boat's communication system. But the ever-growing services available to boaters through these systems make them more than just safety devices. You can stay in constant touch with a fellow cruiser, keep up-to-date on weather reports, call home when you're in the middle of the ocean, or use a cellular phone dockside as you would at a vacation home.

It's important that you know the proper procedures for using communications aboard your boat. The primary vehicles for distress calling are **VHF** and **single sideband (SSB)** marine radio systems. **CBs** and **cellular phones** are more appropriate for personal use.

The following is meant to clarify the use of communications aboard your boat.

VHF Marine Radio

The Federal Communications Commission (FCC) requires a ship station license for **VHF marine radios**. In some cases, an operator license is required, but this is not the case for recreational boats or those carrying six or less passengers for hire. You can obtain license applications by writing: FCC Private Radio Bureau Licensing Division, 1270 Fairfield Road, Gettysburg, PA 17325-7245.

Each of the channels on your VHF has a specific use, as authorized by the FCC. A partial listing of the 48 marine channels follows:

Ch. 06 156.300 MHz
Intership Safety Frequency
Mandatory on all VHF sets. Only use for communicating with the Coast Guard and other rescue vessels.

Ch. 09 156.450 MHz
Ship-to-Ship and Ship-to-Shore

Ch. 13 156.650 MHz
Commercial, Ship-to-Ship, and Bridge and Lock Tenders.
Reserved for navigational communication. Commercial vessels must monitor it at all times.

Ch. 16 156.800 MHz
National Distress Frequency
Monitored continuously by the Coast Guard. All boaters are required to monitor channel 16 when the unit is on.

Ch. 22 157.100 MHz
Primary Coast Guard Communication Channel
Used for communications with the Coast Guard only after establishing contact on channel **16**. Weather reports and Notices to Mariners are broadcast with schedule announced on channel **16**.

Ch. 68 156.425 MHz
General Calling, Ship-to-Ship, and Ship-to-Shore
For general communication. Limited to pleasure craft, shore stations, and harbormasters. Use channels **09**, (also used by commercial vessels) **69**, **71**, and **78** as alternates.

Ch. 72 156.625 MHz
General Calling, Recreational, and Ship-to-Ship
Limited to pleasure craft contacting other pleasure craft.

Ch. WX1 through **WX7** 162.550 MHz through 162.525 MHz
NOAA Weather Reports
Forecasts are broadcast on these channels 24 hours a day. (See **Weathering The Atlantic Coast** on pages 59-61)

SSB Radio

Single sideband radio equipment operates on **medium frequencies (MF)** or **high frequencies (HF)** between **2MHz and 30 MHz**. It is useful for long-distance communications, distances beyond 20 miles.

You cannot obtain a ship station license for an SSB installation alone; you must also have a licensed VHF radiotelephone system. Because of the necessity of a good grounding system, the SSB must be installed by an FCC licensed technician.

The use of the **2 to 3 MHz** band when in harbors, ponds, lakes, or rivers is prohibited for intership communication. It may still be used for distress communications or to contact a public correspondence station if beyond VHF range.

VHF Radio

The international distress and calling frequency for those with MF radiotelephone is **2182 kHz**. Some of the more popular carrier frequencies in the MF band follow:

2082.5 kHz and **2638 kHz**
Intership Safety. May be used in all areas. May also be used for operational and business communications provided there is no interference to safety communications.

2670 kHz
Coast Guard frequency. May be used to receive Coast Guard broadcasts of navigational warning and coastal

weather forecasts and warnings. May also be used to communicate with Coast Guard ship and coast stations after first establishing contact on 2182 kHz.

2738 kHz
Intership Safety. May be used in all areas except Great Lakes and Gulf of Mexico. May also be used for operational and business communications provided there is no interference to safety communications.

Making Ship-to-Shore Calls

You may use your VHF or SSB radio by way of public coast stations to make ship-to-shore radiotelephone calls. For distances up to **20 miles**, you should use **VHF radio**. You will find a listing of channels to use in your area in the **Maritime Radio Users Handbook**. (The address listed later in this section). It's best to make your calls collect and to avoid giving out credit card numbers over the airwaves.

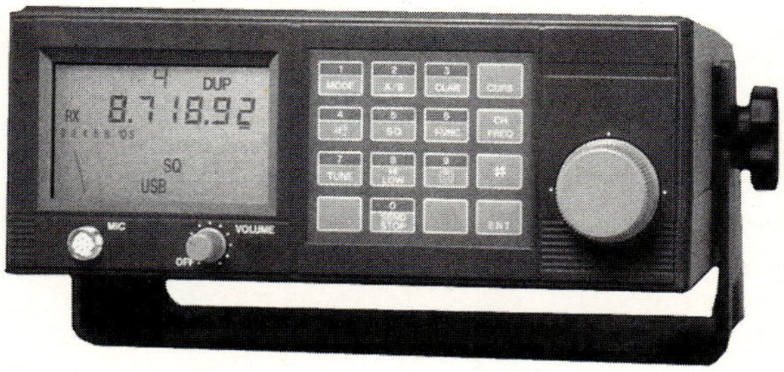
Single Sideband Radio

When making ship-to-shore calls:

- Select the channel assigned to the desired shore station ***(Do not use Ch. 16 or 2182 kHz.)***
- Make sure the channel is not busy (you will hear speech, signaling tones, or a busy signal). If it is, switch to another channel or wait.
- If not busy, transmit the name of the coast station and your call sign. Listen for a reply.
- When the coast station operator answers, transmit the name of your vessel, call sign, ship's telephone number or billing number if assigned, and the city and telephone number you are calling. If billing information for your vessel has not been registered, the operator will ask for additional billing information.

Shore-to-ship calls can be made on VHF radio. The on-shore party should dial 0 and ask for the marine operator.

High Seas Radiotelephone

With SSB equipment you can make ship-to-shore calls from long distances offshore. Two AT&T high seas radiotelephone service coast stations are located on the East Coast: **WOO** in Manahawkin, N.J., and **WOM** in Fort Lauderdale, Fla. Each monitors a line of channels and frequencies. Your proximity to the stations is not neccesarily a factor in which one you should use, although it is best to try the closest station first. You can also get technical information on using the service by calling one of the stations: **WOM 1-800-538-5936**; **WOO** (Call Collect) **1-609-597-2201.**

When making a high seas call to a shore operator, it's important that you transmit:

- The station you are calling
- The channel you are calling on
- Your boat's name
- Call sign
- Approximate location

Once you have given all this information, wait for the response. If you don't hear it within 20 seconds, try again. If you still don't get a response, try another channel.

For high seas shore-to-ship calls to the Atlantic Ocean, the on-shore party should call 1-800-SEA-CALL. Be sure to monitor your radio if you expect a call. For more information on VHF or SSB, please consult the Maritime Radio Users Handbook, which may be purchased from the Radio Technical Commission for Maritime Services (RTCM), Box 19087, Washington, DC. 20036; or write to the FCC, 201 Varick Street, New York, New York 10014.

Citizens Band (CB) Radio

CB Radios are intended primarily for conducting personal or business communications over a short range. You may use CB equipment aboard your boat on **40 channels** on a shared basis to communicate with other CB stations.

Channel 9 in the Citizens Band has been designated as an emergency channel. Certain Coast Guard stations monitor channel 9 on a secondary basis. CB provides an alternative means of alerting the Coast Guard for boaters not equipped with other systems. But remember, CB service is not a substitute for the marine distress system.

For all radio systems, you should consult the owner's manual supplied with your system for particular specifications and operation.

Cellular Telephones

The advent of cellular technology has enabled boaters to have all the advantages of personal telephones aboard boats. A cellular telephone emits a signal to on-shore cell sites, maintained by private companies. The cell site then transmits the signal to a switching office where your call is put through the telephone lines. Because cellular calls are dependent on shore cell sites, their range will vary. They are generally a near-coastal means of communication.

Networks have been developed that enable you to make or receive calls from different regions. When you purchase equipment, you should be provided with detailed information on how to use the cellular phone. Consult a local communications dealer for more information.

Clearing Customs on the East Coast

Who needs to clear Customs?

The master of any American pleasure boat must report arrival from a foreign port or place to U.S. Customs immediately and must also report foreign merchandise that is subject to duty. The report may be made by any means of communication and should include the name of the boat, name of the master, place of docking, and arrival time. American boats need not clear Customs before departing for a foreign port.

The master of a foreign-flagged or undocumented foreign pleasure boat must report arrivals to U.S. Customs immediately and make formal entry into the country within 48 hours. In the absence of a cruising license, vessels in this category must obtain a permit before proceeding to each subsequent U.S. port. The boat must also clear Customs prior to departing for a foreign port.

Cruising licenses exempt pleasure boats of certain countries from formal entry and clearance procedures at all but the first port of entry. They can be obtained from the District Director of Customs at the first port of arrival in the United States.

When should you contact Customs?

You must notify U.S. Customs immediately upon docking. In order to expedite the process, you may want to contact Customs by way of the marine operator several hours before you plan to dock, but be sure to call again after you actually dock. Customs will inform you if an inspection is required.

What is the procedure for an inspection?

For foreign-flagged pleasure boats, and American pleasure boats where an inspection is necessary, everyone must **stay on board**. No person should board or leave the boat prior to Customs processing without permission of the Customs officer, except to report arrival. Violations may result in substantial penalties and forfeiture of the boat.

What documents are required?

For American pleasure boats, everyone on board will need proof of citizenship such as a valid passport, birth certificate, or voter registration card. If the boat is documented with the U.S. Coast Guard, you must have a copy of the document; otherwise a state registration. Forms for individual declarations will be provided by the inspecting officer.

Foreign-flagged vessels are required to present the following documents: original *Customs Form 1300,* masters oath on entry of vessel, general declaration, *Customs Form 1301*, ship stores declaration, *Customs Form 1303*, crew's effects declaration, *Customs Form 1304,* clearance from last foreign port, original crew list, original passenger list, and cargo declaration. Resident aliens must present their "green card" (*Immigration Form I 151 or I 551*). Non-resident aliens should present a valid passport with a non-immigration visa in it. Contact the U.S. consulate or embassy in your country for further information on this subject.

How do you contact U.S. Customs?

Report your arrival in the United States to the U.S. Customs office nearest your point of entry as follows:

Maine	Portland	(207) 775-3131 (800) 343-2840*
Massachusetts	Boston	(617) 565-4657 (800) 343-2840*
Connecticut	Bridgeport	(203) 579-5606 (800) 343-2840*
New York	New York City	(212) 466-3456 (800) 522-5270*
New Jersey**	Newark	(201) 645-3760 (800) 221-4265*
	Perth Amboy	(201) 442-0415 (800) 221-4265*
Pennsylvania	Philadelphia	(215) 597-4605 (800) 343-2840*
Maryland	Baltimore	(301) 962-2666 (800) 343-2840*
Virginia	Alexandria	(703) 557-1950 (301) 953-7454 (202) 566-2321
	Newport News	(804) 245-6470
	Norfolk	(804) 441-6741*
North Carolina	Morehead City	(919) 726-5845 (919) 726-3651 (919) 726-2034
	Wilmington	(919) 343-4616
South Carolina	Charleston	(803) 723-1272
Georgia	Brunswick	(912) 267-2803 (912) 638-4294 (912) 264-4468
	Savannah	(912) 944-4452 (912) 232-7507 (912) 944-8018
Florida	Jacksonville	(904) 791-2775 (800) 432-1216*
	Pt. Canaveral	(305) 783-2066 (800) 432-1216*

The operator of a pleasure boat whose intended destination is a point within the jurisdiction of the Miami, Fla. Customs District, must make the report immediately upon arrival at one of the 23 locations designated by the Miami Customs District. These 23 locations are listed below:

Florida: Ft. Pierce Area

Sebastian Inlet Marina
2580 U.S. 1
Sebastian

Sailfish Marina
3565 Southeast Street
Stuart

Pelican Yacht Club
1120 Seaway Drive
Ft. Pierce

Florida: West Palm Beach Area

Jupiter Marina
97 Lake Drive
Jupiter

Lake Worth Boating Center
7848 S. Federal Hwy.
Hypoluxo

Riviera Beach
Municipal Marina
200 E. 13th Street
Riviera Beach

Florida: Ft. Lauderdale Area

Sands Harbor Marina
125 N. Riverside Dr.
Pompano Beach

Lighthouse Point Marina
2830 N.E. 29th Ave.
Pompano Beach

Lauderdale Marina
1800 S.E. 15th Street
Ft. Lauderdale

The Cove Marina
1755 S.E. 3rd Court
Deerfield Beach

Pier 66 Marina
2301 S.E. 17th Street
Ft. Lauderdale

Delray Harbor Club Marina
1035 S. Federal Hwy.
Delray

Bahia Mar
801 Seabreeze Blvd.
Ft. Lauderdale

Florida: Miami Area

Matheson
Hammock Marina
Matheson Hammock Park
Miami
(sunrise to sunset)

Phillips 66 Marina
1050 MacArthur Causeway
Watson Island
Miami

Crandon Park Marina
Virginia Key
Miami

Sunset Harbor Marina
1982 Purdy Ave.
Miami Beach

Bakers Haulover Marina
10800 Collins Ave.
Miami Beach

Florida: Key West Area

A&B Marina
700 Marina Street
Key West

Ocean Reef Club
Key Largo

Tavernier Creek Marina
Tavernier

Oceanside Marina
5950 Maloney Ave.
Key West

Faro Blanco Marina Resort
1996 Overseas Hwy.
Marathon

* 24 hours-day
** If you arrive in New Jersey south of the Manasquan Inlet contact U.S. Custom's Philadelphia, Pa., office.

Figure at intersection of columns opposite ports in question is the nautical mileage between the two.

Example:
Portland, ME is 100 nautical miles from Boston, MA.

Gulf of Maine Distances
Calais, ME to Cape Cod, MA
(Nautical Miles)

	Calais, Maine 45°11.4'N., 67°16.7'W	Eastport, Maine 44°54.3'N., 66°59.0'W	Lubec, Maine 44°51.7'N., 66°59.0'W	Machiasport, Maine 44°41.9'N., 67°23.6'W	Jonesport, Maine 44°31.6'N., 67°37.0'W	Bar Harbor, Maine 44°23.5'N., 68°12.0'W	Buck Harbor, Maine 44°20.3'N., 68°44.2'W	Stonington, Maine 44°09.2'N., 68°39.8'W	Rockland, Maine 44°06.0'N., 69°05.5'W	Searsport, Maine 44°27.0'N., 68°54.0'W	Bucksport, Maine 44°34.3'N., 68°48.0'W	Bangor, Maine 44°47.7'N., 68°46.3'W	Boothbay Harbor, Maine 43°51.0'N., 69°37.6'W	Wiscasset, Maine 43°59.5'N., 69°40.1'W	Bath, Maine 43°54.5'N., 69°48.7'W	Augusta, Maine 44°18.9'N., 69°46.4'W	Portland, Maine 43°39.4'N., 70°14.7'W	York Harbor, Maine 43°07.9'N., 70°38.6'W	Portsmouth, N.H. 43°04.6'N., 70°44.5'W	Newburyport, Mass 42°48.8'N., 70°52.4'W	Rockport, Mass 42°40.0'N., 70°36.5'W	Gloucester, Mass 42°36.6'N., 70°39.6'W	Salem, Mass 42°31.3'N., 70°52.5'W	Marblehead, Mass 42°30.2'N., 70°50.7'W	Lynn, Mass 42°27.3'N., 70°56.6'W	Boston, Mass 42°22.0'N., 71°03.0'W	Scituate, Mass 42°11.9'N., 70°43.5'W	Plymouth, Mass 41°57.6'N., 70°39.8'W	CAPE COD CANAL E ENT. 41°46.8'N., 70°29.0'W.	Provincetown, Mass 42°02.5'N., 70°10.0'W
Eastport	24																													
Lubec	26	3																												
Machiasport	61	46	35																											
Jonesport	66	42	40	20																										
Bar Harbor	98	83	72	52	34																									
Buck Harbor	125	109	98	80	60	39																								
Stonington	118	102	91	73	53	33	16																							
Rockland	145	130	118	100	82	62	22	20																						
Searsport	152	137	126	107	89	70	16	24	23																					
Bucksport	159	145	133	115	96	77	22	30	33	13																				
Bangor	176	162	150	132	113	94	39	47	50	30	17																			
Boothbay Harbor	168	153	142	123	105	86	57	53	42	61	70	87																		
Wiscasset	189	175	162	144	124	105	68	64	62	80	89	106	21																	
Bath	187	173	161	142	122	103	67	63	59	78	86	104	23	30																
Augusta	214	200	188	169	149	130	94	90	86	105	113	131	50	57	27															
Portland	198	183	172	153	135	115	85	81	71	90	98	115	36	44	40	66														
York Harbor	222	208	196	178	159	140	114	109	99	118	126	144	64	72	67	94	43													
Portsmouth	230	216	204	186	167	148	123	119	107	126	135	151	74	82	78	104	56	11												
Newburyport	241	226	215	197	178	159	137	132	118	137	145	162	86	94	89	116	67	25	22											
Rockport	236	221	210	192	173	154	131	126	115	134	143	160	84	92	88	115	66	29	27	16										
Gloucester	245	230	219	200	182	163	138	133	126	145	153	170	95	103	98	125	79	42	40	31	17									
Salem	252	237	226	208	189	170	149	145	133	152	160	177	102	110	105	132	87	49	47	38	24	12								
Marblehead	251	236	225	207	188	169	148	143	132	151	159	176	101	109	104	131	86	48	46	37	23	11	5							
Lynn	261	246	235	217	198	179	157	153	142	161	169	186	110	118	114	141	95	58	56	47	33	22	18	14						
Boston	265	250	239	221	203	184	162	158	147	166	174	191	115	123	119	146	100	63	61	51	37	26	27	19	13					
Scituate	259	244	233	215	196	177	159	155	142	161	169	186	112	120	115	142	97	60	58	48	34	26	24	22	22	21				
Plymouth	268	253	242	224	206	188	174	163	155	174	182	199	125	133	129	156	112	75	73	64	50	43	45	39	40	40	20			
CAPE COD CANAL E ENT.	270	255	244	226	209	190	176	165	160	178	187	204	130	139	137	164	118	83	81	72	58	52	53	48	48	52	29	20		
Provincetown	258	243	232	214	197	179	165	155	148	166	175	192	119	127	123	150	107	75	73	63	49	45	49	45	47	49	29	26	22	
NANTUCKET SHOALS 40°30.0'N., 69°25.0'W	312	297	286	271	257	243	237	226	223	242	250	267	207	217	213	240	203	182	180	171	157	155	159	156	159	163	143	144	144	132

Coastwise Distances
Cape Cod, MA to New York, NY
(Nautical Miles)

	CAPE COD CANAL E ENT. 41°46. 8'N., 70°29. 0'W.	NANTUCKET SHOALS 40°30. 0'N., 69°25. 0'W.	Nantucket, Mass. 41°17. 2'N., 70°05. 7'W.	Vineyard Haven, Mass. 41°27. 3'N., 70°35. 8'W.	Woods Hole, Mass. 41°31. 4'N., 70°40. 4'W	New Bedford, Mass. 41°38. 1'N., 70°55. 1'W	Newport, R. I. 41°29. 8'N., 71°19. 8'W	Fall River, Mass. 41°42. 4'N., 71°09. 8'W	Providence, R. I. 41°48. 5'N., 71°24. 0'W	Great Salt Pond, R. I. 41°11. 1'N., 71°34. 9'W	Stonington, Conn. 41°19. 9'N., 71°54. 6'W	New London, Conn. 41°21. 4'N., 72°05. 4'W	Hartford, Conn. 41°45. 0'N., 72°39. 0'W	New Haven, Conn. 41°17. 4'N., 72°54. 5'W	Stratford, Conn. 41°11. 3'N., 73°07. 3'W	Bridgeport, Conn. 41°10. 3'N., 73°10. 8'W	South Norwalk, Conn. 41°05. 7'N., 73°24. 7'W	Stamford, Conn. 41°01. 8'N., 73°32. 3'W	Montauk, N. Y. 41°02. 8'N., 71°57. 5'W	Sag Harbor, N. Y. 41°00. 2'N., 72°17. 7'W	Greenport, N. Y. 41°06. 0'N., 72°21. 5'W	Port Jefferson, N. Y. 40°57. 0'N., 73°04. 5'W	MONTAUK POINT, N. Y. 41°01. 7'N., 71°47. 3'W	NEW YORK, N. Y. 40°42. 0'N., 74°01. 0'W	Perth Amboy, N. J. 40°30. 3'N., 74°15. 7'W	Elizabethport, N. J. 40°38. 8'N., 74°11. 2'W
NANTUCKET SHOALS	144																									
Nantucket	69	85																								
Vineyard Haven	43	88	29																							
Woods Hole	22	92	33	7																						
New Bedford	31	111	53	28	15																					
Newport	54	111	71	45	38	38																				
Fall River	69	126	77	51	44	54	16																			
Providence	74	131	91	65	57	58	21	21																		
Great Salt Pond	66	114	80	54	50	48	23	38	43																	
Stonington	77	126	92	67	61	58	34	49	55	19																
New London	89	136	103	77	72	74	48	63	68	29	12															
Hartford	140	187	155	129	125	124	98	113	118	80	66	62														
New Haven	127	171	140	114	109	111	84	100	105	65	52	49	74													
Stratford	132	176	147	121	117	113	90	105	110	72	59	54	81	15												
Bridgeport	138	182	153	127	123	122	95	110	115	78	64	60	86	25	10											
South Norwalk	151	195	166	140	135	135	108	124	129	91	77	73	101	36	26	21										
Stamford	152	196	167	141	136	136	109	125	130	92	79	74	102	37	27	22	11									
Montauk	85	127	99	74	69	66	42	58	63	23	18	20	66	51	56	62	75	76								
Sag Harbor	100	143	114	88	83	81	57	72	77	39	29	27	64	49	54	60	73	74	21							
Greenport	99	142	113	87	82	80	56	71	76	37	28	25	62	47	52	58	71	72	22	11						
Port Jefferson	134	178	149	123	118	118	91	107	112	74	61	56	84	23	15	15	23	24	58	56	54					
MONTAUK POINT	76	113	89	63	59	60	35	51	56	15	19	28	75	62	65	74	84	85	16	32	30	68				
NEW YORK	182	223	196	171	166	166	139	154	159	121	109	104	131	68	57	52	40	33	105	103	102	52	122			
Perth Amboy	202	223	216	191	186	186	159	174	179	141	129	124	151	88	77	72	60	53	125	123	122	72	123	20		
Elizabethport	191	225	206	181	176	176	149	164	169	131	119	114	141	78	67	62	50	43	115	113	112	62	124	10	10	
Port Newark, N. J. 40°41. 8'N., 74°09. 0'W	193	227	208	183	178	178	151	166	171	133	121	116	143	80	69	64	52	45	117	115	114	64	126	12	15	5

Inside-Route Distances

South Side of Long Island
Greenport, NY to East Rockaway Inlet, NY

(Nautical Miles)

	Greenport 41°06.0'N., 72°21.5'W.	Sag Harbor 41°00.2'N., 72°17.7'W.	Riverside 40°55.0'N., 72°39.4'W.	Shinnecock Canal, N. End 40°53.9'N., 72°30.3'W.	Shinnecock Inlet 40°50.3'N., 72°28.6'W.	Westhampton Beach 40°48.2'N., 72°38.4'W.	Moriches Inlet 40°45.8'N., 72°45.3'W.	Bellport 40°45.1'N., 72°56.0'W.	Patchogue 40°45.5'N., 73°01.2'W.	Bay Shore 40°42.8'N., 73°14.2'W.	Fire Island Inlet 40°37.8'N., 73°18.6'W.	Babylon 40°41.2'N., 73°18.9'W.	Amityville 40°39.6'N., 73°24.8'W.	Jones Beach 40°36.2'N., 73°30.8'W.	Jones Inlet 40°34.4'N., 73°34.9'W.	Freeport 40°37.6'N., 73°34.9'W.	Long Beach 40°35.7'N., 73°39.4'W.	East Rockaway Inlet 40°34.9'N., 73°45.4'W.	Rockaway Point * 40°32.4'N., 73°56.5'W.	NEW YORK (The Battery) * 40°42.0'N., 74°01.0'W.	Manasquan Inlet, N. J. * 40°06.1'N., 74°01.9'W.
Sag Harbor	11																				
Riverside	21	22																			
Shinnecock Canal, N. End	16	17	8																		
Shinnecock Inlet	21	22	13	5																	
Westhampton Beach	28	29	20	12	9																
Moriches Inlet	34	35	26	18	15	7															
Bellport	42	43	34	26	23	15	11														
Patchogue	48	49	40	32	29	21	17	6													
Bay Shore	57	58	49	41	39	30	27	16	13												
Fire Island Inlet	62	63	54	46	44	35	32	21	18	9											
Babylon	61	62	53	45	42	34	30	19	17	5	8										
Amityville	66	67	58	50	47	39	35	24	22	10	12	6									
Jones Beach	72	73	64	56	54	45	42	31	28	17	16	13	7								
Jones Inlet	76	77	68	60	58	49	45	35	32	21	20	17	11	4							
Freeport	77	78	69	61	58	49	46	35	32	21	21	18	12	4	4						
Long Beach	80	81	72	64	61	53	49	38	36	24	24	21	15	8	5	6					
East Rockaway Inlet	85	86	77	69	66	58	54	44	41	29	29	26	20	13	10	11	5				
Rockaway Point	94	95	86	78	75	67	63	53	50	38	38	35	29	22	19	20	14	9			
NEW YORK (The Battery)	107	108	99	91	88	80	76	66	63	51	51	48	42	35	32	33	27	22	13		
Manasquan Inlet, N. J.	116	117	108	100	97	89	85	75	72	60	60	57	51	44	41	42	36	31	27	40	

Distances on Hudson River

New York, NY to Troy Lock, NY

(Nautical Miles)

	NEW YORK (The Battery) 40°42.0'N., 74°01.0'W.	Yonkers 40°56.1'N., 73°54.3'W.	Tarrytown 41°04.7'N., 73°52.2'W.	Nyack 41°05.4'N., 73°54.9'W.	Ossining 41°09.6'N., 73°52.3'W.	Haverstraw 41°11.8'N., 73°57.5'W.	Peekskill 41°17.3'N., 73°56.0'W.	West Point 41°23.1'N., 73°57.3'W.	Newburgh 41°30.1'N., 74°00.3'W.	Poughkeepsie 41°42.3'N., 73°56.5'W.	Hyde Park 41°47.3'N., 73°56.9'W.	Kingston 41°55.1'N., 73°59.0'W.	Saugerties 42°04.4'N., 73°56.7'W.	Catskill 42°13.0'N., 73°52.1'W.	Hudson 42°15.3'N., 73°48.1'W.	Athens 42°15.6'N., 73°48.5'W.	Coxsackie 42°21.0'N., 73°47.6'W.	Coeymans 42°28.5'N., 73°47.4'W.	Albany 42°37.9'N., 73°45.3'W.	Rensselaer 42°37.9'N., 73°45.1'W.	Troy 42°43.7'N., 73°41.8'W.	Watervliet 42°43.7'N., 73°41.9'W.	Troy Lock 41°45.1'N., 73°41.1'W.
Yonkers	16																						
Tarrytown	24	9																					
Nyack	25	10	2																				
Ossining	29	14	6	6																			
Haverstraw	33	18	9	10	5																		
Peekskill	38	23	15	15	11	6																	
West Point	45	29	21	22	17	13	8																
Newburgh	53	37	29	29	25	21	15	8															
Poughkeepsie	66	50	42	43	38	34	29	21	13														
Hyde Park	71	55	47	48	43	39	34	26	18	5													
Kingston	80	64	56	57	52	48	43	35	27	14	9												
Saugerties	89	74	66	66	62	58	52	45	37	24	19	12											
Catskill	99	83	75	75	71	67	61	54	46	33	28	21	11										
Hudson	102	86	78	78	74	70	64	57	49	36	31	24	14	5									
Athens	102	86	78	78	74	70	64	57	49	36	31	24	14	5	1								
Coxsackie	108	93	85	85	80	76	71	63	55	42	37	30	21	11	7	6							
Coeymans	115	100	92	92	88	84	78	70	62	49	44	38	28	19	14	14	7						
Albany	126	110	102	102	98	94	88	81	73	60	55	48	38	29	24	24	18	10					
Rensselaer	126	110	102	102	98	94	88	81	73	60	55	48	38	29	24	24	18	10	0				
Troy	132	116	108	108	104	100	94	87	79	66	61	54	44	35	30	30	24	16	6	6			
Watervliet	132	116	108	108	104	100	94	87	79	66	61	54	44	35	30	30	24	16	6	6	0		
Troy Lock	134	118	110	110	106	102	96	89	81	68	63	56	46	37	32	32	26	18	8	8	2	2	

Coastwise Distances

New York, NY to Chesapeake Bay Entrance, VA

(Nautical Miles)

	NANTUCKET SHOALS 40°30.0'N, 69°25.0'W	MONTAUK POINT, N.Y. 41°01.7'N, 71°47.3'W	NEW YORK, N.Y. 40°42.0'N, 74°01.0'W	Manasquan Inlet, N.J. 40°06.1'N, 74°01.9'W	Barnegat Inlet, N.J. 39°46.0'N, 74°06.3'W	Atlantic City, N.J. 39°22.6'N, 74°24.9'W	Cape May Harbor, N.J. 38°57.1'N, 74°52.6'W	DELAWARE BAY ENTRANCE 38°50.5'N, 75°03.3'W	Harbor of Refuge, Del. 38°49.0'N, 75°05.2'W	CHES. & DEL. CANAL E. ENT. 39°33.8'N, 75°32.8'W	Wilmington, Del. 39°43.2'N, 75°31.5'W	Marcus Hook, Pa. 39°48.2'N, 75°25.2'W	Chester, Pa. 39°50.0'N, 75°22.0'W	Philadelphia, Pa. 39°56.8'N, 75°08.3'W	U.S. Steel Basin, Pa. 40°08.2'N, 74°45.3'W	Trenton, N.J. 40°11.4'N, 74°45.4'W	Indian River Inlet, Del. 38°36.5'N, 75°03.6'W	Ocean City, Md. 38°19.6'N, 75°05.6'W	Chincoteague, Va. 37°56.1'N, 75°22.8'W
MONTAUK POINT, N.Y.	113																		
NEW YORK, N.Y.	223	122																	
Manasquan Inlet, N.J.	212	117	40																
Barnegat Inlet, N.J.	221	131	63	22															
Atlantic City, N.J.	242	159	94	52	32														
Cape May Harbor, N.J.	271	192	128	85	65	37													
DELAWARE BAY ENTRANCE	285	212	153	97	78	49	16												
Harbor of Refuge, Del.	285	212	153	98	79	50	17	2											
CHES. & DEL. CANAL E. ENT.	336	263	204	148	129	100	67	51	52										
Wilmington, Del.	347	274	215	159	140	111	78	62	63	11									
Marcus Hook, Pa.	353	280	221	165	146	117	84	68	69	17	8								
Chester, Pa.	356	283	224	169	150	121	88	72	73	21	11	3							
Philadelphia, Pa.	372	299	240	184	165	136	103	87	88	36	26	18	15						
U.S. Steel Basin, Pa.	395	322	263	207	188	159	126	110	111	59	49	41	38	23					
Trenton, N.J.	400	327	268	212	193	164	131	115	116	64	54	46	43	28	5				
Indian River Inlet, Del.	285	209	145	105	86	57	24	15	14	66	77	83	86	101	124	129			
Ocean City, Md.	295	227	161	121	101	73	40	32	31	83	95	101	104	119	142	147	20		
Chincoteague, Va.	328	262	201	161	141	113	80	72	71	123	134	140	144	159	182	187	60	41	
CHESAPEAKE BAY ENT. 36°56.3'N, 75°58.6'W	381	322	267	219	199	171	141	155	155	206	218	224	227	242	265	270	118	100	69

Distances by Intracoastal Waterway

Manasquan Inlet, NJ to Cape May Canal, NJ

(Nautical Miles)

	NEW YORK, N.Y. (The Battery)* 40°42.0'N, 74°01.0'W	Shark River Inlet* 40°11.2'N, 74°00.5'W	Manasquan Inlet* 40°06.1'N, 74°01.9'W	Bay Head 40°03.8'N, 74°03.1'W	Mantoloking 40°02.2'N, 74°03.4'W	Toms River (town) 39°56.9'N, 74°11.8'W	Seaside Park 39°55.3'N, 74°05.0'W	Forked River (town) 39°50.1'N, 74°11.7'W	Barnegat Inlet 39°46.0'N, 74°06.3'W	Beach Haven 39°34.0'N, 74°14.8'W	Atlantic City 39°22.6'N, 74°24.9'W	Mays Landing 39°26.9'N, 74°43.4'W	Ocean City 39°17.3'N, 74°34.4'W	Sea Isle City 39°09.4'N, 74°42.0'W	Avalon 39°06.6'N, 74°44.0'W	Stone Harbor 39°03.4'N, 74°46.0'W	Wildwood 39°00.5'N, 74°49.8'W	Cape May Harbor 38°57.1'N, 74°52.6'W	Cape May Canal W. Ent. 38°58.0'N, 74°58.0'W
Shark River Inlet*	34																		
Manasquan Inlet*	40	6																	
Bay Head	44	9	4																
Mantoloking	46	11	6	2															
Toms River (town)	58	23	18	14	12														
Seaside Park	54	20	14	10	9	7													
Forked River (town)	63	29	23	19	17	15	10												
Barnegat Inlet	66	32	26	22	20	18	13	8											
Beach Haven	79	45	39	35	33	31	26	21	20										
Atlantic City	97	62	57	53	51	49	44	39	38	18									
Mays Landing	124	90	84	80	79	77	71	66	65	45	30								
Ocean City	108	74	68	64	63	60	55	50	49	29	13	18							
Sea Isle City	119	85	79	76	74	72	66	61	60	40	25	29	11						
Avalon	123	89	83	79	77	75	70	65	64	44	28	33	15	4					
Stone Harbor	126	94	88	85	83	81	75	70	69	49	34	38	20	9	5				
Wildwood	133	99	93	89	88	86	80	75	74	54	39	43	25	14	10	5			
Cape May Harbor	138	103	98	94	92	90	85	80	79	59	43	47	30	18	15	9	5		
Cape May Canal W. Ent.	142	108	102	98	96	94	89	84	83	63	47	52	34	23	19	14	9	4	
CHES. & DEL. CANAL E. ENT. 39°33.8'N, 75°32.8'W	190	156	150	146	144	142	137	132	131	111	95	100	82	71	67	62	57	52	48

Distances on Delaware Bay and River

(Nautical Miles)

DELAWARE BAY ENT. 38°50. 5'N., 75°03. 3'W.	Roosevelt Inlet, Del. 38°47. 7'N., 75°09. 4'W.	Cape May Canal W. Ent., N.J. 38°58. 0'N., 74°58. 0'W.	St. Jones River Mouth, Del. 39°04. 0'N., 75°22. 5'W.	Mauricetown, N.J. 39°17. 1'N., 74°59. 5'W.	Bridgeton, N.J. 39°25. 5'N., 75°14. 2'W.	Smyrna River Mouth, Del. 39°22. 2'N., 75°30. 2'W.	Salem, N.J. 39°34. 6'N., 75°28. 7'W.	CHES. & DEL. CANAL E. ENT. 39°33. 8'N., 75°32. 8'W.	New Castle, Del 39°39. 4'N., 75°33. 6'W	Wilmington, Del 39°43. 2'N., 75°31. 5'W	Marcus Hook, Pa 39°48. 2'N., 75°25. 2'W	Bridgeport, N.J 39°48. 0'N., 75°21. 3'W	Chester, Pa 39°50. 0'N., 75°22. 0'W	Schuylkill River Mouth, Pa 39°52. 8'N., 75°11. 9'W	Philadelphia, Pa 39°56. 8'N., 75°08. 3'W	Burlington, N.J. 40°04. 9'N., 74°51. 8'W	U.S. Steel Basin, Pa 40°08. 2'N., 74°45. 3'W	Bordentown, N.J. 40°09. 1'N., 74°43. 0'W	Trenton, N.J. 40°11. 4'N., 74°45. 4'W.
6																			
9	14																		
20	20	21																	
33	37	26	30																
51	52	47	35	51															
39	40	36	21	39	25														
54	55	51	36	54	39	16													
51	52	48	34	51	36	13	5												
58	59	55	40	58	43	20	12	7											
62	63	59	45	62	47	24	16	11	5										
68	69	65	51	68	53	30	22	17	12	8									
72	73	69	55	72	57	34	26	21	15	11	4								
72	73	69	55	72	57	34	26	21	15	11	3	6							
80	81	77	63	80	65	42	34	29	23	19	12	14	9						
87	88	84	69	87	72	49	41	36	30	26	18	22	15	7					
103	104	100	85	103	88	65	57	52	46	42	34	37	31	23	16				
110	111	107	92	110	95	72	64	59	53	49	41	44	38	29	23	7			
112	113	109	94	112	97	74	66	61	55	51	43	46	40	31	25	9	2		
115	116	112	97	115	100	77	69	64	58	54	46	49	43	34	28	12	5	4	

Chesapeake Bay Distances

(Nautical Miles)

CHES. & DEL. CANAL E. ENT. 39°33. 8'N., 75°32. 8'W.	Chesapeake City, Md 39°31 8'N., 75°48 9'W	Havre de Grace, Md 39°32. 7'N., 76°05 0'W	Baltimore, Md 39°16 0'N., 76°34 5'W	Chestertown, Md 39°12 4'N., 76°03 8'W	Annapolis, Md 38°59 0'N., 76°28. 6'W	St Michaels, Md 38°47 2'N., 76°13. 2'W	Cambridge, Md 38°34 4'N., 76°04 3'W	Solomons, Md 38°19. 2'N., 76°27 4'W	Salisbury, Md 38°21 9'N., 75°36 3'W	Potomac River Mouth 37°57 7'N., 76°16 7'W	Washington, DC 38°52 4'N., 77°01 4'W	Crisfield, Md 37°58 6'N., 75°51 9'W	Fredericksburg, Va 38°17 8'N., 77°27 2'W	Cape Charles, Va 37°15 9'N., 76°01 4'W	Yorktown, Va 37°14 4'N., 76°30 5'W	West Point, Va 37°31 6'N., 76°48 1'W	Newport News, Va 36°58 0'N., 76°26 0'W	Suffolk, Va 36°44 3'N., 76°35 0'W	Hopewell, Va 37°19 0'N., 77°16 4'W	Petersburg, Va 37°14 1'N., 77°24 0'W	Richmond, Va 37°31 4'N., 77°25 2'W	Norfolk, Va 36°50 9'N., 76°17 9'W	CHESAPEAKE BAY ENT 36°56 3'N., 75°58 6'W
13																							
33	20																						
62	49	41																					
78	65	61	45																				
65	52	45	28	40																			
83	70	62	45	59	25																		
98	85	78	60	72	39	36																	
103	90	84	66	78	45	48	39																
143	130	124	107	119	86	89	81	51															
122	109	101	84	96	64	65	58	27	49														
213	200	192	175	187	155	156	149	118	141	96													
134	121	115	98	110	77	80	72	42	43	27	121												
243	230	224	206	219	186	190	182	150	165	125	221	129											
169	156	149	132	146	112	116	107	76	87	50	146	51	122										
192	179	163	155	156	130	126	117	87	100	68	164	64	132	28									
214	201	185	174	178	152	149	140	109	122	90	186	86	154	50	22								
207	194	172	170	167	136	138	129	97	110	86	182	74	143	29	55	63							
209	196	190	172	185	152	156	147	115	128	89	185	92	161	48	55	78	21						
265	252	230	228	225	194	196	187	155	168	144	240	132	201	88	114	122	58	79					
259	246	240	222	235	204	206	197	165	178	137	233	142	211	97	101	123	68	89	10				
284	271	249	247	244	213	215	206	174	187	163	259	151	220	106	132	140	77	98	19	28			
209	196	175	173	170	140	141	132	100	113	89	185	77	146	32	58	66	12	29	70	80	90		
187	174	166	150	162	129	132	123	92	103	67	163	67	136	21	34	56	24	42	82	92	101	27	

Coastwise Distances
Norfolk, VA to Key West, FL
(Nautical Miles)

	CHESAPEAKE BAY ENTRANCE	Norfolk, Va.	DIAMOND SHOALS	Morehead City, N.C.	Southport, N.C.	Wilmington, N.C.	Georgetown, S.C.	Charleston, S.C.	Port Royal, S.C.	Savannah, Ga.	Brunswick, Ga.	Fernandina Beach, Fla.	Jacksonville, Fla.	St Augustine, Fla.	Cape Canaveral, Fla.	Fort Pierce, Fla.	Stuart, Fla.	Port of Palm Beach, Fla.	Port Everglades, Fla.	Miami, Fla.	Key West, Fla.
CHESAPEAKE BAY ENTRANCE 36°56.3'N., 75°58.6'W.																					
Norfolk, Va. 46°50.9'N., 76°17.9'W.	27																				
DIAMOND SHOALS 35°08.0'N., 75°15.0'W.	117	144																			
Morehead City, N.C. 34°42.8'N., 76°41.8'W.	222	249	105																		
Southport, N.C. 33°54.8'N., 78°01.0'W.	315	342	198	133																	
Wilmington, N.C. 34°14.0'N., 77°57.0'W.	336	363	219	154	21																
Georgetown, S.C. 33°21.4'N., 79°16.9'W.	365	392	248	184	87	108															
Charleston, S.C. 32°47.2'N., 79°55.2'W.	402	429	285	220	130	151	79														
Port Royal, S.C. 32°22.3'N., 80°41.6'W.	465	492	348	284	191	212	141	90													
Savannah, Ga. 32°05.0'N., 81°05.7'W.	476	503	359	295	206	227	154	102	51												
Brunswick, Ga. 31°08.0'N., 81°29.7'W.	527	554	410	346	260	281	210	156	110	104											
Fernandina Beach, Fla. 30°40.3'N., 81°28.0'W.	533	560	416	352	265	286	216	166	120	115	50										
Jacksonville, Fla. 30°19.2'N., 81°39.0'W.	560	587	443	379	294	315	247	197	152	145	82	53									
St Augustine, Fla. 29°53.6'N., 81°18.5'W.	557	584	440	377	296	317	246	199	157	152	90	61	56								
Cape Canaveral, Fla. 28°24.6'N., 80°36.5'W.	612	639	495	438	367	388	324	283	251	251	195	169	167	120							
Fort Pierce, Fla. 27°27.5'N., 80°19.3'W.	647	674	530	476	407	428	368	329	298	298	242	216	214	167	69						
Stuart, Fla. 27°12.2'N., 80°15.6'W.	666	693	549	497	423	444	391	353	324	324	268	242	240	192	91	32					
Port of Palm Beach, Fla. 26°46.1'N., 80°03.0'W.	678	705	561	509	443	464	407	369	341	340	285	262	259	211	110	52	36				
Port Everglades, Fla. 26°05.6'N., 80°07.0'W.	720	747	603	550	485	506	449	411	383	382	327	304	301	253	152	94	78	46			
Miami, Fla. 25°47.0'N., 80°11.0'W.	743	770	626	573	508	529	472	434	406	405	350	327	324	276	175	117	101	68	27		
Key West, Fla. 24°33.7'N., 81°48.5'W.	881	908	764	711	646	667	610	572	544	543	488	465	462	414	313	255	239	207	165	151	
STRAITS OF FLORIDA 24°25.0'N., 83°00.0'W.	942	969	825	772	707	728	671	633	605	604	549	526	523	475	374	316	300	267	226	211	73

Inside-Route Distances
Norfolk, VA to Fernandina Beach, FL

Nautical miles

	Norfolk, Va.	Elizabeth City, N.C.	Hertford, N.C.	Columbia, N.C.	Edenton, N.C.	Plymouth, N.C.	Manteo, N.C.	Belhaven, N.C.	Washington, N.C.	Ocracoke, N.C.	Oriental, N.C.	New Bern, N.C.	Beaufort, N.C.	Morehead City, N.C.	Swansboro, N.C.	Jacksonville, N.C.	Wrightsville, N.C.	Wilmington, N.C.	Southport, N.C.	Little River, S.C.	Bucksport, S.C.	Georgetown, S.C.	McClellanville, S.C.	Charleston, S.C.	Beaufort, S.C.	Savannah, Ga.	Thunderbolt, Ga.	Brunswick, Ga.
Norfolk, Va. 36°50.9'N., 76°17.9'W.																												
Elizabeth City, N.C. 36°18.1'N., 76°13.0'W.	77*																											
Hertford, N.C. 36°11.6'N., 76°28.0'W.	89	39																										
Columbia, N.C. 35°55.0'N., 76°15.4'W.	89	39	26																									
Edenton, N.C. 36°03.3'N., 76°36.6'W.	98	48	35	25																								
Plymouth, N.C. 35°51.8'N., 76°45.6'W.	105	56	42	33	14																							
Manteo, N.C. 35°54.6'N., 75°40.2'W.	80	39	46	45	54	61																						
Belhaven, N.C. 35°32.1'N., 76°37.4'W.	120	79	79	78	87	95	70																					
Washington, N.C. 35°32.6'N., 77°03.7'W.	156	114	115	114	123	130	105	39																				
Ocracoke, N.C. 35°06.8'N., 75°59.1'W.	131	91	97	96	104	113	61	43	60																			
Oriental, N.C. 35°01.5'N., 76°41.8'W.	160	118	119	118	126	134	91	43	56	41																		
New Bern, N.C. 35°06.1'N., 77°02.1'W.	180	139	139	138	146	155	110	63	76	62	23																	
Beaufort, N.C. 34°43.1'N., 76°40.2'W.	177	134	137	136	144	152	110	61	74	59	22	38																
Morehead City, N.C. 34°42.8'N., 76°41.8'W.	178	136	137	136	145	152	110	61	74	59	22	38	3															
Swansboro, N.C. 34°41.0'N., 77°07.3'W.	200	158	160	158	166	175	132	83	96	81	44	60	24	22														
Jacksonville, N.C. 34°44.7'N., 77°26.3'W.	231	189	190	189	197	206	162	114	127	112	75	91	55	53	31													
Wrightsville, N.C. 34°13.1'N., 77°48.8'W.	246	204	205	204	213	221	178	130	142	128	91	107	70	69	46	48												
Wilmington, N.C. 34°14.0'N., 77°57.0'W.	273	231	232	231	240	248	205	157	169	155	118	134	97	96	73	75	27											
Southport, N.C. 33°54.8'N., 78°01.0'W.	268	226	227	226	235	243	201	152	164	150	113	129	93	91	69	70	22	21										
Little River, S.C. 33°52.2'N., 78°36.6'W.	299	257	259	257	266	274	232	183	195	181	144	160	124	122	100	101	53	52	31									
Bucksport, S.C. 33°39.0'N., 79°05.6'W.	328	286	287	286	294	303	260	211	224	209	172	188	152	150	128	129	82	81	60	28								
Georgetown, S.C. 33°21.4'N., 79°16.9'W.	352	311	311	310	318	327	283	235	248	233	196	213	177	174	152	154	106	105	84	53	24							
McClellanville, S.C. 33°04.7'N., 79°27.6'W.	374	332	333	332	340	349	306	257	270	255	218	235	199	196	174	176	128	127	106	75	46	24						
Charleston, S.C. 32°47.2'N., 79°55.2'W.	406	364	366	364	373	381	337	290	302	288	251	267	231	229	207	208	160	159	138	107	79	56	32					
Beaufort, S.C. 32°25.6'N., 80°40.2'W.	466	424	425	424	432	441	397	349	362	347	310	327	290	288	266	268	220	219	198	167	138	116	92	60				
Savannah, Ga. 32°05.0'N., 81°05.7'W.	508	466	467	466	475	483	440	392	404	390	353	369	333	331	309	310	262	261	240	209	181	159	135	104	43			
Thunderbolt, Ga. 32°01.5'N., 81°02.8'W.	507	465	466	465	474	482	439	391	403	389	352	368	332	330	307	309	261	260	239	208	179	157	133	103	41	14		
Brunswick, Ga. 31°08.0'N., 81°29.7'W.	595	553	554	553	562	570	527	479	491	477	440	456	420	418	396	397	349	349	327	296	268	246	222	191	130	102	88	
Fernandina Beach, Fla. 30°40.3'N., 81°28.0'W.	623	581	582	581	589	598	555	506	519	504	467	484	447	445	423	425	377	376	355	324	295	273	249	218	157	130	116	35

Statute miles

	Elizabeth City, N.C.	Hertford, N.C.	Columbia, N.C.	Edenton, N.C.	Plymouth, N.C.	Manteo, N.C.	Belhaven, N.C.	Washington, N.C.	Ocracoke, N.C.	Oriental, N.C.	New Bern, N.C.	Beaufort, N.C.	Morehead City, N.C.	Swansboro, N.C.	Jacksonville, N.C.	Wrightsville, N.C.	Wilmington, N.C.	Southport, N.C.	Little River, S.C.	Bucksport, S.C.	Georgetown, S.C.	McClellanville, S.C.	Charleston, S.C.	Beaufort, S.C.	Savannah, Ga.	Thunderbolt, Ga.	Brunswick, Ga.	Fernandina Beach, Fla.
Norfolk, Va.	89*	102	102	113	121	92	138	180	151	184	207	204	205	230	266	283	314	308	344	377	405	430	467	536	585	583	685	717
Elizabeth City, N.C.		45	45	55	64	45	91	131	105	136	160	154	157	182	217	235	266	260	296	329	358	382	419	488	536	535	636	669
Hertford, N.C.			30	40	48	53	91	132	112	137	160	158	158	184	219	236	267	261	298	330	358	383	421	489	537	536	638	670
Columbia, N.C.				29	38	52	90	131	110	136	159	157	157	182	217	235	266	260	296	329	357	382	419	488	536	535	636	669
Edenton, N.C.					16	62	100	142	120	145	168	166	167	191	227	245	276	270	306	338	366	391	429	497	547	545	647	678
Plymouth, N.C.						70	109	150	130	154	178	175	175	201	237	254	285	280	315	349	376	402	438	508	556	555	656	688
Manteo, N.C.							81	121	70	105	127	127	127	152	186	205	236	231	267	299	326	352	388	457	506	505	606	639
Belhaven, N.C.								45	49	49	72	70	70	96	131	150	181	175	211	243	270	296	334	402	451	450	551	582
Washington, N.C.									69	64	87	85	85	110	146	163	194	189	224	258	285	311	348	417	465	464	565	597
Ocracoke, N.C.										47	71	68	68	93	129	147	178	173	208	241	268	293	331	399	449	448	549	580
Oriental, N.C.											26	25	25	51	86	105	136	130	166	198	226	251	289	357	406	405	506	537
New Bern, N.C.												44	44	69	105	123	154	148	184	216	245	270	307	376	425	423	525	557
Beaufort, N.C.													3	28	63	81	112	107	143	175	203	229	266	334	383	382	483	514
Morehead City, N.C.														25	61	79	110	105	140	173	200	226	264	331	381	380	481	512
Swansboro, N.C.															36	53	84	79	115	147	175	200	238	306	356	353	456	487
Jacksonville, N.C.																55	86	81	116	148	177	203	239	308	357	356	457	489
Wrightsville, N.C.																	31	25	61	94	122	147	184	253	302	300	402	434
Wilmington, N.C.																		24	60	93	121	146	183	252	300	299	402	433
Southport, N.C.																			36	69	97	122	159	228	276	275	376	409
Little River, S.C.																				32	61	86	123	192	241	239	341	373
Bucksport, S.C.																					28	53	91	159	208	206	308	339
Georgetown, S.C.																						28	64	133	183	181	283	314
McClellanville, S.C.																							37	106	155	153	255	287
Charleston, S.C.																								69	120	119	220	251
Beaufort, S.C.																									49	47	150	181
Savannah, Ga.																										16	117	150
Thunderbolt, Ga.																											101	133
Brunswick, Ga.																												40

Inside-Route Distances

Fernandina Beach, FL to Key West, FL

Statute miles (above the diagonal)

Nautical miles (below the diagonal)

	Norfolk, Va.	Fernandina Beach, Fla.	Jacksonville, Fla.	St. Augustine, Fla.	Marineland, Fla.	Daytona Beach, Fla.	New Smyrna Beach, Fla.	Titusville, Fla.	Cocoa, Fla.	Eau Gallie, Fla.	Melbourne, Fla.	Vero Beach, Fla.	Fort Pierce, Fla.	Salerno, Fla.	Stuart, Fla.	Port Mayaca, Fla.	Clewiston, Fla.	Moore Haven, Fla.	Fort Myers, Fla.	Jupiter, Fla.	Port of Palm Beach, Fla.	Fort Lauderdale, Fla.	Port Everglades, Fla.	Miami, Fla.	Tavernier, Fla.	Matecumbe Harbor, Fla.	Marathon, Fla.	Flamingo, Fla.	Key West, Fla.
Norfolk, Va. 36°50.9'N., 76°17.9'W.		717	758	778	796	831	846	879	898	915	918	952	966	990	995	1026	1053	1066	1122	1005	1018	1064	1067	1090	1150	1170	1203	1208	1244
Fernandina Beach, Fla. 30°40.3'N., 81°28.0'W.	623		41	60	79	114	130	162	181	198	201	235	249	273	278	310	336	349	406	289	302	348	350	373	435	453	486	491	527
Jacksonville, Fla. 30°19.2'N., 81°39.0'W.	659	36		56	75	109	124	158	176	193	197	230	244	268	274	305	331	344	402	284	297	344	345	368	430	449	481	486	522
St. Augustine, Fla. 29°53.6'N., 81°18.5'W.	676	53	49		18	53	68	101	120	137	140	174	188	212	217	248	275	288	344	227	241	287	289	312	373	392	425	430	466
Marineland, Fla. 29°40.1'N., 81°13.0'W.	692	69	65	16		35	51	83	102	119	122	157	170	194	199	230	257	270	327	209	222	269	270	293	356	374	407	412	448
Daytona Beach, Fla. 29°12.6'N., 81°00.7'W.	722	99	95	46	30		16	48	67	84	87	121	135	159	165	196	222	235	292	175	188	235	236	259	321	339	372	377	413
New Smyrna Beach, Fla. 29°01.7'N., 80°55.1'W.	735	113	108	59	44	14		32	52	68	72	106	120	144	150	181	207	220	276	159	173	219	220	243	305	323	356	362	397
Titusville, Fla. 28°37.3'N., 80°47.9'W.	764	141	137	88	72	42	28		20	36	39	74	87	112	116	147	174	188	244	127	139	186	188	211	273	291	323	329	365
Cocoa, Fla. 28°21.3'N., 80°43.1'W.	780	157	153	104	89	58	45	17		16	21	54	68	92	98	129	155	168	224	107	121	167	168	191	253	273	304	311	346
Eau Gallie, Fla. 28°07.9'N., 80°37.1'W.	795	172	168	119	103	73	59	31	14		3	38	52	76	81	112	138	152	208	91	104	151	152	175	237	257	288	293	329
Melbourne, Fla. 28°05.0'N., 80°35.5'W.	798	175	171	122	106	76	63	34	18	3		34	47	71	77	108	135	148	205	87	100	147	148	171	234	252	284	290	326
Vero Beach, Fla. 27°39.0'N., 80°22.4'W.	827	204	200	151	136	105	92	64	47	33	29		14	38	44	75	101	114	170	53	67	113	114	137	199	219	250	257	292
Fort Pierce, Fla. 27°27.5'N., 80°19.3'W.	839	216	212	163	148	117	104	76	59	45	41	12		24	30	61	87	100	157	39	53	99	100	123	185	205	236	243	278
Salerno, Fla. 27°08.8'N., 80°11.6'W.	860	237	233	184	169	138	125	97	80	66	62	33	21		9	40	67	81	137	20	32	79	80	104	166	185	216	222	258
Stuart, Fla. 27°12.2'N., 80°15.6'W.	865	242	238	189	173	143	130	101	85	70	67	38	26	8		31	58	70	128	25	38	85	86	109	171	190	222	228	264
Port Mayaca, Fla. 26°59.1'N., 80°36.8'W.	892	269	265	216	200	170	157	128	112	97	94	65	53	35	27		26	40	96	56	69	116	117	141	203	221	253	259	295
Clewiston, Fla. 26°45.6'N., 80°55.2'W.	915	292	288	239	223	193	180	151	135	120	117	88	76	58	50	23		13	70	83	96	143	144	167	229	247	280	285	321
Moore Haven, Fla. 26°50.0'N., 81°05.3'W.	926	303	299	250	235	204	191	163	146	132	128	99	87	70	61	34	11		56	96	108	155	156	180	242	261	292	298	334
Fort Myers, Fla. 26°38.9'N., 81°52.3'W.	975	353	349	299	284	254	240	212	195	181	178	148	136	119	111	83	61	49		152	166	212	213	236	298	318	349	356	390
Jupiter, Fla. 26°56.8'N., 80°05.4'W.	873	251	247	197	182	152	138	110	93	79	76	46	34	17	22	49	72	83	132		14	60	61	84	146	166	197	204	239
Port of Palm Beach, Fla. 26°46.1'N., 80°03.0'W.	885	262	258	209	193	163	150	121	105	90	87	58	46	28	33	60	83	94	144	12		47	48	71	133	152	184	190	226
Fort Lauderdale, Fla. 26°06.8'N., 80°07.2'W.	925	303	299	249	234	204	190	162	145	131	128	98	86	69	74	101	124	135	184	52	41		1	24	86	106	137	144	178
Port Everglades, Fla. 26°05.6'N., 80°07.0'W.	927	304	300	251	235	205	191	163	146	132	129	99	87	70	75	102	125	136	185	53	42	1		23	85	105	136	142	177
Miami, Fla. 25°47.0'N., 80°11.0'W.	947	324	320	271	255	225	211	183	166	152	149	119	107	90	95	122	145	156	205	73	62	21	20		62	82	113	120	154
Tavernier, Fla. 25°00.7'N., 80°31.3'W.	999	378	374	324	309	279	265	237	220	206	203	173	161	144	149	176	199	210	259	127	116	75	74	54		22	54	60	96
Matecumbe Harbor, Fla. 24°51.1'N., 80°44.5'W.	1017	394	390	341	325	295	282	253	237	223	219	190	178	161	165	192	215	227	276	144	132	92	91	71	19		33	40	75
Marathon, Fla. 24°42.2'N., 81°06.7'W.	1045	422	418	369	353	323	309	281	264	250	247	217	205	188	193	220	243	254	303	171	160	119	118	98	47	29		39	48
Flamingo, Fla. 25°08.5'N., 80°55.4'W.	1050	427	423	374	358	328	315	286	270	255	252	223	211	193	198	225	248	259	309	177	165	125	123	104	52	35	34		84
Key West, Fla. 24°33.7'N., 81°48.5'W.	1081	458	454	405	389	359	345	317	301	286	283	254	242	224	229	256	279	290	339	207	196	155	154	134	83	65	42	73	

Acknowledgements

Embassy Marine Publishing would like to thank the following individuals and organizations for their invaluable assistance and expertise in producing *Embassy's Coastal Cruising Guide: The Atlantic Coast.*

Ken Textor
Laura Pope
Ellen Schneider
Joseph P. Cavaluzzi
Ronald L. Speer
Cy Liberman
Eddie Jones
Gloria A. Shearin
Joseph Devenney
Becky Bovell
Kimberly Kennedy
Mike Beringer
Richard Gilbert
Gary Stogner
Ann Kennard
Gigi Windley
Rebecca Moore
Karen G. Shaffer
John Johnson
Karin Koser
Thomas Dimock
Isabelle Becker
Joan Carlson
Margaret Partner
Paul Spezzaferri
Mike Rounds

Custom Navigation Systems, Westbrook, CT
Duke University, Durham, NC
Joseph Kelley, University of Maine
Francis McQueeney, United States Customs Service
Arlene Dietz, Army Corps of Engineers Navigational Data Center
National Hurricane Center
Ann B. Deaton, United States Coast Guard
Joel A. Johnson Sr., North Carolina Department of Transportation
Maine Office of Tourism
New Hampshire Office of Travel & Tourism Development
Massachusetts Office of Travel & Tourism
New Jersey Division of Travel & Tourism
Delaware Development Office
Maryland Office of Tourism Development
Virginia Division of Tourism
North Carolina Division of Travel and Tourism
South Carolina Department of Parks, Recreation, & Tourism
Georgia Department of Industry, Trade, & Tourism
Florida Division of Tourism

EMBASSY'S COASTAL NAVIGATION GUIDE®

How to Navigate

Navigational buoys, markers and lights come in all shapes, sizes and colors; but if you follow these three basic rules, coastal navigation becomes like "connect the dots" — follow from one buoy to another, and each marker will tell you what to do next. For instance:

Always keep the **Red** buoys on your **RIGHT** when **RETURNING** to land or going up a river.

Always keep the **GREEN** buoys on your **LEFT** when returning to land or going up a river.

(If you are going out to sea, or down river — do the reverse.)

All buoys are either rectangles or cones. Rectangles are always painted **GREEN** and should be kept on your **LEFT** when returning to land or going up a river. Cones are always painted **RED** and should be kept on your **RIGHT** when **RETURNING** to land or going up a river, as shown below:

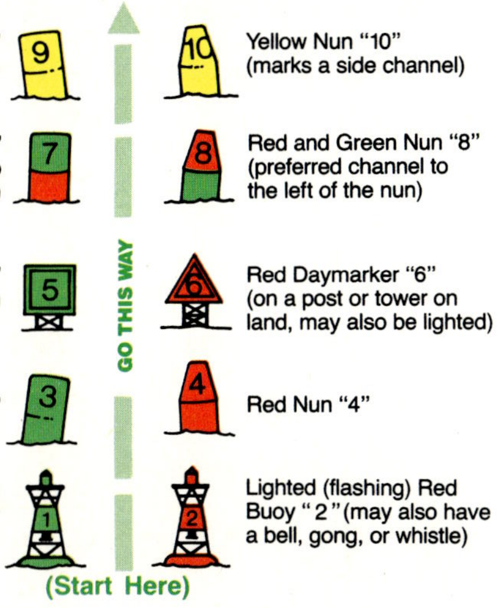

GREEN buoys are **ODD** numbers. **LEFT** is **PORT**

RED buoys are **EVEN** numbers. **RIGHT** is **STARBOARD**

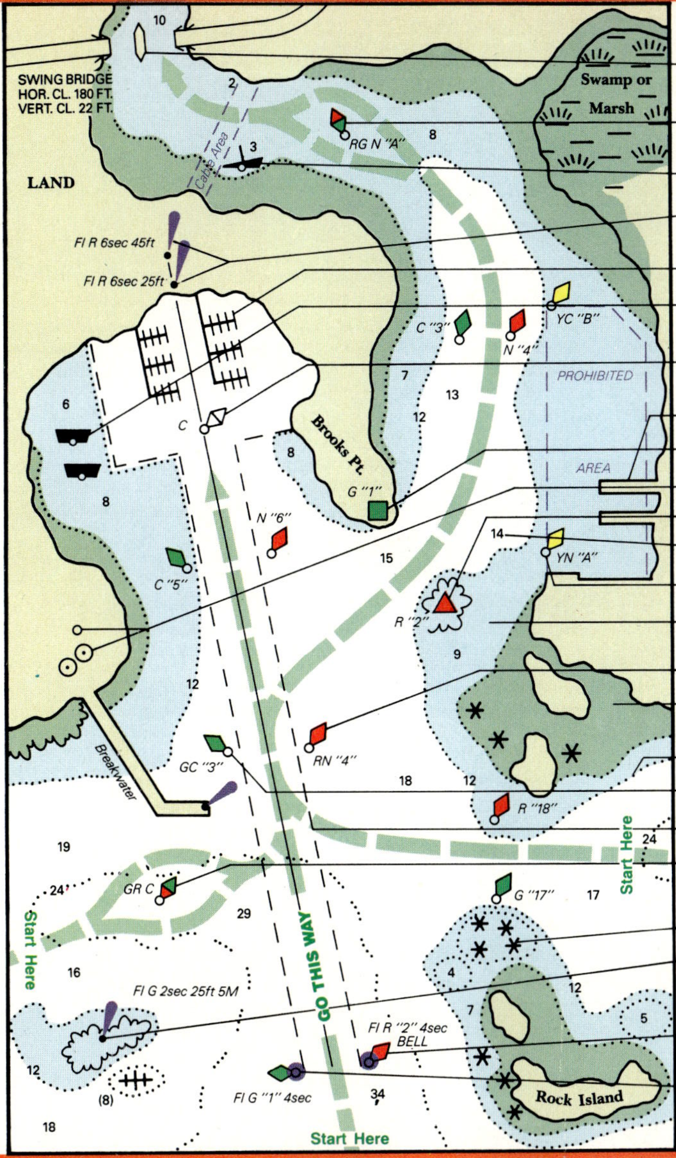

Legend

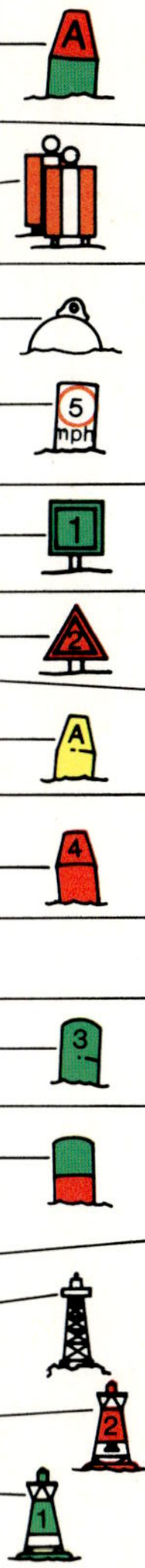

- Turnstile or swing bridge with vertical clearance (in feet) at high tide, and horizontal clearance.
- Red and green nun "A" — preferred channel is to the left of the nun.
- Partially submerged wreck.
- Range Markers — line up both markers for a safe approach to harbor.
- Marina docks, slips, and finger piers.
- Commercial mooring buoys — usually large, unlighted, steel balls of various colors.
- White, orange, and black caution buoy — may warn of speed limits, dangers, or other restrictions.
- Piers, commercial docks, wharfs, etc.
- Green daymark "1" on a post.
- Towers, steeples, water tanks, etc.
- Red daymark "2" on a tower.
- Sounding or depth of water (in feet) at low tide.
- Yellow nun "A" marking a side channel to commercial dock.
- Shallow water is shown in blue, deeper water is shown in white.
- Red nun "4".
- Shoals, shallows, sand bars, and mud flats which are visible at low tide but covered by water at high tide.
- Water depth contour line.
- Green can "3".
- Shipping channel or fairway. Always keep to the right.
- Green and red can — preferred channel is to the right of the can.
- Rocks awash at low tide.
- 25-foot-tall tower with a green light flashing every 2 seconds, visible for 5 miles, and surrounded by protective riprap or rocks.
- Lighted red bell buoy "2" with a red light flashing every 4 seconds.
- Lighted green buoy "1" with a green light flashing every 4 seconds.

How to get HELP

Wave arms, hold brightly colored towel or white shirt.

Sound Signals: S.O.S.
● ● ● — — — ● ● ●
or repeated long blasts

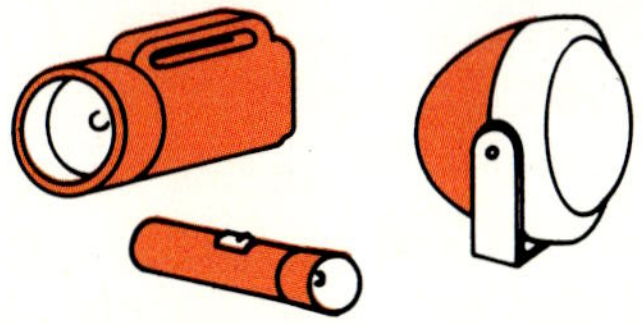

Light Signals: S.O.S.
● ● ● — — — ● ● ●

Flares or Smoke Signals

VHF Radio Channel 16
CB Radio Channel 9
Call "MAYDAY," Give Name, Location, Describe Emergency

Bridges of the Intracoastal Waterway

Virginia to Florida ICW Mile: 2.2 to 1134.0

Sound Signals to open bridges: One long and one short blast of the horn. All movable bridges will open on demand except as noted below.

Statute Mile	Bridge Name	Bridge Type	Clearance @ Center	VHF	Restricted Period	Scheduled Openings
			Virginia			
2.2	N & PB L Railroad	Lift	6'			Open except for traffic
2.8	Jordan Highway (Rte. 337)	Lift	15'	13–16	6:30–7:30am & 3:30–5pm	No openings
3.6	Norfolk & Western RR	Lift	10'			Open except for traffic
5.8	Gilmerton Highway	Bascule	11'	13–16		
7.1	I64 Highway	Bascule	65'			
7.8	N & PBL RR	Swing	7'	13–16		
8.8	Rte. 104	Bascule	12'	13–16	7am–8am & 4pm–6pm, M–F	
					(except Federal holidays)	Opens at 5pm
11.6	Great Bridge Lock	Lock			6am–7pm	On the hour
12.0	Great Bridge	Swing	6'	13–16	6am–7pm	On the hour
12.4	Rte. 168 bypass	Fixed	65'			
14.0	N S RR	Bascule	7'			Open except for traffic
15.2	Rte. 604	Swing	4'	13–16	7am–7pm	On the hour & half hour
22.0	North Landing	Swing	6'	13–16	6am–7pm	On the hour & half hour
28.0	Pungo Ferry	Fixed	65'			
			North Carolina			
50.1	Coinjock	Fixed	65'			
84.3	Alligator River	Swing	14'	13–16		
113.7	Fairfield	Swing	7'	13–16	Apr. 1 to Nov. 30, 7am–7pm	On the hour & half hour
125.7	Wilkerson	Fixed	65'			
157.2	Hobucken	Swing	6'	13–16	Apr. 1 to Nov. 30, 7am–7pm	On the hour & half hour
195.8	Core Creek	Swing	16'	13–16	Apr. 1 to Nov. 30, 6am–7pm	On the hour & half hour
203.0	Morehead City	Fixed	65'			
203.0	Morehead City RR	Bascule	4'			
206.5	Atlantic Beach	Fixed	65'			
225.7	Cedar Point	Fixed	65'			
240.6	Onslow Beach	Swing	12'	13–16		
252.0	Near Alligator Bay	Fixed	65'	13–16		
260.7	Surf City	Swing	12'	13–16	7am–7pm	On the hour
278.5	Figure Eight Island	Swing	20'	13–16		
283.1	Wrightsville	Bascule	20'	13–16	7am–7pm	On the hour
295.8	Snows Cut	Fixed	65'			
311.8	Fort Caswell	Fixed	65'			
323.6	Holden Beach	Fixed	65'			
333.8	Ocean Isle	Fixed	65'			
337.9	Sunset Beach	Pontoon		13–16	Apr. 1 to Oct. 31, 7am–7pm	On the hour
347.1	Little River	Fixed	65'			
347.3	Little River	Swing		13–16		
365.5	Highway and RR	Fixed	65'			
371.0	Socastee (Rte. 544)	Swing	10'	13–16	Apr 1–Jun 30 & Oct 1–Nov 30,	
					7am–10am, 2pm–6pm, M–F	
					(except Federal holidays)	On the 1/4 & 3/4 hour
					May 1–Jun 30 & Oct 1–Oct 30,	
					10am–2pm, Sat, Sun & Federal holidays	On the 1/4 & 3/4 hour
371.0	Socastee	Fixed	65'			
401.1	Lafayette	Fixed	65'			
462.0	Highway	Fixed	65'			
462.2	Ben Sawyer	Swing	31'	13–16	7am–9am & 4pm–6pm, M–F,	
					(except Federal holidays)	No openings
					9am–7pm, Sat & Federal holidays	On the hour
470.8	Wappoo Creek (Rte. 171)	Bascule	33'	13–16	6:30am–9am & 4pm–6:30pm, M–F	
					(except Federal holidays)	No openings
					Apr 1–Nov 30, 9am–4pm, M–F,	
					(except Federal holidays)	On the hour & half hour
					Apr 1–Nov 30 9am–7pm,	
					Sat & Sun & Federal holidays	On the hour & half hour
479.3	John Limehouse	Swing	12'	13–16	6:30am–9am & 4pm–6:30pm, M–F	
					(except Federal holidays)	On the hour & half hour
					Mar 15–Jun15 & Sep 15–Nov 15,	
					9am–4pm, M–F (except Federal holidays)	Every 20 minutes

Continued on next page . . .

Statute Mile	Bridge Name	Bridge Type	Clearance @ Center	VHF	Restricted Period	Scheduled Openings
			North Carolina continued			
501.1	Whooping Island (Rte. 174)	Swing	8'	13–16		
536.0	Lady's Island	Swing	30'	13–16	7am–9am & 4pm–6pm, Mon–Sat	
					(except Federal holidays)	On the hour
					Apr, May, Jun, Sep, Oct & Nov,	
					9am–4pm, Mon–Sat	
					(except Federal holidays)	Every 20 minutes
539.6	Burkmeyer Beach	Fixed	65'			
557.5	Hilton Head	Fixed	65'			
			Georgia			
579.9	Causton Bluff	Bascule	26'	13–16	6:30am–9am & 4:30pm–6:30pm,	
					M-F (except Federal holidays)	7am, 8am & 5:30pm
583.0	Thunderbolt	Fixed	65'			
592.5	Skidway Narrows	Bascule	22'	13–16		
675.0	St. Simons	Fixed	65'			
684.5	Jekyll Creek	Lift	9'	13–16		
			Florida			
720.6	SS RR	Swing	5'	13–16		
720.7	Highway	Fixed	65'			
739.1	Sisters Creek	Bascule	24'	13–16		
744.7	Salamander Ldg	Fixed	65'			
747.5	McCormick Bridge	Bascule	37'	13–16	Apr–May & Oct–Nov, 7am–8:30am,	
					4:30pm–6pm, M–F (except Federal holidays)	On the hour & half hour
					Apr–May & Oct–Nov, 12noon–6pm,	
					Sat, Sun & Federal holidays	On the hour & half hour
749.4	Pablo Creek	Fixed	65'			
759.0	Palm Valley	Bascule	9'	13–16		
775.8	Vilano Beach	Lift	5'	13–16	Mar 15–Dec 15, 7am–6pm, M–F	
					(except Federal Holidays)	Every 20 minutes
					Mar 15–Dec 15, 9am to sunset	
					Sat, Sun & Federal holidays	Every 20 minutes
777.9	Bridge of Lions	Bascule	25'	13–16	7am–6pm, except 8am, noon & 5pm, M–F	On the hour & half hour
					7am–6pm	
					Sat, Sun & Federal holidays	On the hour & half hour
780.0	Highway	Fixed	65'			
789.0	Crescent Beach	Bascule	25'			
802.7	Palm Coast	Fixed	65'			
810.6	Flagler Beach	Bascule	14'	13–16		
815.9	Highbridge Road	Bascule	15'	13–16		
824.9	Ormond Beach	Fixed	65'			
829.1	Seabreeze Blvd.	Bascule	20'	13–16	7:30–8:30am & 4:30–5:30pm, Mon–Sat	
					(except Federal holidays)	8am & 5pm
829.7	Main Street	Bascule	22'	13–16		
830.1	Broadway	Bascule	20'	13–16		
830.6	Memorial Bridge	Bascule	21'	13–16	7:45–8:45am & 4:45–5:45pm, Mon–Sat	
					(except Federal holidays)	8:15am & 5:15pm
834.5	Port Orange	Fixed	65'			
845.0	Coronado Beach	Bascule	14'	13–16	7am–6pm	Every 15 minutes
846.3	Harris Saxon	Fixed	65'			
868.2	Haulover Canal	Bascule	27'			
876.6	NASA RR	Bascule	7'	13–16		Open except for traffic
878.9	Rte. 402	Swing	9'	13–16	M–F, 6:15–7:15am & 3pm to 4:30pm	
					(except Federal holidays)	No openings
885.0	JFK Space Center	Bascule	27'	13–16	M–F, 6:30–8am & 3:30–5pm,	
					(except Federal holidays)	No openings
894.2	City Point	Fixed	65'			
897.0	Merritt Island	Fixed	65'			
909.1	Palm Shores	Fixed	65'			
914.0	Eau Gallie	Fixed	65'			
918.4	Melbourne	Fixed	65'			
943.5	Near Wabasso Beach	Fixed	65'			
951.9	Merrill Barber	Bascule	22'	13–16	M–F, 7:45–9am, 12–1:15pm,4–5:15pm	
					(except Federal holidays)	8:30am, 12:30 & 4:30pm
					Oct 1–May 31, M–F, 7am–6pm & as above	
					(except Federal holidays)	Every 15 minutes
953.4	Dark Point	Fixed	65'			
965.0	Ft. Pierce N.	Bascule	26'	13–16		
966.0	Ft. Pierce S.	Fixed	65'			

Continued on next page . . .

Statute Mile	Bridge Name	Bridge Type	Clearance @ Center	VHF	Restricted Period	Scheduled Openings
			Florida continued			
981.4	Jensen Beach	Bascule	24'	13–16	Dec 1–May 1, 7am–6pm, M–F	
					(except Federal holidays)	On the hour & half hour
984.9	Ernest Lyons	Bascule	28'	13–16	Dec 1–May 1, 7am–6pm, M–F	
					(except Federal holidays)	On the hour & half hour
995.8	Rte. 708	Bascule	21'	13–16		
1004.2	Jupiter	Bascule	25'	13–16		
1004.8	Jupiter (Rte. 1)	Bascule	26'	13–16		
1006.2	Indiantown Road	Bascule	15'	13–16	7am–6pm	Every 20 minutes
1009.4	Donald Ross	Bascule	14'	13–16		
1012.6	PGA Blvd.	Bascule	24'	13–16	7am–9am & 4am–7pm, M–F	
					(except Federal holidays)	On the 1/4 & 3/4 hour
					Nov 1–Apr 30, 9am–4pm, M–F	
					(except Federal holidays)	Every 20 minutes
					8am–6pm, Sat, Sun & Federal holidays	Every 20 minutes
1013.7	Parker (US1)	Bascule	25'	13–16	7–9am & 4–7pm, M–F	
					(except Federal holidays)	On the hour & half hour
					8am–6pm, Sat, Sun & Federal holidays	Every 20 minutes
1016.9	Riviera Beach	Fixed	65'			
1021.9	Flagler Memorial	Bascule	25'	13–16	Nov 1–May 31, 8am–9:30am, 4pm–5:45pm,	
					8:30am & 4:45pm, M–F (except Federal holidays)	8:30am & 4:45pm
					Nov 1–May 31, 9:30am–4pm, M–F	
					(except Federal holidays)	On the hour & half hour
1022.6	Royal Park	Bascule	14'	13–16	Nov 1–May 31, 8am–9:30am, 3:30–5:45pm, M–F	8:45am, 4:15pm & 5pm
					Nov 1–May 31, 9:30am–3:30pm, M–F	
					(except Federal holidays)	On the 1/4 & 3/4 hour
1024.7	Southern Blvd.	Bascule	14'	13–16	Nov 1–May 31, 7:30am–9am & 4:30pm–6:30pm	
					M–F, (except Federal holidays)	8:15am & 5:30pm
1029.0	Lake Worth	Bascule	35'	13–16		
1031.0	Lantana Ave.	Bascule	13'	13–16	Dec 1–Apr 30, 10am–6pm	
					Sat, Sun & Federal holidays	Every 15 minutes
1035.0	Boynton Beach	Bascule	10'	13–16		
1035.7	Ocean Ridge	Bascule	25'	13–16		
1038.7	NE 8th St. Delray Beach	Bascule	9'	13–16	Nov 1–May 31, 11am–6pm	
					Sat, Sun & Federal holidays	Every 15 minutes
1039.6	Atlantic Ave.	Bascule	12'	13–16	Nov 1–May 31, 10am–6pm, M–F	On the hour & half hour
1041.0	12th Street	Bascule	30'	13–16		
1044.9	NE 40th St (Boca Raton)	Bascule	25'	13–16		
1047.8	Palmetto	Bascule	19'	13–16		
1048.2	Camino Real	Bascule	9'	13–16	7am–6pm	Every 15 minutes
1050.0	Hillsboro Blvd.	Bascule	21'	13–16	Oct 1–May 31, 7am–6pm	Every 15 minutes
1055.0	N.E. 14th Street	Bascule	15'	13–16	7am–6pm	On the 1/4 & 3/4 hour
1056.0	Atlantic Blvd. (Rte. 814)	Bascule	15'	13–16	7am–6pm	On the hour and half hour
1059.0	Commercial Blvd.	Bascule	15'	13–16	Nov 1–May 15, 8am–6pm	Every 15 minutes
1060.5	Oakland Park Blvd.	Bascule	22'	13–16	Nov 15–May 15, 7am–10pm, M–F	Every 20 minutes
					Nov 15–May 15, Sat, Sun & Federal holidays	Every 15 minutes
1062.4	Sunrise Blvd.	Bascule	25'	13–16		
1064.3	Los Alos Blvd.	Bascule	31'	13–16		
1065.9	Brooks Memorial	Bascule	25'	13–16	7am–7pm	15 min. after last closure
					(A clock will display time until next opening)	
1069.5	Dania Town Canal	Bascule	22'	13–16		
1070.6	N. Hollywood	Bascule	22'	13–16		
1072.2	Hollywood Beach Blvd.	Bascule	25'	13–16	Nov 15–May 15, 10am–6pm,	
					(except Federal holidays)	On the hour & half hour
					May 16–Nov 14, 9am–7pm,	
					Sat, Sun & Federal holidays	On the hour & half hour
1074.1	Hallandale Beach Blvd	Bascule	22'	13–16	7:15am–6:15pm	On the 1/4 & 3/4 hour
1076.1	Dumfoundling Bay	Fixed	65'			
1078.0	Rte. 826	Bascule	19'	13–16	7am–6pm, M–F (except Federal holidays)	On the 1/4 & 3/4 hour
					10am–6pm, Sat, Sun & Federal holidays	On the 1/4 & 3/4 hour
1081.4	Broad Causeway	Bascule	16'	13–16	8am–6pm	On the 1/4 & 3/4 hour
1084.6	79th Street Causeway	Bascule	25'	13–16		
1086.9	Julia Tuttle Causeway	Fixed	56'			
1088.6	Venetian Causeway	Bascule	8'	13–16	Nov 1–Apr 30, 7am–9am & 4:30–6:30pm	
					M–F, (except Federal holidays)	On the hour & half hour
1088.8	MacArthur Causeway	Bascule	35'	13–16	Nov 1–Apr 30, 7–9am & 4:30–6:30pm	On the hour & half hour
1089.4	Dodge Island	Fixed	65'			
1089.4	Dodge Island RR	Bascule	22'			Open except for traffic
1091.5	Rickenbocker Causeway	Fixed	76'			
1134.0	Jewfish Creek	Bascule	14'	13–16	10pm–sunset, Thur–Sun & Federal holidays	On the hour & half hour

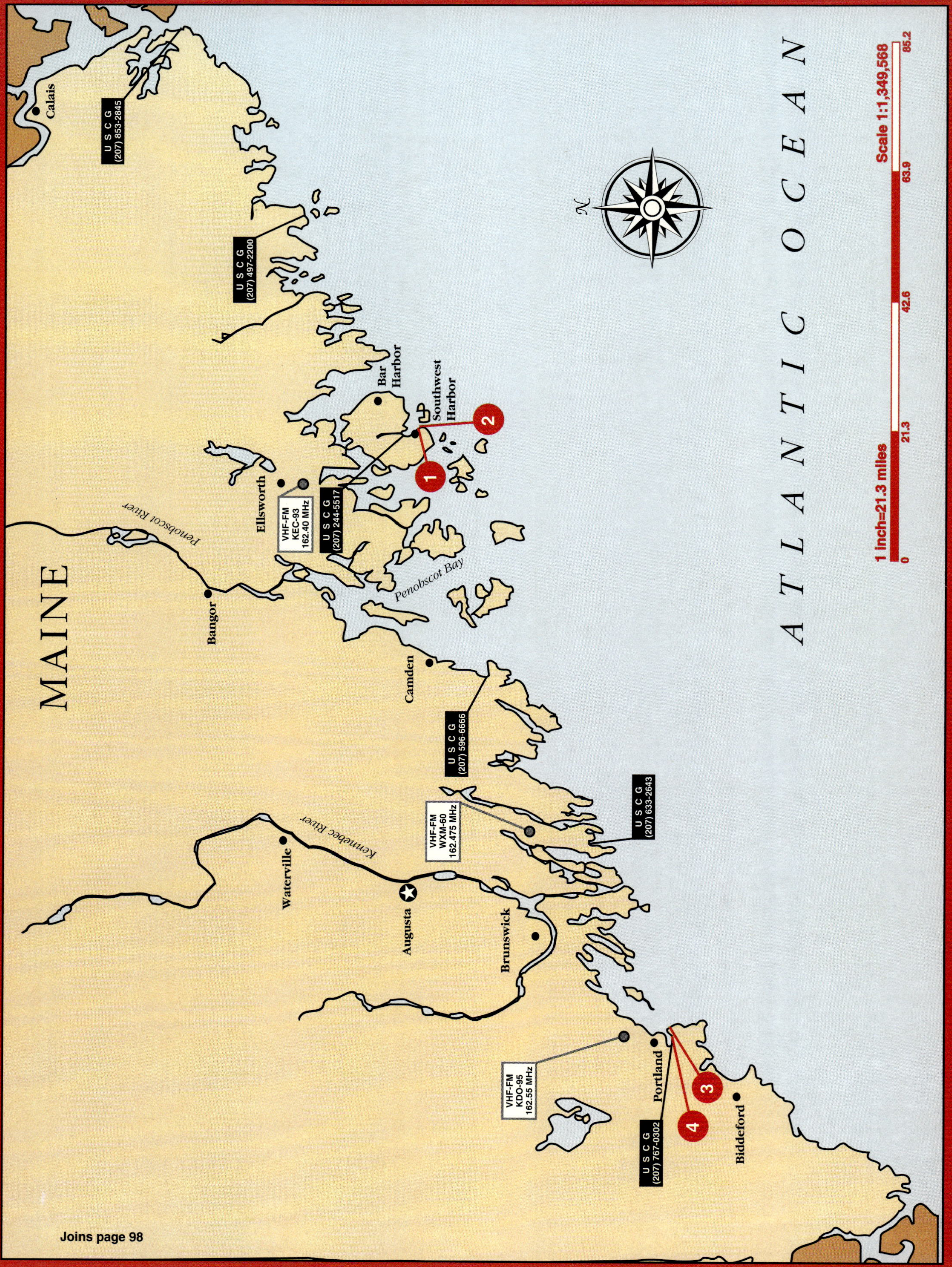
MAINE
ATLANTIC OCEAN
N
Calais
U S C G (207) 853-2845
U S C G (207) 497-2200
Bar Harbor
Southwest Harbor
1
2
Ellsworth
VHF-FM KEC-93 162.40 MHz
U S C G (207) 244-5517
Penobscot River
Bangor
Penobscot Bay
Camden
U S C G (207) 596 6666
U S C G (207) 633-2643
VHF-FM WXM-60 162.475 MHz
Kennebec River
Waterville
Augusta
Brunswick
VHF-FM KDO-95 162.55 MHz
Portland
3
4
Biddeford
U S C G (207) 767-0302
Scale 1:1,349,568
1 inch=21.3 miles
0
21.3
42.6
63.9
85.2
Joins page 98

The Pine Tree State

Exploring the coast of Maine could take a lifetime. And considering the nearly 2,500 miles of shoreline with its unspoiled natural beauty, myriad of safe harbors, long list of things to do, and abundance of the state's patron saint—the boiled lobster, it would be a life well spent. Maine is one of those cruising areas where the nautical dream matches the reality.

The reality does vary, though. There are coastal areas where you will wonder if anyone other than a couple of lobstermen live there. Then, there are other areas that bustle with activities from weekly regattas to world-famous shopping opportunities. There are sandy beaches and rocky shores, frigid waters and tepid swimming holes, ranges of mountains and plains of salt marsh. But mostly, Maine's coast is a quiet place to get away from it all and enjoy an anchorage to yourself.

© Joseph Devenney/Maine Office of Tourism

A classic Maine windjammer sails past Rockland breakwater.

The limitations on this dream come in the form of numerous navigational hazards, thick fog, and capricious tidal currents. The weather is fickle, too, keeping the wise captain always alert for subtle wind shifts and cloud formations. In general, there are few problems a seasoned skipper can't overcome, but planning and preparation are essential.

Personal preparations as well as nautical ones are necessary. Maine's mid-summer weather can vary from shorts and T-shirt to sweater and jacket in the space of a few hours. A well-stocked galley is important because some sections of the coast, especially the islands, have little more than basic provisions. And a good book or deck of cards makes waiting out a fog in a sleepy little harbor much easier to bear.

Your planning should include a thorough understanding of the tidal currents and heights in this state, which vary greatly from north to south. While the tide range in Eastport in the north is up to 25 feet, south in Kittery it's around 9 feet. Both ports, and many others in between, have tidal currents up to 6 knots. So timing entrances to and departures from harbors with a favorable current, and anchoring accordingly, is often crucial.

Along with the currents, keep an eye on the wind. Southwesterlies are the most common summer wind. But when wind and tide run hard against each other, even in enclosed bays, the resultant chop can thoroughly rattle most boats, overwhelm some. Add a thick fog into the mixture and the scene becomes truly challenging.

The prudent mariner will find Maine's benefits far exceed the few inconveniences now and then. Most of the coast is so empty, so elemental, so filled with nature's glory, that you easily fall under its spell and forget you ever had any cares back home. This is particularly true of the stretch from Eastport to Schoodic Point.

Marine Facilities and Services

Marina	Phone	Monitors VHF Channel	Working Channel	Number of Transient Berths	Number of Moorings	Seasonal / Year-round	Maximum Length Over-All	Approach Depth in Feet at MLW	Dockside Depth in Feet at MLW	Hookups: Fresh Water / Phone / Cable TV	110V★ 220V▲ 3 Phase ■ Maximum Amps	Ramp / Dinghy Dock / Launch Service	Railway / Lift / Crane: Capacity (tons)	Repairs: Propeller / Sail / Electronics / Rigging / Gas / Diesel / Wood / Fiberglass	SCUBA: Air / Rentals / Equipment / Charters	Gas / Diesel Fuel	Fuel Brand	Marine / Groceries / Bait / Ice / LPG / CNG	Showers / Laundry / Pumpout / Restrooms	Hotel / Pool / Tennis / Golf	Public Phone / Restaurant / Snack Bar	MasterCard / Visa / Discover / AmEx
		DOCKAGE								SERVICES						SUPPLIES			AMENITIES			
1 Hinckley Great Harbor Marina (p. 90)	207-244-5531	9		150		S	150	10	10	FC	★▲■ 50	All				D		All	SLR		PRS	All
2 Hinckley Service (p. 90)	207-244-5531	9			30	S	85	8	7	F		All	L70	All		D		All	SLR		PR	All
3 Spring Point Marina (p. 96)	207-767-3254	9	78	25		Y	150	9.5	9.5	F	★▲■ 50	D	L35	All	N	GD	Mob	MGBI	All		PR	All
4 Channel Crossing Marina (p. 96)	207-767-4729	9	68	20		Y	250	15	10	F	★▲■ 50	RD				GD		I	SLR		PR	MVD

COAST GUARD: See Adjacent Chart for Coast Guard Information

Information in these listings is provided by the facilities themselves.

The Sunrise Coast: Eastport to Schoodic Point

The standard joke in **Eastport** is: "This may not be the end of the earth, but you can see it from here." Eastport is the principal port of the **Cobscook Bay** area. And as the first bay on the United States East Coast, Cobscook is essentially where it all begins. Cobscook is an isolated area with strong currents, back eddies, whirlpools, and tidal fluctuations that can leave you high and dry. Seeing another pleasure boat here is unusual so be careful when transiting the narrows around Falls Island and investigating the nether reaches of Dennys and Whiting Bays. Eastport is an important provisioning stop for anyone exploring this desolate area.

Nearby, **Lubec** is to the south of Eastport at the eastern U.S. entrance to the Cobscook area. Located on Lubec Narrows, it is a quiet town struggling to survive the demise of the sardine canneries that once ruled the waterfront.

You must plan your use of this U.S. entrance to the Cobscook Bay area very carefully. The currents are very strong and you must pass under a fixed bridge with a 47-foot clearance. If you are traveling from Canada, you would be better off to investigate the Head Harbor or Western Passage approaches. If you do use this approach, you will enter from Grand Manan Channel. The approach is via the West Quoddy Head Light (83-foot, 15sec group flashing), which also has a loud fog horn on it. The flashing green whistle buoy "1" off West Quoddy marks the beginning of the Quoddy Narrows entrance. From here, run north to the red and white bell buoy "WQ." The increasingly narrow channel heads northerly with red buoys to starboard, green to port, until you reach the bridge (vert. clearance 47 feet) at Lubec Narrows. Once through the bridge, the town of Lubec is to port, and Eastport roughly straight ahead just beyond Treat Island.

Lubec has no marinas, but there is a town dock. The streets make for a quiet walk, and minor provisions are available. In Eastport, new marine facilities are inside the prominent breakwater.

From the Eastport area, your trip down the coast is south and west along a barren stretch of classic rocky Maine coast for nearly 20 nautical miles without a harbor to duck into. These are beautiful, but lonely, shores until **Cutler**, which is the first good harbor (though it is crowded and dedicated to working lobstermen). Salmon farming operations are carried on in the harbor and are fascinating to watch. The hatchery for the farming is in the basement of the town hall, which is a short walk from the docks. Being isolated and self-sufficient by necessity, you'll find the townspeople that run the luncheonette, village market, and other establishments remarkably chatty and witty.

The leading buoy is red whistle "2LR" well offshore from the entrance, then red and white bell "LR" near the east side of the entrance. You can anchor and find basic provisions ashore. There is a busy town dock for dinghies only.

On the coast below Cutler, there are literally dozens of empty harbors worthy of exploration. In Machias Bay you'll find **Starboard Cove, Bucks Harbor, Holmes Bay,** and the **Machias River,** as well as **Little Kennebec Bay** just around the corner. Many pleasure boats head straight for renowned **Roque Island**, with its huge sand beaches and half a dozen gunkholes to explore. Good anchorages can be found at **Lakeman Harbor,** in the northeast, **Bunker Hole**, in the south between Great Spruce and Little Spruce islands, and **Shorey Cove**, on the north shore. These are beautiful areas to explore, but none have shoreside facilities.

Shorey Cove can be reached by proceeding in a northwest direction up Englishman Bay. Leave green cans "1" and "3" to port, then swing around Great Head into the cove with 10 feet of water.

All this solitude and beauty may suit you just fine, but you will have to provision at some point. To do so, you can continue on to the fishing village of **Jonesport,** on the north shore of Moosabec Reach. Maine lobsterboats fill this area and there's even a lobsterboat race every summer to determine "the fastest lobsterboat in the world." You can also catch sightseeing boat trips to the islands in the area should you not want to take the time to navigate the area in your own boat. Aside from the fish trade, boatbuilding is part of the activity at Jonesport.

Entrance from the east is from the green bell buoy "1" north of Mark Island. Though the channel is narrow, it's well marked. Unless visibility has fallen below a quarter of a mile, you should be able to proceed from marker to marker, red buoys beginning with the nun "2" to starboard, green buoys beginning with the daymarker on Gilchrist Rock "3" to port. Jonesport itself may be clear when it is thick outside. The hot air rising off Beals and Great Wass islands produces this effect. It's probably why another Coast Guard station is located in Jonesport rather than on the outer coast. The town's marina and a private marina are both in Sawyer Cove, from which it's an easy walk to grocery, hardware, and other stores.

South of the marina is a small channel that leads to **Eastern Bay** between Head Harbor Island and Great Wass Island, an area with another dozen small Maine gunkholes to explore. Tides run swiftly between anchorages, so navigate with care and only travel when visibility is good.

Basic fuels and some amenities can be found at the fishing community of **South Addison** in Eastern Harbor between Cape Split and Moose Neck.

To reach Eastern Harbor coming from Moosabec Reach, pass between red nun "28" and flashing green 4sec "27" at Tibbett Narrows. Leave red bell "30" to starboard; enter the harbor between can "1" and nun "2."

As you travel west of Moosabec Reach, you're seeing some of the many examples of how Maine's coast was etched by glacial activity thousands of years ago. Pleasant, Harrington, and Narraguagus bays, extending into rivers of

© Mary Jane Hayes

A cirrus sundown over Southwest Harbor.

the same names, are carved into the coast. **Pleasant River** is a good example of what you'll find at many such estuaries all along the Maine coast. Although water temperatures downriver rarely rise much above 50 degrees, upriver near **Addison** the water is often warmed to 70 degrees or more. Swimming is not only possible, it's pleasant.

Addison, about 5 miles above the mouth of the Pleasant River, is used mostly by a few fishing vessels. Should you decide to make the trip up the river, approaches are well marked and straightforward, but keep in mind that the river is reported to have shoaled in many places.

Just when you think there can't be any more of this geologic wonder, an explorer's feast is served between Nash Island Light and Schoodic Point. **Pigeon Hill Bay, Dyer Bay, Gouldsboro Bay,** and **Prospect Harbor** all offer superb gunkholing. If the weather is calm and clear, be sure to stop at Petit Manan Island's National Wildlife Refuge. Nesting terns, puffins, great auks and other rare birds are often at the refuge and easily spotted from a passing boat. Be careful of the waters south and north of the island. They are normally choppy, and wind running against tide can create an awe-inspiring chop in very short order.

Down East: Bar Harbor to Rockland

Your "return to civilization" is heralded by entering **Frenchman Bay.** Here, the mountains of Acadia National Park (207) 288-5262 on Mount Desert Island are to port, granite-clad Schoodic Peninsula to starboard. Eye-opening in its own right, this landscape is reminiscent of Scandinavia. **Bar Harbor** is where most cruising boats congregate. There are dozens of shoreside facilities in addition to easy access.

Because you haven't had the luxury of dockside water and electricity since Eastport, you may want to tie up at one of the marinas, which offer fuel and other dockside necessities. There is also a huge town pier with floats. Ashore, necessities are available in a town that positively hums with summer visitor activities. Dozens of restaurants, from plain to fancy, compete for your attention. There are tour buses to Acadia's miles of footpaths, lakes, and memorable mountain-top views. Once in the park, you can hike to the top of well-known Cadillac Mountain or take a bus for a spectacular view of the park. Or you can take one of the many less-populated trails for all hiking levels. Along the "loop road" that circles through the park, stop at Thunder Hole where waves crash with fury into a cavelike formation; have lunch at Jordan Pond House overlooking flower gardens and forest; rent a bike and peddle through miles of old carriage paths; visit the Indian museum or lighthouses. That's just a taste of what Acadia has to offer. The park rivals any national park in the country for beauty.

In Bar Harbor, whale-watching boats and big old-fashioned schooners come and go under the gaze of verandas of 19th century hotels and guest houses that seem unchanged since they were built.

To get to Bar Harbor, you'll enter Frenchman Bay, the eastern border of Mount Desert Island. The only major obstruction while entering Frenchman Bay, Egg Rock, has a good light (64 foot, flashing red 5sec) and loud fog horn. Be careful of the breakwater extending southward from Bald Porcupine Island. During high tide, it's submerged—only the white beacon shows. If in doubt of your position, go east around Bald Porcupine.

As you approach Bar Harbor, you'll see **Winter Harbor** on the east side of the bay. You'll see grand and beautiful summer houses along Grindstone Neck in this area. If you're in the area during the second week in August, you can attend the annual lobster festival. The harbor is a decent anchorage with a marina and adequate shoreside facilities in Henry Cove, but it is rolly in southerlies.

Exploring the upper reaches of Frenchman Bay can easily take a pleasant week. Eastern Bay to **Jordan River**, **Skillings River**, **Sullivan Harbor**, and **Flanders Bay** are just some of the areas readily accessible and offering good to excellent anchorages. There are few if any facilities outside **Sorrento Harbor**, south of Waukeag Neck, which offers a shutter-snapping view of Cadillac Mountain as well as moorings and a hospitable yacht club.

The approach is from the red and white bell "SH" just south of Dram Island. Pass close east or west around Dram, and work your way up into the eastern end of the harbor.

There are many places to stay on the south side of Mount Desert Island. The three main harbors are Northeast Harbor, Southwest Harbor, and Bass Harbor. They all offer easy access and are safe in almost any weather.

Northeast Harbor is a sedate summer haven. Municipal facilities for boats range from dockage to showers and laundromats. There are also two private marinas and several handsome old hotels overlooking it all. It is the home port of the only continuous sea-going ministry in the United States. The vessel *Sunbeam IV* welcomes visitors aboard when she's in port. There are also well-narrated sightseeing tours and whale-watching trips available, plus self-guided walking tours of the surrounding mansions.

From Frenchman Bay, head around the south side of Mount Desert Island through Eastern Way. Watch out for East Bunker Ledge, marked by green can "3A." Follow the channel keeping the red nuns to starboard until you're past Bear Island where you can make your entrance into Northeast Harbor.

At **Southwest Harbor**, the Mount Desert Oceanarium (207) 244-7330 is one of the most popular attractions in town, especially for children. There is also the Wendell Gilley Museum (207) 244-7555 about a mile outside of town on Route 102. It features hand-carved sea birds, prints, photos, and exhibits, all beautifully presented by a local plumber who pursued his true interest. Bike or car rentals are available if you'd rather not walk.

Southwest Harbor is a short hop from Northeast Harbor. Head for red nun "6" at the south side of Greening Island. Leaving the nun to starboard, you'll be able to make your approach into the harbor. Southwest Harbor is headquarters for all the Coast Guard stations of eastern Maine. There is also a hospital. Stores and restaurants of all types abound and numerous summer visitors populate the otherwise quiet streets. From **Hinckley Great Harbor Marina** in Southwest Harbor you can enjoy all that this wonderful area has to offer. Run by the Hinckley Company, Maine's famous builder and charterer of yachts, the full-service marina can accommodate vessels up to 100 feet.

The magnificent fiord of Somes Sound is just a short sail from Northeast or Southwest harbors. The Cranberry Islands about a mile to the southeast are also worth a few days exploration.

The villages of **Bernard** and **Bass Harbor** both offer a quiet down east town atmosphere. There are several antique shops on the Bernard side of the harbor. Naturalists' walking paths to Bass Harbor Light and along the rocky southern shores of Acadia National Park are also readily accessible from Bernard. Car rentals offer access to all of Acadia's attractions.

Bass Harbor is the southernmost and westernmost anchorage, clearly marked by the Bass Harbor Head Light (56-foot, 4sec occulting red) just east of the entrance. You must cross the Bass Harbor Bar, which is clearly marked with red and white buoys "EB" and "WB" just south of the light. The only obstruction, Weaver Ledge, is marked by buoys on its east and west end. There are several private marinas that offer varying degrees of services, as well as a town dock on the east side of the harbor.

Blue Hill Bay borders the southwest side of Mount Desert Island. It is deep and wide with no major navigational obstructions. In its northwest corner is Blue Hill Harbor with the town of **Blue Hill**. Just walking the streets of Blue Hill is a pleasant experience, with its well-tended homes, antique stores, and the shops of excellent potters. The town was named for the 940-foot "hill," which you can hike for a view of Blue Hill Harbor. Among many activities are the Blue Hill Fair, a true country fair, and regular concerts by chamber musicians at Kneisel Hall (207) 374-2811.

Blue Hill Harbor is slightly more than 2nm northwest of Long Island. Be sure to stay on the starboard side of green cans "1," "3," and "5" when entering the inner harbor via Sculpin Point. Stay close to the can and nun on the inside and anchor in a perfect basin next to the town of Blue Hill. There is an accommodating yacht club here and a quiet, but entertaining and useful, town ashore. **Morgan**, **Union River**, **and Patten Bays** are northeast of Blue Hill.

South of Blue Hill Bay are islands, lots of them. Geologic forces have made **Swan's Island** and **Long Island** (don't get this confused with the Long Island inside Blue Hill Bay) the most hospitable to passing vessels. Swan's Island is largely a lobstermen's island, with a growing summer community. It does have a small marina in Burnt Coat Harbor and some stores ashore for basic provisions. On Long Island, Lunt Harbor is the only suitable anchorage. At **Frenchboro,** there are miles of walking paths and tame deer that keep them beaten down. All else is silent isolation.

Swan's boasts two good harbors, **Mackerel Cove** on the north side and **Burnt Coat Harbor** on the south. Mackerel Cove is best approached between the green 4sec gong "1" and close to nun "2" north and west of North Point. Run up to Roderick Head and turn south until you like the soundings. You can go ashore at the car ferry dock nearby. Burnt Coat Harbor on the south side of Swan's requires a little more attention when approached from the southeast. Be sure to pick up the red gong "2" and the cans "3" and "5." In fog, this narrow entrance actually can be better than the southwest entrance via the red and white bell "HI" and Hockamock Head.

Approach Lunt Harbor from the northeast. Pick up red and white gong "LI" east of Crow Island. Head southwest 1nm staying south of Crow and Harbor islands. There is good holding ground in the cove, but watch out for the cable area.

Working westward, you have the option of running up Toothacher Bay to the south or through Casco Passage to the north of Swan's. Toothacher is more straightforward, but it's apt to be the more foggy of the two. Casco

© Mary Jane Hayes

Mount Desert Island dwarfs a lonely sailboat.

Passage is more tide scoured, setting east on the flood, west on the ebb. Casco's northern channel has less current than the southern York Narrows channel, but both have adequate depths for most vessels.

Both passages bring you to **Jericho Bay**, where three new passage options are presented. You can continue northwest via Eggemoggin Reach to the north, or west through Deer Island Thorofare straight ahead, or a little south of west through Merchant's Row to the south. The area is so varied, you may want to poke around all three passages. In that case, the best base is **Stonington**.

This active seaport is on the south end of Deer Isle near the west end of the Deer Island Thorofare. It is protected by dozens of granite islands to the south and east and the town itself has a rich history of quarrying granite. You can look out at Crotch Island, another quarrying site. And if you're tired of navigating yourself, you can take a sightseeing boat out of Stonington to Isle au Haut.

To get to Stonington from Toothacher Bay, proceed via Halibut Rock Light (25-foot 6sec flashing white) 2nm nearly northwest to the red and white bell "SI" just east of Saddleback Island. Then pick up the red nun "2" northwest of Saddleback, followed by two green cans "1" and "3" on the north side of Bold Island. The channel is obvious from here. A similar route can be followed from York Narrows. The town has virtually all provisions.

The Merchant Row passage is reached via previously mentioned Halibut Rock Light. Head southwest about

2nm to the north side of Southern Mark Island and then pick up the green cans beginning with "1." Once past can "9," however, you may want to follow the north shore of Merchant Island around to the south side and then on to **Isle au Haut.** This huge island is also part of Acadia National Park, but without the crowds. Anchor east of **Kimball Island** and go ashore for a feast of natural beauty. The small village ashore can provide a few provisions. Take a bicycle if you're intent on seeing it all.

Accessing the Eggemoggin Reach to the northeast is simply a matter of finding Egg Rock beacon and red bell buoy "8" at the west end of Casco Passage, and then proceeding to the red and white bell "EE" at the south end of the reach. On the north shore of the reach, **Center Harbor** and **West Brooklin** in the Benjamin River are both worthwhile stops. Though there is little of interest ashore, the array of moored boats—both classic and otherwise—is incomparable.

All three of the foregoing passages lead into Penobscot Bay, Maine's largest single body of salt water. You'll see more large old touring schooners here than anywhere else in the world. It's likely you will spend an evening anchored with one, particularly if you lay over at **Buck's Harbor** in **South Brooksville**. This is located at the northwest extremity of Eggemoggin Reach, 1nm north-northeast of the red and white bell buoy "EG." Even without schooners, it's an excellent stop, with good boat facilities and a sleepy village with adequate provisions ashore.

Emerging from the reach you can follow the shores of Cape Rosier northerly for about 5nm until you find **Castine**, which is—among other things—home to Maine's Maritime Academy (207) 326-4311. The *State of Maine* training ship dominates the waterfront and is usually open to the public. There is also a town dock, full-service private boat yard, and a yacht club. The town ashore is full of history, having been a hot spot of naval activity during the Revolution; there are some 100 historical markers. Many amenities are ashore.

Located in the mouth of the Bagaduce River, entrance is made easy by Dice Head Light (27-foot, 6sec flashing white) on the north shore and a deep, wide, and obvious channel due east of red and white bell "CH." From Castine, you can continue up the bay to the **Penobscot River**. It's a gorgeous 20nm winding trip, particularly in the autumn when trees on the high shores are in their full glory. The channel is well buoyed because oil barges still use the river as far as the city of **Bangor**. The city makes an effort to please passing pleasure boats with a good public landing, complete with water, electricity, and showers. Along the way, too, are **Bucksport**, **Winterport**, and **Hampden**. All three are a bit industrial in appearance but do provide either public or private facilities for passing boats. All have adequate shoreside attractions and necessities. In Bucksport, for example, visit the museum located in the old railroad station and learn about this small Maine town's history. You can cross the Bucksport Bridge and visit Fort Knox State Park—not the gold depository—where you can wander through the still-intact fortification from the Revolutionary War.

Back at the mouth of the river, you continue west across the northern part of the bay toward **Belfast**, a town that is making a decided effort to attract pleasure boats. There are art galleries and shady streets lined with handsome architecture on the way to shops and restaurants.

After leaving Fort Point on Cape Jellison, find the flashing green 4sec gong south of Sears Island. Use it to take a bearing north of west to the light on Steels Ledge Light (4sec flashing red 20-foot) at the mouth of the harbor. Note the oil transfer areas on the chart—these are fairly active zones, with large seagoing barges and/or tankers potentially crossing your route, so stay alert. Once in Belfast Harbor, though, you're in good hands again. There are both public and private boating facilities, which will fill most of your needs.

Farther south, **Penobscot Bay** is divided in half by Islesboro Island. This is 10nm of varying topography and home to a few hundred souls in the winter, and hundreds more in the summer. Aside from the baker's dozen of coves worthy of overnight anchoring, the primary boating center is in **Gilkey Harbor**. This is reached via the Grindel Point Light (39-foot 4sec flashing green) just northeast of Warren Island. The channel is plainly marked, and moorings can be found off the village of **Dark Harbor**. There is a marina on Seven Hundred Acre Island, the shore opposite Dark Harbor.

Many of the boats in Gilkey Harbor arrive when traveling northeast from **Camden**, a short 4nm hop across West Penobscot Bay. Camden used to be combined as a township with **Rockport** to the south, but the two split 1891. They both still share an oft-quoted theme, though, as the area "where the mountains meet the sea." With complete outfitters, bookstores, fish markets, restaurants and numerous other stores, there are few places that can compete with Camden. Along the waterfront you'll see the windjammers at the head of the harbor, the Chestnut Street Baptist Church staring out true New England style. Visit the outdoor Bok amphitheater for the arts and crafts show in July and October. The Camden Civic Theatre has live performances year-round. But this is only the tip of the iceberg. Your best bet is to stop at the Chamber of Commerce (207) 236-4404 at the public landing for the list of activities going on here and in Rockport.

Departing Gilkey Harbor via the green bell buoy south of Ensign Islands, there is a clear course to Camden's Curtis Island Light (52-foot flashing green). Be sure to pick up the red bell buoy off Curtis to avoid the ledges protecting the outer harbor. Traffic is also heavy here, so keep a sharp lookout. There's a yacht club and public landing which are often filled to capacity. Once inside the harbor, get ready for one of Maine's premier harbors. Transportation connections to the rest of the world are readily available. Beautiful architecture, lobster dinners, and summer visitors abound. In short—if it's not in Camden, it's not in Maine.

Just south of Camden is **Rockport,** one of the two "rocks of Penobscot." Here, some of the options include watching wooden boatbuilding by builder William Cannell and students at the Artisans School (207) 236-6071, visiting the Maine Photographic Workshop for seminars on still photography and film or just looking at a variety of photography and art at the Maine Coast Artist in the village. There are also performances held year-round at the Rockport Opera House (207) 236-4885.

The approach to Rockport from Camden is largely a matter of following the western shore. Be careful of The Graves (26-foot flashing green 6sec "5") just northeast of Deadman Point. After the red and white bell "RO" southeast of Lowell Rock, you have a clear shot into Rockport. This is a good anchorage in all but southerly winds, and it's rarely crowded.

Beyond Rockport to the south is the other rock of Penobscot—**Rockland**. The Owls Head Transportation Museum (207) 594-4418 is probably the most popular diversion in the Rockland area. About two miles out of town, it features vintage cars and planes from all over the country. There's also the Shore Village Museum (207) 594-0311 in town, with the largest collection of lighthouse memorabilia in the country. And for those attracted to the art of the Wyeths and other famous American artists, there's the Farnsworth Art Museum (207) 596-6457 a couple of blocks from the water.

Rockland Harbor is another 5nm south of Rockport. Porterfield Ledge is the only obstruction along the way and it's marked with a green and white day beacon. From here you can head south until the Rockland Breakwater Light (39-foot 5sec flashing white) is on the starboard beam and then run in south of the breakwater. This is where you'll see another big schooner fleet similar to the one in Camden, as well as a Coast Guard station and scores of offshore fishing vessels. The portion of the harbor north of Crockett Point is where the best facilities are located. There is also a town dock. Boats are still built in Rockland, and it is worth watching amid all the other activities going on ashore. If in need of anything nautical, from sailmaker to seine mender, it's in Rockland.

Camden, Rockland, and Rockport often end up being a base for boats exploring the lower end of Penobscot Bay. The most-frequented area is the **Vinalhaven/North Haven** Island group, which could take a week to explore in itself. A visit to Vinalhaven Island will show you a true offshore fishing island at work. Ashore is the slow pace of a summer island community, broken only by the arrival of the small mainland car ferry every few hours. The Camden Hills are visible to the northwest. On North Haven Island, you can purchase wool items made from island sheep or visit one of the other shops along the waterfront with quality items.

© Donna Caruso

Lobster pot buoys decorate home port in many fishing villages.

The two islands require two different approaches. The approach to North Haven from the west is via the red and white bell "FT" just 1nm southwest of Stand-in Point. From there it's 2nm east-northeast to noisy Brown's Head Light (39-foot flashing white with horn) and a safe entrance into Fox Island Thorofare. The town of North Haven is located on the north shore. There is a town float and a marina. For shopping, you may want to take a taxi to Pulpit Harbor.

Approaching **Carver's Harbor** on Vinalhaven is a bit more treacherous. Pick up the same "FT" leading buoy as for North Haven, but turn southeast toward the red and white bell "GI." Then head through Lairey's Narrows, down Hurricane Sound and into The Reach north of Green's Island. The Reach is marked, but it is a narrow, much obstructed channel, only 100 feet wide at its narrowest, and should only be transited with great care. This is obviously no place to be in heavy fog. In addition to all the obstructions, there's heavy commercial fishing traffic and another regular car ferry.

Mid-Coast: Tenants Harbor to Casco Bay

Leaving Penobscot Bay takes you past **Tenants Harbor**. The art of Andrew Wyeth, particularly his famous painting "Christina," makes you feel as if you've been to Tenants before. Wyeth used much of this scenery in his art. A marina and quiet village ashore will add to the pleasures of this small harbor.

Whether you cruise through the big deep main shipping channel west of Vinalhaven, or down the more picturesque Muscle Ridge Channel beginning at Owl's Head Light (100-foot flashing white with horn) just outside of Rockland Harbor, the abandoned lighthouse on Southern Island and the lighted green bell "1" clearly mark the entrance to Tenants.

Moving south and west, the first port in Muscongus Bay is **Port Clyde**. From here, you can take a ferry to Monhegan Island. Although you can go there by boat (described shortly), it is difficult to find a good anchoring area and there are no marinas. You don't want to miss this beautiful island and the way to go may be by ferry.

The approach to Port Clyde is via Marshall Point Light (30-foot flashing white with horn). From the east, pick up the red bell buoy "2" immediately south of Mosquito Island and follow the channel in. From the south or west, enter via the Davis Strait nun "12" just north of Davis Island and follow the channel. The harbor is rarely crowded, but the town dock is often busy. The freight/tour boat for Monhegan Island departs from here, and it's usually filled with scores of fishermen.

There are four other ports worth visiting in Muscongus Bay—**Thomaston, Round Pond, Friendship,** and **New Harbor.** The latter two are primarily lobstermen's harbors, but there's easy access to all four and they all make transient boats welcome. The channels to them are clear from the chart.

As mentioned, **Monhegan Island** is a fabulous place to visit. It is the outer guardian of Muscongus Bay and is Maine's best-known and most-visited offshore island. There are only a hundred or so year-round residents here. But tour boats from the mainland arrive daily during the summer, doubling and tripling the population as they dock. The pleasures of the island are all elemental. Walking paths radiate north and east from the village, eventually taking you to the cliffs on the island's east side. If a heavy easterly swell is running, the picture is dramatic. Be careful of your footing, though. More than one person has not returned from this remote end of the universe. The island's other entertainment is watching the boats come and go in the harbor.

The island is easy to find, the approach easy in anything short of pea-soup fog. The crowded harbor is on the southwest end. The easiest approach is from the southwest after you find the red whistle buoy "14M." Once again, the holding ground at Monhegan is dicey. The best bet for your dinghy is one of the beaches near the town wharf.

Pemaquid Point marks your arrival in Johns Bay, in which the harbor of most significance is **Pemaquid Harbor**. There's the William Henry Memorial Fort to visit here with exhibits and a wonderful view, and there's a museum at the pier.

To enter the harbor, leave Johns Island to port or starboard, being sure of the location of Knowles Rocks if using the eastern channel. Inside the harbor and up the river are a marina and adequate shoreside facilities.

If you need more facilities, go directly across the bay into **South Bristol**. Access to this area is via the north side of Witch Island and through a bridge that opens on signal. Still more facilities are found directly across the Damariscotta River in **East Boothbay** or about 1nm down river in **Christmas Cove**. The latter is apt to be crowded, although the amenities are the most upscale of the four harbors mentioned. The entrance is clear of unmarked obstructions.

The harbor that calls itself "The Boating Capital of New England" is the next stop westward. During the high season, **Boothbay Harbor** is certainly a busy anchorage, with tour boats, yachts, and commercial fishermen coming and going nearly 24 hours a day. Like Camden, **Boothbay** is another old boatbuilding town. Today, it is filled with shops and restaurants for visitors. If you leave Boothbay bored, there's probably not much anyone can do for you. In late June, attend the annual Windjammer Festival; in late July, you can see Friendship Sloop Days, a celebration of the sloop-rigged boat well-known in Maine. Call the Chamber of Commerce (207) 633-2353 for information. Just a few miles outside of town you'll see old resorts. And it's one of the many great places to enjoy the "Maine-stay" of the state—tasty lobster.

From the east, access is through Fisherman Island Passage just south of Linekin Neck. Run up the east shore of Squirrel Island and into the inner harbor. From the south or west, follow a similar route via the Cuckolds Light (59-foot 6sec group flashing white) and the east or west side of Squirrel. Marinas, a yacht club, and a town dock are all in the inner harbor. Facilities ashore are more than complete, including a Coast Guard station and hospital.

A handy sail away from Boothbay's hubbub is Sheepscot Bay. The easiest access to this is back around the Cuckolds and a short run north. Avoiding the red sector and leaving Hendricks Head Light (43-foot, flashing white) to starboard, run up the east shore as far as Dogfish Head. A turn to starboard will bring you into the quiet waters of **Ebenecook Harbor**. You can drop your anchor anywhere, or pick up a mooring in the southwesternmost cove. This is a full-service marina that offers tranquillity ashore.

From Sheepscot, you head southwest again toward Seguin Island. It marks the mouth of the Kennebec River, which runs nearly forty miles inland and offers a dozen ports to visit. It's important to catch the tide right here. Most Kennebec ports are small and have limited facilities, generally catering to small, high-speed powerboats. The city of **Bath** is the major port on the river, featuring the construction of the U.S. Navy's new destroyer fleet at Bath Iron Works. The best moorings are at the Maine Maritime Museum (207) 443-1316, which has an excellent exhibit of 300 years of shipbuilding on the Kennebec River. It is a working museum with boatbuilders and exhibits on sea life and fishing.

Beyond Seguin, you enter **Casco Bay**, which has been a recreational boating mecca for 150 years. Sebasco Harbor, just east and south of Harbor Island, is one of the better-known layovers here. There is a nine-hole golf course and two old-fashioned lodges known for exquisite cuisine and quiet ambiance. Farther north, there is a spectacular tidal race at Gurnet Strait, just west of Bombazine Island. Kayakers often use it as training for white-water contests.

The New Meadows River is the easternmost attraction. The entrance is marked by Fuller Rock Light (25-foot 4sec flashing white) off Cape Small. From it, you find red bell "2BH" and then red bell "4" about 2nm to the north. From here, it's straightforward. **Cundy's Harbor** is the best place for a few supplies and a good lobster dinner with local ice cream for dessert. It doesn't make a very good anchorage, so try **The Basin** or **Brigham's Cove** up in Winnegance Bay. Both are secure and the latter the least crowded. The rest of the river is perfect for exploring to the headwaters, where there is a full-service marina.

Making your way across Casco Bay, head up Broad Sound, the entrance to which is clearly marked by Eagle Island. The best destination is the Harraseeket River and **South Freeport**. This is your access to the world's shopping mecca: Freeport, Maine. The attraction is L.L. Bean and the scores of spin-off retailers. There are two marinas on the Harraseeket, plus an active yacht club and town dock. Between the two towns is shopper's heaven.

The smoothest water to western Casco Bay is found via the channel out of the Harraseeket and just south of Moshier's and Littlejohn Islands. Stick close to these islands and you'll stay out of trouble.

As you're traveling this way, you may want to stop at **Falmouth Foreside** on the mainland. Here you will see the several hundred boats of the Portland Yacht Club and a full-service marina. The principal entertainment ashore is trading cruising stories at PYC. If you catch the right mid-August weekend, the annual running of the Monhegan Regatta (207) 781-9820 will be more than absorbing. And, of course, many yacht and cruising clubs from all over the East Coast make Falmouth a "must-stop" destination. At any given time during the summer, dozens of cruising boats could swarm into the anchorage.

To get to Falmouth Foreside, head for the green beacon 2nm southwest on Lower Basket Ledge. Then turn west toward the red beacon off the northeast side of Clapboard Island.

South Coast: Portland to Kittery

It is a short, easy sail to **Portland**, a truly fun city in Maine's more southern reaches. If you dock in Portland, spending time will be easy. You can casually wander the streets of this city, window-shopping and watching street musicians here and there to get the "feel" of Portland. The Portland Museum of Art (207) 775-6148, the Portland Symphony (207) 773-8191, the retired lightship *Nantucket,* and the Maine Historical Society (207) 774-1822 are just some of the ongoing diversions. The Children's Museum of Maine (207) 828-1234 features "hands-on" exhibits in the arts, sciences, and natural history. Look around for the Maine Sunday Telegram's *Audience* flyers, free in many of the shops near the waterfront. It's always loaded with listings of the latest plays, concerts, festivals, and other activities. For those committed to seagoing attractions, there is the *Scotia Prince* ferry (207) 775-5616 to Nova Scotia, or the Casco Bay Lines ferries (207) 774-7871 to various island communities just outside Portland Harbor. If you're in Portland in mid-March, don't miss the Maine Boat Show (207) 775-3458 and the Maine Boatbuilders Show (207) 774-3458.

© Mary Jane Hayes

Harvesting pogies in Boothbay Harbor.

The entrance is easy for small boat sailors and ocean-going ships alike. Once behind Great and Little Diamond Islands, the city's lights will provide the best navigational aid. Diamond Island Ledge is so well marked that avoiding it is merely a matter of heading for Fish Point.

The harbor itself is bounded by Portland on the north and **South Portland** on the south. South Portland has excellent restaurants and is within easy reach of all the attractions of Portland itself. There is excellent public bus transportation and inexpensive taxis.

Most small boat facilities for transient boats are in South Portland. We have two recommendations. The first full-service yard you come to is **Spring Point Marina** near the Spring Point Light. More in the middle of the inner harbor is **Channel Crossing Marina**. Both put you within easy reach of Portland and the Casco Bay Islands.

From Portland south, the Maine coast once again changes personalities. Gone are the hundreds of choices in harbors, anchorages, and potentially confusing channels. The coast trends southward with well-defined coastal harbors and inlets, beginning with the Saco River and Biddeford Pool 10nm southwest of Cape Elizabeth at the entrance to Portland Harbor.

The architecture of these towns will remind you of the less genteel side of New England. The sprawling wool mills along the banks of the river are a testament to fortunes made by the sweat of innumerable poor immigrants. Those fortunes often went into the huge, ornate 19th century houses that line the better streets in town, not into the poorer tenements that still stand in some areas.

The approach to these harbors is by prominent Wood Island Light (71-foot 10sec alternating flashes white and green) which also has a loud horn on it for foggy weather. If bound for the Saco River and the towns of **Biddeford** and **Saco**, head for the light until you see the red and white bell "SA" off Ram Island Ledge. Follow the channel to a somewhat industrial pair of river towns with adequate marina and standard shoreside amenities. Alternatively, you could head southwest from "SA" between Negro and Stage Islands to the uneasy anchorage of **Wood Island Harbor**. Better yet, continue into The Pool and pick up a mooring in perfect protection. There is a basic restaurant ashore and a very accommodating yacht club.

Another 8nm southwest down the coast will take you to **Cape Porpoise**, a good stopover where the town, harbor, and cape all share the same name. You needn't run to the offshore buoy here. Just pick up red bell "2" off Old Prince Ledge and follow the buoys in. There are a lot of lobster trap buoys at this entrance, so keep a sharp lookout. Once inside, you may find a vacant mooring or, more likely, drop the anchor. This is primarily a lobstermen's harbor, but there is town dock at which you are welcome to land. Some basic restaurants and a grocery store are ashore. But the chief advantage is being in a safe harbor right amid the breakers of the outer coast.

If you need more facilities, Cape Arundel and the Kennebunkport River are another 2nm southwest. There used to be restrictions when approaching **Kennebunkport** in a boat strange to the area. But since President Bush has retired, you won't have any problems entering this area. Even more than its famous resident, you may especially appreciate the tours of handsome architecture, whale-watching excursions, and the town's relaxed personality.

The green can "3" is probably the first mark you'll see if running close inshore. From it, the next green can "5"

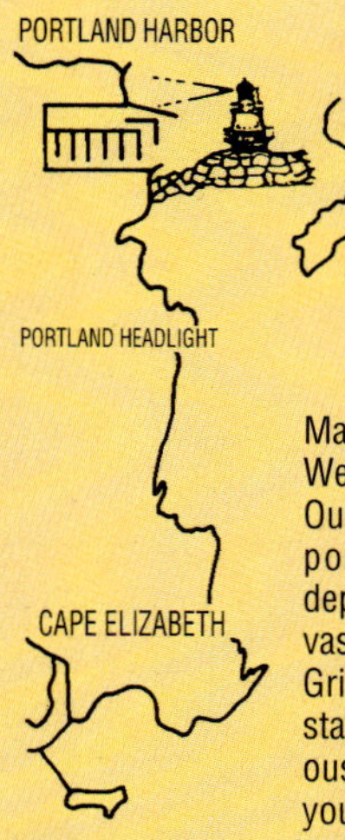

Spring Point Marina

3 p. 86

P.O. Box 2350 1 Spring Point Dr.
S. Portland, Maine 04106
(207) 767-3254

Maine's largest marina welcomes you to Casco Bay! We can provide you with 100% of your boating needs. Our facility has 250 slips, a 35-ton travel lift, transporters, Maine's largest marine store and parts department, award-winning service technicians, canvas, fiberglass, engine repair shops, and the Mariner's Grille just steps away from your dock. If you plan to stay with us or are just passing through, our courteous and professional staff will be glad to help make your stay comfortable and trouble-free.

- Portland Tides; High: -0:02 Low: -0:02, Mean 8.9 feet
- $ 1.00 / ft per day 25+ Berths, 150+' LOA
- Prop., Sail, Elec., Rigging, Gas, Diesel, Wood, Fiberglass
- Mobil Gas, Diesel
- Fresh Water 110V, 240V, 50 Max Amps
- Nearby
- Mariner's Grille 207-767-1200 Casual $11
- MasterCard, VISA, Discover American Express

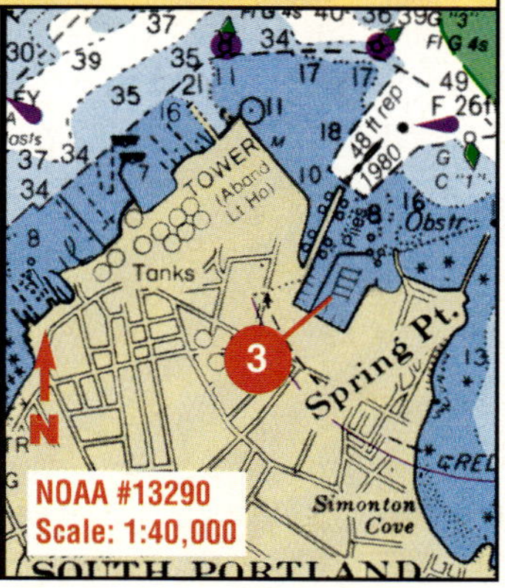

231 Front St. S. Portland, ME
207-767-4729

Channel Crossing

4 p. 86

Observe the breathtaking view of Portland's skyline from the deck of Channel Crossing's own restaurant, or simply relax and enjoy the sunset from the comfort of your own vessel. Portland's old port offers many gift shops, restaurants, pubs, taverns, and Casco Bay's own ferry boat service to the beautiful Casco Bay Islands. Channel Crossing is located in South Portland's historic Ferry Village, walking distance from scenic Portland Headlight, 10 minutes from the Portland International Airport, the Maine Mall, and downtown Portland. Portland sights include the Wyeth & Homer exhibits at the Museum of Art, the Wadsworth-Longfellow House, the Victorian Mansion, and the Portland Observatory.
A rental car can put you into the spectacular Maine countryside or in its spectacular beaches in just minutes. There's no better place to try a Maine lobster!

- Portland Tides; Daily Predictions, Mean 9.1 feet
- $1.25 / ft per day 20 Berths, 250' LOA
- Locally available
- Gas, Diesel
- Fresh Water 110V, 220V, 50 Max Amps.
- Legion Square Market (207) 799-3374
- Channel Crossing, Spectacular view, $10-20 (207) 799-5552
- MasterCard, Visa, Discover

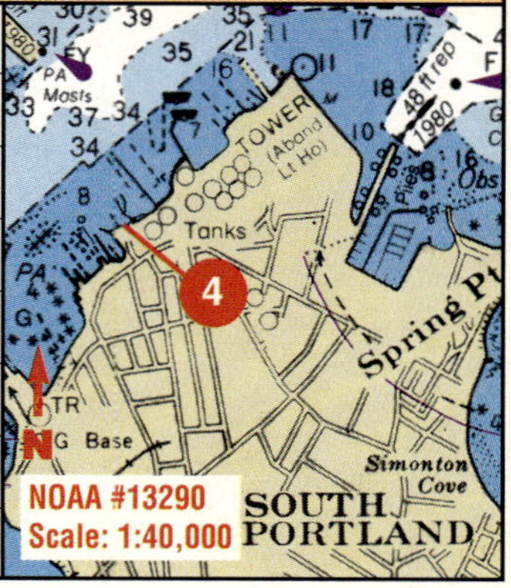

© Mary Jane Hayes

Sportfishing boat departing Kennebunkport.

and the flashing red tower on the breakwater are readily visible. At low tide it may be difficult to carry much more than five feet to Kennebunkport itself.

The next reasonable harbor is **York** just beyond Cape Neddick. There are deep sea charter fishing boats here and also a red trolley "bus" that stops at the harbor and proceeds to other points of interest. The seaside "cottages" on the way to the beach are reminiscent of Newport, R.I., mansions, though less ornate. There are also half a dozen inns with fine dining, the most popular of which is a screened deck affair on Harris Island. Here you can watch boats coming and going in the nearby York Harbor channel or on the distant ocean.

It's 15nm directly southwest of Kennebunkport, making it a spray-filled passage on an average summer day. But there are no obstructions along the route until you pick up the red and white bell "YH" just off East Point. The channel is clearly marked on the way in, although you will probably notice a strong tidal current. Anchoring is nearly impossible in the currents, but the harbor is full of moorings and there are marinas. There is a town dock at the north end and plenty going on ashore.

The end of the Maine coast is marked by the town of **Kittery**, which shares the Piscataqua River with the city of Portsmouth, N.H. The channel here is deep, wide, and tide-scoured, making it advisable for low-powered auxiliaries to enter with a favorable current. Also be wary of ocean-going ships arriving and departing. There is a full-service marina beyond Fort Point.

While you can see the end of the Maine coast on a chart, many boaters see it as a change of mind. Since few ever see all the nooks and crannies of the entire coast, it's probably best to concede there is no end, just a look forward to "the next time."

Shoreside and Emergency Services

Ambulance, Fire, Police: 911*

Coast Guard:

South Portland (207) 767-0302 or VHF 16
Boothbay Harbor (207) 633-2643 or VHF 16
Rockland (207) 596-6666 or VHF 16
Southwest Harbor (207) 244-5517 or VHF 16
Jonesport (207) 497-2200 or VHF 16
Eastport (207) 853-2845 or VHF 16

Radio Telephone:

VHF 24, 28; I.D.: Portland Marine Operator (207) 846-0700
VHF 26, 84; I.D.: Camden Marine Operator (207) 236-3605

Tow Service:

Contact the nearest Coast Guard facility for information on local commercial towing businesses.

*** Note:**

911 only works at mainland locations, generally west of Schoodic Peninsula. At other locations, contact the nearest Coast Guard facility. ✦

Local Services

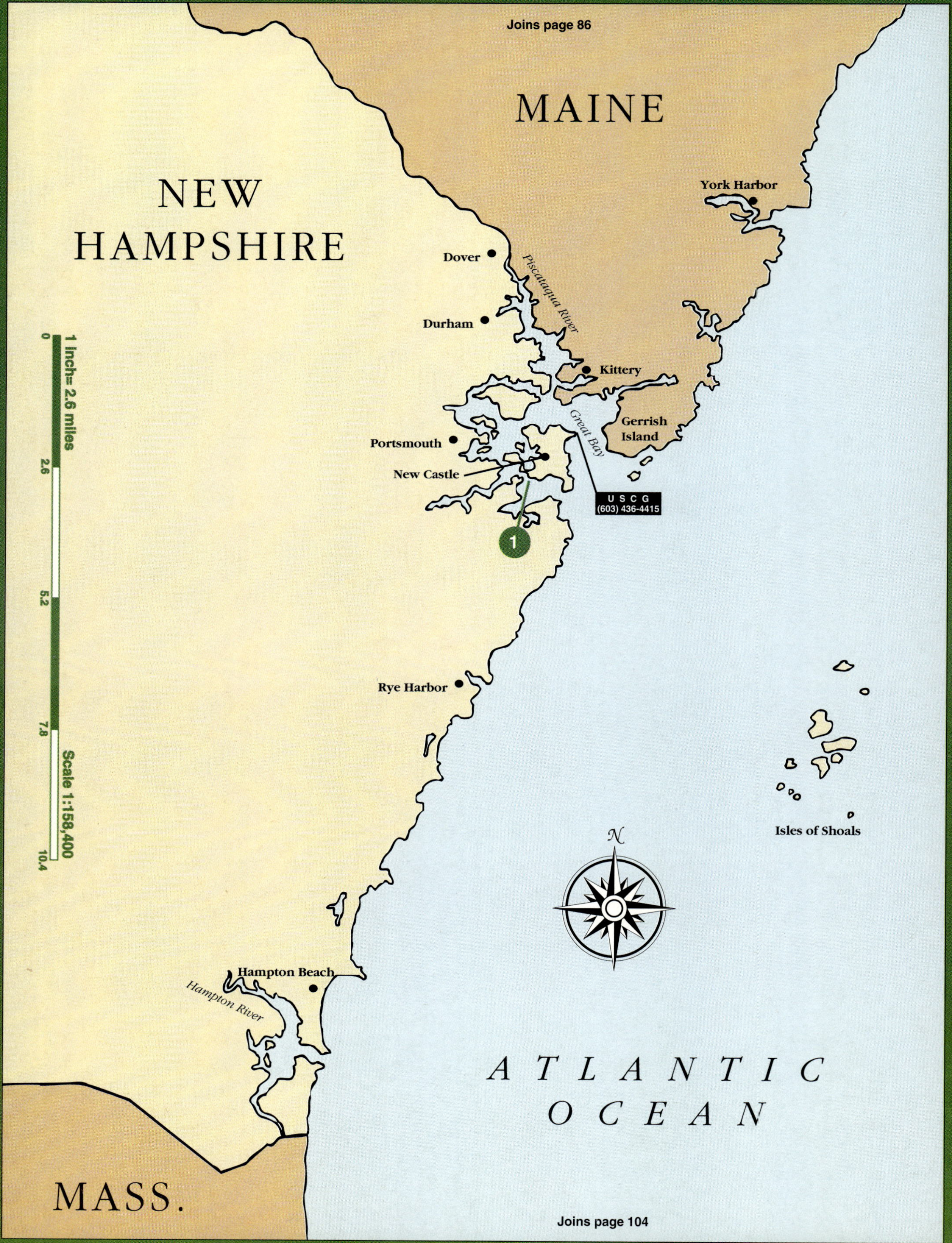

Joins page 86
MAINE
NEW HAMPSHIRE
York Harbor
Dover
Piscataqua River
Durham
Kittery
Great Bay
Gerrish Island
Portsmouth
New Castle
U S C G
(603) 436-4415
1
1 Inch= 2.6 miles
0
2.6
5.2
7.8
10.4
Scale 1:158,400
Rye Harbor
Isles of Shoals
N
Hampton Beach
Hampton River
ATLANTIC OCEAN
MASS.
Joins page 104

The Granite State

As the old proverb goes: "Good things come in small packages." And so it is with New Hampshire boating. With 18 miles or so of coastline, it comes in a very small package. But when you unwrap it, you'll find that the sizable and inviting port of Portsmouth, pristine Great Bay, the smaller harbors of Rye and Little Harbor, and the Isles of Shoals all make a stop along these shores a must for boaters in search of history and diverse landscape.

There are few "harbors" along the New Hampshire coast. The three main ports are at Rye, New Castle, and Portsmouth, which offer a variety of services and land activities. Plenty of good boating also exists in and around the Isles of Shoals. The parts of the coast between harbors are dotted with beaches, private homes, rocky coast, and rich wooded areas, all of which makes for a rather scenic backdrop. Like the short coastline of New Hampshire, the state's history often gets unduly overlooked. The years leading up to the American Revolution were filled with political strife in Portsmouth, then the capital of the state. A prime example is that of tavern owner John Stavers: In 1766, Stavers was jailed as a Tory sympathizer after a patriot mob attacked his establishment. Tearing down his tavern sign emblazoned with its Tory name—The Earl of Halifax—the mob smashed windows and later caught Stavers in Greenland, N.H. Thinking over the situation, Stavers renamed his pub (the William Pitt Tavern) and signed an oath of loyalty. George Washington visited the tavern to pay tribute to officials of the state and hosted principal war figures in Portsmouth. The tavern has since been restored and converted into a museum.

The months of May through October are the best months to visit New Hampshire by boat. At this time, the communities tied to the Atlantic are in an especially festive mood, hosting many celebrations and outdoor festivals. But you may want to start out in the beautiful natural environment of Great Bay.

© Mary Jane Hayes

Star Island in the Isles of Shoals.

Piscataqua River: Great Bay to Little Harbor

You get to **Great Bay** by rounding Gerrish Island into the Piscataqua River. At two miles long and three miles wide, Great Bay is the second largest estuarine bay on the East Coast. The area is constantly teeming with recreational craft of all kinds taking advantage of the fantastic boating conditions as well as an abundance of bluefish and striped bass. Vast areas of unspoiled woods, private homes, and docks line the banks of this body of water. It's a favorite among naturalists and photographers who can enjoy the wildlife amidst the pastoral settings of northern New England.

At the mouth of the Piscataqua, make for flashing red 2.5sec "2KR" whistle at Kitts Rocks, then bear northwest past 59-foot Whaleback Lighthouse to flashing red "2." From the 52-foot New Castle Lighthouse on Fort Point, continue west to flashing green 4sec "5" off Salamander Point. It is especially important to stay mid-channel when passing the Naval Shipyard on Seavey Island as it is a restricted military area, and traffic can get rather heavy.

As you pass several industrial complexes to port, look for flashing green 4sec "11" at South Beacon Shoal. Then

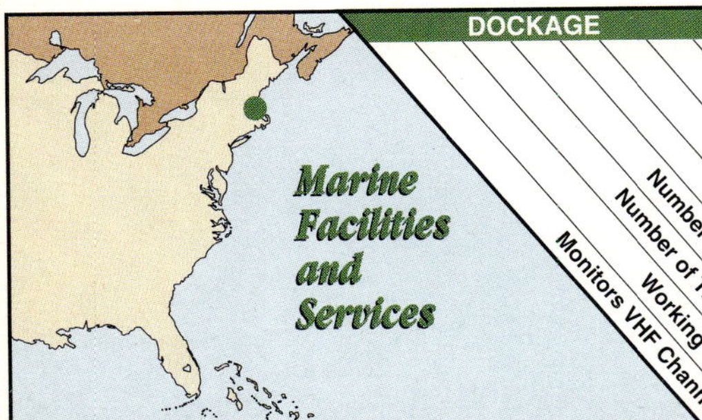

		DOCKAGE							SERVICES						SUPPLIES			AMENITIES			COAST GUARD
Marina	Phone	Monitors VHF Channel	Working Channel	Number of Transient Berths	Number of Transient Moorings	Seasonal / Year-round	Maximum Length Over-All	Approach Depth in Feet at MLW	Dockside Depth in Feet at MLW	Hookups: Fresh Water / Phone / Cable TV	110V★ 220V▲ 3 Phase ■ Maximum Amps	Ramp / Dinghy Dock / Launch Service	Railway / Lift / Crane: Capacity (tons)	Rigging / Gas / Diesel / Wood / Fiberglass	Repairs: Propeller / Sail / Electronics	SCUBA: Air / Rentals / Equipment / Charters	Gas / Diesel Fuel	Fuel Brand	Marine / Groceries / Bait / Ice / LPG / CNG	Showers / Laundry / Pumpout / Restrooms	Hotel / Pool / Tennis / Golf	Public Phone / Restaurant / Snack Bar	MasterCard / Visa / Discover / AmEx
1 Wentworth by the Sea Marina (p. 103)	603-433-5050	16	71	50		Y	250	8	10	All	▲■ 100	L		ERGDFP			GD	Ind	GBI	All	TG	All	MVA

See Adjacent Chart for Coast Guard Information

Information in these listings is provided by the facilities themselves.

© Embassy

The lighthouse at Portsmouth.

continue along the channel to green can "13." Here the river forks; take the left-hand fork under two fixed bridges, and head to Great Bay almost four miles upriver. While the fringes of the bay are lined mostly with private homes and docks, there are marinas and services available.

A center of maritime commerce and shipbuilding in the 17th and 18th centuries, **Portsmouth** is a small, cosmopolitan city with a flourishing waterfront district and reputation for exceptional restaurants and entertainment. All are within walking distance from the central downtown square to the old waterfront. Prescott Park (603) 431-8748 is worth a visit with its formal gardens replete with fountains, brick walkways, and sculptures making it a peaceful and beautiful little bit of New England. In July and August, the park hosts an arts festival with a musical play performed every night, Wednesday through Sunday, on its outdoor stage. Annual events at the park include a sock hop, big-band gala, and jazz Sunday brunch.

Across the street from the park is Strawbery Banke (603) 433-1100, a 10-acre maritime museum with nine furnished period houses reflecting the endeavors of craftsmen and sea captains who lived here. The oldest house dates to 1783; the newest is a 1950s-style restoration. Several exhibits trace traditional occupations such as architecture, early colonial house construction, and boatbuilding. There's also a superb archaeology exhibit. On-site craftsmen include a potter, barrel maker, and boatbuilder. Some noteworthy events held each year are Boatbuilders Day in July, Carvers and Cabinet Makers Day in August, New England Gardening Day, and a reading of the Declaration of Independence on July 4th. And don't forget about the William Pitt Tavern, now a museum at Strawbery Banke.

Just inside the harbor entrance, look for green 52-foot New Castle Light, located on Fort Point at the northeast end of New Castle Island. Other landmarks include Fort William and Mary, also called Fort Constitution. Next you'll see Fort McClary on Kittery Point, followed by Seavey Island where the U.S. Navy has a working naval shipyard that repairs submarines.

There are several marinas in Portsmouth just past the Memorial Bridge. Tides determine whether a request to open the bridge is necessary. During summer months, the bridge opens on the half hour from 7 a.m. to 7 p.m.; thereafter the bridge will open on request. Located just before the Memorial Bridge are the Prescott Park Public Docks. Accommodations for up to 25 vessels can be arranged with the dockmaster by calling (603) 431-8748 or hailing the marina on VHF 09.

The best local anchorage lies across the river in Maine's **Pepperell Cove**, just across the river from New Castle Lighthouse. A launch service is available out of the Pepperell Cove Yacht Club. To make arrangements, contact the Kittery Point Harbormaster on VHF 68.

If activities ashore don't float your boat (so to speak), two ferry services offer a variety of excursions (603) 436-8084. Harbor cruises sail through channels and coves, around the 14 small islands in the harbor and past tankers, sailboats, and submarines. Inland river excursions travel upriver to Great Bay.

As you get to **New Castle Island,** you'll begin to notice the sense of history that permeates the New Hampshire coast. Here, trade and commerce yielded an 18th century urban landscape, leaving a tradition of shipbuilding and extensive Georgian and Federal-era architecture in its wake. Preceding political furor in Boston, the first colonial uprising against the British took place at New Castle's Fort Constitution in 1774 when colonists sacked the fort and took its stores of gunpowder. Today, the fort is open to the public and is next door to the New Castle Coast Guard Station.

Providing protection for Portsmouth, New Castle is a quiet and picturesque village filled with traditional homes huddled close together and numerous ocean and river vistas. A small beach and park complex at the New Castle Commons is a favorite with locals for its splendid view of the lighthouse and river traffic.

Little Harbor is the best port from which to access the island. Once out of Portsmouth Harbor, head southwest from Whaleback Lighthouse to green can "1" and red nun "2" at the harbor entrance. From here, a winding, marked, and dredged channel heads north to Portsmouth

© Paul Spezzaferri / New Hampshire Travel and Tourism

The peaceful banks of Portsmouth Harbor near Strawbery Banke.

Harbor, giving you the opportunity to explore a few of the smaller islands that dot the area. In the quiet inlet of Little Harbor, just after nun "8" and before the bascule bridge, lies **Wentworth By The Sea Marina**, a luxurious full-service facility with two restaurants, tennis courts, a nearby 18-hole golf course, and shuttle service into Portsmouth. You'll also find some nice gunkholes and great exploring grounds if you head southwest from the bridge into Sagamore Creek.

Rye Harbor to Hampton Harbor, including The Isles of Shoals

A quintessential New England port, the few attractions to be found in **Rye Harbor** cater to transients and locals alike, which makes Rye very inviting. Two state-operated beaches, Wallis Sands and Jenness Beach, are not far from the harbor and are extremely popular with locals as well as transients. An annual tuna tournament and a bluefish tournament are held every August. To arrange a stay at Rye Harbor, contact the harbormaster at (603) 431-1779.

From red nun "2" outside the Little Harbor entrance, head to flashing green 4sec bell "1," then southwest to lighted whistle buoy "RH." Green can "3" marks the entrance to Rye Harbor, protected by two breakwaters.

The Isles of Shoals, nine islands running in a northeast-southwest direction 5 miles east of Rye Harbor, are among New Hampshire's best cruising grounds as well as sure landmarks from which to approach mainland ports. Unfortunately, the islands offer no marine services, so be prepared when you venture out. Island visits are available by steamship from Portsmouth; call (603) 431-5500.

The islands served an important role in early maritime history—New England's largest fishery and fish processing center peaked here in the late 18th and early 19th centuries. Austere and lovely, the islands (divided between New Hampshire and Maine) sustained a year-round fishing population of up to 600 families that pulled a living from the rich fishing grounds surrounding them. Inventors of a slow fish-drying process called "dunning," the Shoalers prospered, receiving shipments from around the world to be processed in their facilities.

In the mid-1800s, **Appledore Island** became a revered island retreat for the writers of that era. Mark Twain, Nathaniel Hawthorne, and Ralph Waldo Emerson sought refuge here at this wild place. At the center of the retreat was Celia Laighton Thaxter, an author and poet whose father built the Appledore Hotel in 1848. Thaxter's garden and her grave site are still maintained to this day on Appledore, which is home to the Shoals Marine Laboratory (603) 964-9011 of the University of New Hampshire and Cornell University.

As you head out to the islands, take care to avoid the shallows, and keep an eye on your chart as you approach—there's a good reason for the name of these islands. The Isles of Shoals Light, located at the southern point of White Island, has a fog signal. Ledges surround

© Mike Rounds/New Hampshire Travel and Tourism

Competing Hobie Cats glide onto Hampton Central State Beach.

most of the islands, although most are marked. Many cruising vessels navigate **Haley Cove**, formed by the breakwater joining **Smuttynose Island** to **Malaga Island**. Otherwise, there are several deep and well-marked channels that lead into Gosport Harbor on the northwest side.

Appledore Island and **Star Island** attract the most visitors. **Gosport** at Star Island is the site of a large religious and educational conference center run by the Universalist-Unitarian and Congregational churches. Visits to Appledore are restricted, and calling ahead is advisable (603) 964-9011.

The best and safest way to visit the islands is to anchor offshore and take a dinghy, but great care should be taken when selecting an anchorage around the islands—ledges and boulders mark much of the surrounding waters. Locals recommend a spot midway between the northeast corner of Star and Cedar Island, but as the summer rolls through, so do the many fishermen who frequent the area. It can get congested, and if a visit to the islands is on your itinerary, keep a watchful eye on the weather. These ledges wreak havoc when anchoring—they can be treacherous in foul weather and have claimed several vessels. As we are well aware—shipwrecks love company!

Near the southern border of the state, about 10 miles southwest of Portsmouth Harbor, **Hampton** and **Seabrook** share **Hampton Harbor**, a tight little port that offers good protection in foul weather. The harbor offers deep-sea fishing cruises, and a short drive to Hampton Beach will give you the opportunity to visit Hampton Beach Casino (603) 929-4100. On the other side of the inlet is Seabrook Beach—a great beach that is seldom as crowded as Hampton.

From White Island Lighthouse, head southwest to flashing red 4sec bell "2." Then pick up red gong "4" and nun "4A" before heading under the Seabrook-Hampton bridge. Two jetties mark the entrance to the harbor. Part of the south jetty is submerged, but both are marked by daybeacons so you shouldn't have a problem. Half a mile from the entrance are several rocky ledges: Old Cellar Rock, Inner Sunk Rocks, and Outer Sunk Rocks. This area is marked by a buoy to the northeast.

All in all, New Hampshire cruising spans the gamut—from the cosmopolitan diversions of a city port in Portsmouth to the old-time Yankee atmosphere of its quieter ports. And the wonder of it all is that it comes in a package less than 20 miles long.

Shoreside and Emergency Services

Ambulance, Fire, Police: 911

Coast Guard:

Portsmouth Station (603) 436-4415

Emergency (603) 436-4414 or VHF 16 or 23

Radio Telephone:

VHF 26, 84; I.D. Camden Marine Operator

VHF 25; I.D. Gloucester Marine Operator

Tow Service:

Portsmouth Harbor Towing

Portsmouth, NH (603) 436-0915 or VHF 16

Marine Services

Salisbury, MA (508) 465-8876 or VHF 16 ✦

From meeting you at the breakwater to meeting your every need, Wentworth Marina goes a little bit further when it comes to making your stay a memorable one. We can accommodate boats of virtually any length and requests of virtually any kind. Our on-site concierge will see to your dinner reservations, get you a tee time at the Wentworth golf course, or arrange a ride into historic Portsmouth aboard our land shuttle to enjoy the area restaurants, shops and festivals. At Wentworth Marina, you'll find we go to greater lengths to give you the facilities you require and the service you deserve.

- Free Transportation to Historic Portsmouth
- Slips Accommodating Boats cf 250 Feet
- On-site Restaurants
- Full-service Fuel Dock with Bulk Discounts
- 18-Hole Golf Course
- Tennis Courts
- Heads/Showers/Laundry
- Pump-out station
- Full Staff and Concierge
- On-site TV/Telephone Hook-up
- Electric Power to 100 Amp 3 Phase
- Annual Tuna Tournament

For reservations, please call (603) 433-5050 or use Channel 09 VHF FM (we also monitor channels 68 & 71). When cruising the New England coast, look for us on NOAA Chart NR 13283, latitude 43°03' north, longitude 70°43' west. By land, we are located on U.S. Route 1-B in New Castle, New Hampshire.

Portland, ME Tides; High: -0:03, Low: -0:05, Mean: 8.7 feet	Freshwater, TV,Telephone Elec: 100 Amp 3Phase
Under 50':$1.75/ ft, over 51': $2.25/ft — 170 Berths, 250 LOA	
Electronics, Rigging, Gas, Diesel, Fiberglass, Propeller	Pontevecchio's (Italian) 603-433-5019
Gas, Diesel	MasterCard, VISA, American Express

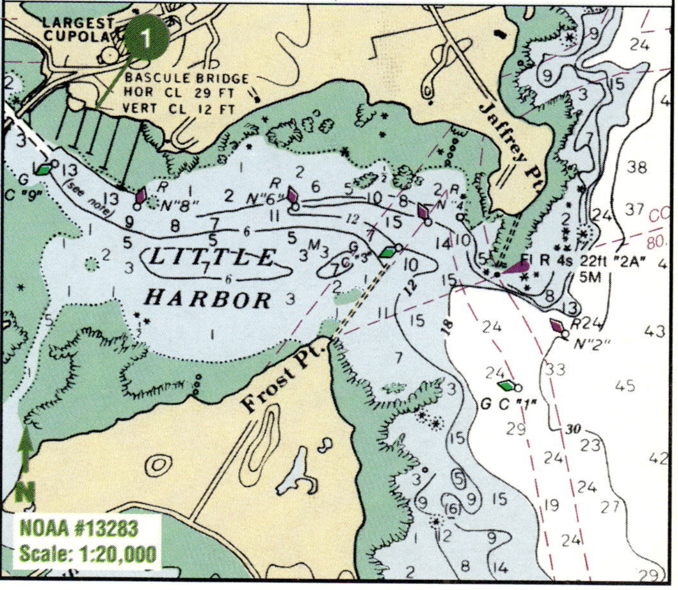

NH
Joins page 98
FOR MORE INFORMATION:
Embassy's Complete Boating Guide & Chartbook
• Rhode Island & Massachusetts (see p. 12)
Embassy's Complete WATERPROOF Charts
• Massachusetts Bay & Cape Cod
• Block Island to Nantucket Sound (see p. 32)
Gloucester
USCG (508) 283-0704
ATLANTIC OCEAN
Salem
Lynn
Cambridge
Boston
USCG (617) 223-3225
USCG (617) 925-0165
USCG (617) 545-3801
Quincy
VHF-FM KHB-35 162.475 MH-z
MASSACHUSETTS
USCG (508) 487-0077
Provincetown
Plymouth
Cape Cod Bay
RI
VHF-FM WXJ-39 162.40 MH-z
Joins page 122
VHF-FM KEC-73 162.55 MH-z
USCG (508) 888-0020
Cataumet
Megansett
Hyannis
S. Yarmouth
Harwichport
USCG (508) 945-3830
Fall River
New Bedford
Westport
USCG (508) 457-3277
Nantucket Sound
Martha's Vineyard
USCG (508) 645-2661
USCG (508) 228-0398
Nantucket
Rhode Island Sound
1 Inch=13 miles
Scale 1:823,680
0 13 26 39 52

The Bay State

Would you like to cruise to an island? On a protected bay? In a good stiff breeze on the ocean? Or maybe you would prefer to pick your way along protected coastline. In Massachusetts, you can choose any or all of the above. Take a jaunt to the beautiful Elizabeth Islands or Martha's Vineyard and Nantucket. See the northern shore of Cape Cod while sailing on protected Cape Cod Bay. Or use the Cape Cod Canal and hop between historic towns without ever leaving the coastline.

With its varied coastal environment, Massachusetts seems to have more than its fair share of geologic diversity. Like most New England states, the Bay State was once home to glaciers, sandy land such as Cape Cod being the result of deposits left by glacial retreat. Exposed to ocean waves and currents for 10,000 years, Cape Cod was pummeled into its current shape. As recently as 1987, a Chatham beach at the Cape's elbow split; one stretch of land became two. It's a perfect example of the forces that are constantly working on the state's coastline.

© Embassy

The south end of scenic Marblehead Harbor.

Other forces traveled to these shores early on as well. The state's leading role in the American Revolution left a bevy of historical towns that greet you as you go ashore. Just about any port in Massachusetts will remind you of our country's austere beginnings—the witch hunts of

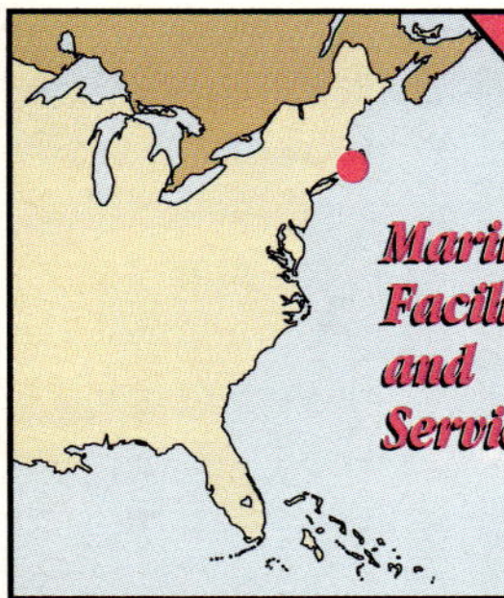

Marine Facilities and Services

See adjacent chart for Coast Guard information.

		DOCKAGE							SERVICES					SUPPLIES			AMENITIES			COAST GUARD
Facility	Phone	Monitors VHF Channel	Working Channel	Number of Transient Berths	Number of Moorings	Seasonal / Year-round	Maximum Length Over-All	Approach Depth in Feet at MLW	Dockside Depth in Feet at MLW	Hookups: Fresh Water / Phone / Cable TV	110V★ 220V▲ 3 Phase ■ Maximum Amps	Ramp / Dinghy Dock / Launch Service	Railway / Lift / Crane: Capacity (tons)	Repairs: Propeller / Sail / Electronics / Rigging / Gas / Diesel / Wood / Fiberglass	SCUBA: Air / Rentals / Equipment / Charters	Gas / Diesel Fuel	Fuel Brand	Marine / Groceries / Bait / Ice / LPG / CNG	Showers / Laundry / Pumpout / Restrooms	Hotel / Pool / Tennis / Golf	Public Phone / Restaurant / Snack Bar	MasterCard / Visa / Discover / AmEx
1 Hawthorne Cove Marina (p.106)	508-740-9890	9	8	10	5	Y	55	7	8	All	★▲50	All	L35	PGDWF				I	SLR			MV
2 Pickering Wharf Marina (p. 107)	508-744-2727	9	72	15	6	Y	125	8	8	F		LD		PSGD				I	SLR		All	MV
3 Fred J. Dion Yacht Yard (p. 107)	508-744-0844	9	8	10	12	Y	100	8	8	F	★▲■ 50		R100 L50	PSRGDF				GLC	SR		P	
4 Boston Boatyard and Marina (p. 109)	617-561-1400	9	68	15		Y	200	30	30	All	★▲■ 50		L50	All		GD	Cit	I	All		PS	All
5 Marina at Rowes Wharf (p. 109)	617-439-3131	9	69	19		S	150	15	15	F	★▲■ 50	D						IB	All	H	PR	MV
6 Boston Waterboat Marina (p. 110)	617-523-1027		8	10	6	Y	140	20	11	F	50	D							SLR	HP	All	MA
7 Boston Yacht Haven (p. 110)	617-523-7352	9	8	13		Y	140	27	27	F	★▲50							GI	SLR	H	All	MV
8 Constitution Marina (p. 111)	617-241-9640	9	69	50+		Y	138	40+	20+	All	★▲■			PEWF				GBI	All	T	All	MV
9 Shipyard Quarters Marina (p. 111)	617-242-2020	9	71	50		Y	200	40	35	All	★▲■ 100	RD		PSERWF				GI	All		PR	MVA
10 Marina Bay (p. 112)	617-847-1800	10	10	50		Y	210	13	13	FP	★▲■ 100		35	All	C	GD	Coas	All	All		All	MV
11 Kingman Marine (p. 114)	508-563-7136	9	71	235	130	Y	100	6	10	All	★▲■ 50	DL	L60 C15	All		GD	Cit	I	All		PR	MVD
12 Moby Dick Wharf Restaurant (p. 115)	508-636-6500	9	74	6	5	Y	55	20	9	F	★▲ 50	All									All	MVA
13 Seaport Inn & Marina (p. 116)	508-997-1281	9	69	10	2	Y	110	15	10	All	★50	All			Avail			All	All	All	All	
14 Harborside Inn (p. 118)	508-627-4321	9	72	10		S	60	25	6	F										HP	All	MVA
15 Nantucket Boat Basin (p. 119)	800-NAN-BOAT	9	11	245		Y	300	15	10	All	★▲■ 100	All		PSRGDF	ARE	GD	Mob	MGBIL	SLR	HTG	All	MVA
16 Fiddler's Cove Marina (p. 120)	508-564-6327	9	68	15		Y	65	7	7	All	★▲■ 50	D	L35	All		GD	Mob	MGI	All		P	MVA
17 Allen Harbor Marine Service (p. 120)	508-432-0353	9	68	3	3	5	50	7	5	F	★50	R	L15	PERGDWF		GD	Tex	MI	R		P	All

Information in these listings is provided by the facilities themselves.

Salem, the landing at Plymouth Rock, the Boston Tea Party, whaling, farming, and shipbuilding. The architecture of different periods—some original, some restored—and pride in the past envelop you as you approach the shores of this great state.

The waters of Massachusetts can be just as hospitable if you simply pay attention to the rules. You can take to your boat as early as May and as late as October, but temperate weather is most guaranteed in the summer season. And that, of course, is when everyone else goes boating. Reservations are a good idea if you plan to visit the more popular harbors at the height of the season.

On the Atlantic side of Cape Cod, you'll find some of the most daunting conditions for boaters. Some 3,000 ships have gone down in the waters between Race Point, at the tip of Cape Cod, and the island of Nantucket. The waters of Nantucket Sound are riddled with long shoals. So while you travel from shore to island to cape, follow the rules. Use the channels, don't try to skip over shallows or shoals if you have a deep draft, watch the weather, and enjoy some of the finest cruising in the Northeast.

Massachusetts Bay: Newburyport to North River

Massachusetts cruising begins in **Newburyport** on the Merrimack River. Wander the streets of this historic town and you'll get the first taste of all you're about to see and feel in Massachusetts. Visit the Old Hill Cemetery, the Old Gaol, the Cushing House Museum (508) 462-2681, and the Custom House Maritime Museum (508) 462-8681.

Red bell "2" and green can "1" mark the channel to the Merrimack River. Newburyport is located on the south bank three miles upriver. You'll spot Newburyport by its large water tank and church spires.

As you head south, you'll see Cape Ann stretched before you. If your boat's draft is shallow enough, you can make straight for the **Annisquam River**, which combines with the Blynman Canal to form a connection between Ipswich Bay and Gloucester Harbor. This connection is used primarily by small craft due to areas as low as five feet. If you choose this route, you'll come out in Gloucester Harbor and miss the port of Rockport on the outside of Cape Ann.

Rockport is a quaint town with winding streets, but it does get inundated with tourists, including boaters, in the summer season. You'll find plenty to do in Rockport. Attend one of the many concerts given by the Rockport Chamber Music Festival (508) 546-7391. Watch or join in the Fourth of July Parade (508) 546-6575. Head to the 54-acre state park at Halibut Point, or sit back and enjoy the picturesque harbor.

Halibut Point is marked by flashing green 4sec gong "1AHP." Rockport Harbor to the south is protected by two breakwaters. A flashing red "6" marks the northern breakwater. The entrance channel is between the breakwaters; moorings and berths are available in the harbor.

On the south shore of Cape Ann sits the fishing village of **Gloucester**, with the statue of the raincoat-clad fisherman staring out to sea. You can stroll through town and feel Gloucester's nautical history. Today, there are less than one hundred commercial ships; the shift has been to recreational boaters and tourists. But you can always hop aboard a whale-watching (508) 283-6995 or fishing boat (800) 942-5464 and see for yourself why the fishermen of Gloucester went to sea.

To approach Gloucester, pass Brace Cove on Eastern Point, and feel free to stay close to the shore as you approach the harbor. A flashing 5sec 57-foot horn marks the shore end of the breakwater; the other end is marked by an occulting red 4sec 45-foot horn. Red nun "2DB" sits just west of this horn. Keep this nun to starboard and its counterpart, green can "1DB," to port to enter the channel, which will give you a clean shot into the harbor.

Farther down the coast in **Manchester-by-the-Sea**, wealthy Bostonians were more the order of the day. Because Manchester is compact and centered around the harbor, a walking tour is a good way to see the sights. You can stroll by elegant homes once belonging to well-known families, such as the Coolidges. On your way, take in some local history on exhibit at the Trask House (508) 526-7230 on Union Street, or inquire in town about the occasional summer concerts in Masconomo Park, located on the harborfront off Beach Street.

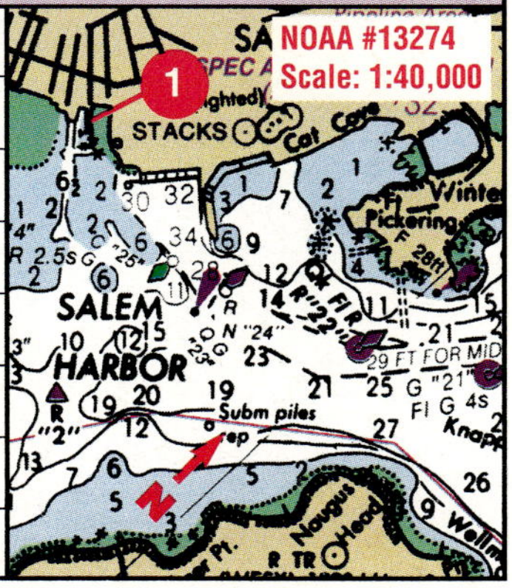

Coming from Gloucester, pick up the flashing red 4sec whistle "2" off Eastern Point and head northwest along Salem Channel to red nun "6" at Gales and Pilgrim Ledges. Make sure to take in the fantastic view of the Boston skyline. Next, head for the red daymarker "8" at Whaleback Rock. Leave it well to starboard, and follow the channel buoys into the harbor. Inside at the large red buildings, you'll find the excellent services and facilities of **Manchester Marine**, close to town and restaurants.

If you feel a spell coming over you, it could be that **Salem Harbor** is around the corner. Your curiosity will undoubtedly be peaked by Salem's legendary history of witchcraft. A short stroll to the Salem Maritime National Historic Site (508) 744-4323, the House of Seven Gables (508) 744-0991, and the Witch House (508) 744-0180 will surely enlighten (or bewitch) you.

Coming from Manchester, the easiest approach is to pass through Salem Channel off Great Misery Island. Keep the lighted flashing red 2.5sec "10" off Great Misery Island to starboard. Continue up Salem Channel, leaving flashing green 4sec "11" to port and then the flashing red 4sec "12" to starboard. You are now in Salem Sound. A straight shot to the flashing green 4sec "13" will bring you to Salem or Beverly.

To approach Salem, look for the stacks of the New England Electric Power Plant on Salem Neck. Enter the main channel at the flashing red 4sec "16," then follow the buoys in. **Hawthorne Cove Marina** is located on the northern shore as you enter the harbor, next to the famous House of Seven Gables. Restaurants, shopping, services, and the historic center of the city are all a short walk away. **Pickering Wharf Marina**, located inside of Derby Wharf in the South River, is well-protected and is a full-service facility. Several shops and boutiques, as well as five restaurants, can be found in the shopping village. In Palmer Cove, you'll find **Fred J. Dion Yacht Yard**, a full service facility with transient dockage for vessels up to 100 feet. A courtesy car is available to explore the historic sites in the area.

For a quieter time, head to **Beverly** with its beautiful historic district overlooking the harbor. Approaching from Salem Sound, the entrance to Beverly Harbor is well-marked, beginning at red nun "2." But be careful, as you come through Beverly Channel to Tuck Point, not to veer too far off to starboard—the bottom shoals more quickly than the chart suggests.

Heading southwest, **Marblehead** is full of little shops tucked away along winding, one-way streets. Old houses dating back to the 1600s and 1700s crowd narrow lanes. History buffs will want to stop in at the Jeremiah Lee Mansion on Washington Street, an impressive Georgian style mansion and home to the Marblehead Historical Society (617) 631-1069. If you're in town for July Fourth, there's a weekend street fair, with crafts, games, and music.

© Embassy

The lightship *Nantucket* rests at the Charlestown Navy Shipyard.

Head south down Marblehead Channel, passing between green can "5" at Coney Ledge and red nun "6" at Eagle Bar. Continuing down the channel, you'll pass tree-covered Eagle Island. Then pass between the flashing white 4sec light atop green can "3" and the flashing red 4sec light at the south end of Eagle Bar. From green "3," turn west into the harbor. Marking Marblehead Neck on the south side of the harbor is a 130-foot fixed green light on a brown and black square skeleton tower.

On the way to Boston, you may want to make a quick stop at **Lynn** or **Nahant**. Tie up in Lynn and enjoy some of the areas attractions: free concerts on Wednesday evenings and Sundays, five miles of beautiful beaches, tennis courts, and bicycle routes. If you come ashore at the Tudor Wharf in Nahant, you can leave your dinghy at the beach and walk the length of town. You can scuba dive at Castle Rock, cool off at Short Beach, or tee-off at Kelley's Green 9-hole golf course (617) 581-0840.

Heading to **Nahant Harbor**, pick up the flashing red 4sec buoy "2" off Shag Rocks. Then swing into the harbor, while keeping an eye on the Spindle, marked by red nun "6," and Joe Beach Ledge, marked by red nun "4." To approach Lynn Harbor, continue past Nahant Harbor until you pick up flashing red 2.5sec bell "2," which will start you into a well-marked channel.

An archipelago consisting of 30 islands at the entrance to **Boston Harbor** awaits you as you head southeast towards the capital city. The one thing you're sure to find at the islands is peace and quiet. Although many are now open to visitors as the Boston Harbor Islands State Park (617) 727-5290, they are undeveloped. You'll want to visit Georges Island which serves as the visitor and transportation hub of the entire park.

Your entrance into the island area is either the North or South Channels, leading you into President Roads. Use the Narrows to reach Georges Island. Follow the channel between Lovell and Gallops before rounding flashing green 4sec horn "7" to the west. You'll see Fort Warren, a pier, and a brick information center on the western shore of Georges Island.

While you're inside the islands you can head straight to **Boston** ("Beantown," if you prefer). The list of things to do in Boston is as long and varied as the list of ports in the state. So if you plan to pull into Boston, make reservations and then let yourself be spontaneous. Sounds like an oxymoron, but there's simply so much to see and do in Boston that it's better to loosen the reins and let the city guide you. Chances are you will miss essential parts of this great city if you plan to adhere to a strict and inflexible time schedule.

Aside from the actual "attractions" in Boston—and they are plentiful—an unshakeable sense of history permeates everything you see and everywhere you go. There's a presence, an almost palpable mystique about the city. It's as if the 350 years since the first settlers arrived in Boston hasn't passed into history, but is lingering in every corner of town. The close-knit, tightly situated streets of the North End—the Italian section of town—will surely make this phenomenon apparent, as will the cobblestone streets of Quincy Market, the architecture in Copley Square, and even a stroll down Newbury Street. The roads alone, situated rather haphazardly and winding through town with no real rhyme or reason, are evidence that time doesn't seem to affect Boston.

Another aspect of the capital city that is sometimes easy to overlook is how relatively clean it is. Of course, as with any city of considerable size, there are bad neighborhoods with graffiti-stained, decrepit old houses and all the telltale signs of poverty. But in comparison to New York, for example, Boston manages to remain fairly clean, and this is especially striking while riding the "T"—Boston's public transportation system similar to a subway.

At the Charlestown Navy Shipyard, you can visit the *USS Constitution* (617) 242-5670, also known as "Old Ironsides." Or you can head next door to the Constitution Museum (617) 426-1812 or the Bunker Hill Pavilion (617) 241-7575. The Bunker Hill Monument and Museum are only a 10-minute walk away. At the downtown waterfront, drop in at the New England Aquarium (617) 973-5200 or hop on one of the trolleys for a sightseeing tour of the city. And, of course, a trip to Boston is not complete without a walk around Faneuil Hall and Quincy Market (617) 523-2980 with its shops, restaurants, musicians, dancers, and artists.

That's only the beginning of seeing this New England city. There's dining galore and museums for a variety of tastes. And because there are several colleges, universities, and specialty schools in town, there's an intellectual atmosphere that accompanies the young crowds and nightlife. Across the river in Cambridge you will be witness to the artistic and intellectual pursuits of Harvard and MIT (Massachusetts Institute of Technology). The Art Institute of Boston and the Berklee School of Music provide the city with a never-ending source of street performers, from sidewalk artists to traveling jazz ensembles. And the co-op program at Northeastern University gives students the opportunity to gain experience, sending thousands into the workplace during the course of their academic programs.

From President Roads, follow the channel to the Inner Harbor at flashing red 4sec "10." **Boston Boatyard & Marina** will be the first marina on the right after the airport, easily identified by the Citgo fuel dock. The marina has transient dockage as well as a host of services. Continuing west and north toward the city, you'll see **The Marina at Rowes Wharf** about 1.3nm from red "10." This is one of the most spectacular waterfront developments you will find anywhere—the walk-through dome is most impressive. Just two piers to the north is **Boston Waterboat Marina**, which offers convenient access to downtown Boston. Moorings are available. Another excel-

Enjoy the best of what Boston has to offer from centrally located Boston Boatyard & Marina! We boast the best view of the Boston skyline as well as complete support services including yacht hauling, yacht repair and chartering.

BOSTON BOAT YARD MARINA

With 24-hour security, plenty of free parking, and transient dockage always available, Boston Boatyard and Marina strives to take care of all your needs. Relax and enjoy Boston! We'll take care of everything else.

- See Boston Tides
- $1.50 / foot per day, 160 Berths, 200 LOA
- Prop., Sail, Elec., Rigging, Gas, Diesel, Wood, Fiberglass
- Citgo, Premium Gas, Diesel, Oil
- Fresh Water, Phone, Cable TV, 30V, 30/50 Max Amps
- Liberty Market
- Several restaurants in area
- MasterCard, VISA, American Express

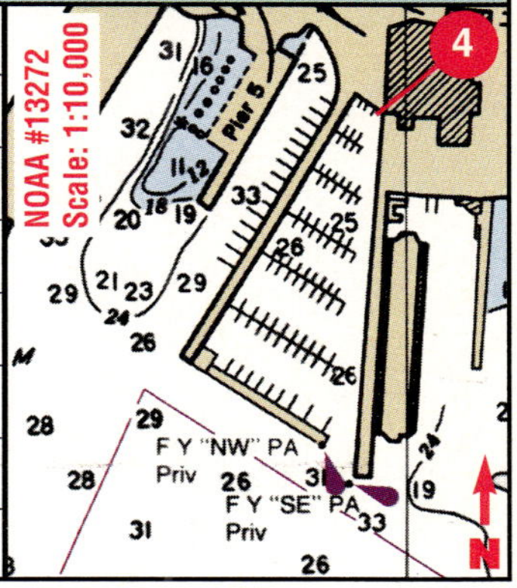

In the heart of Boston, the Marina at Rowes Wharf is the gateway to the city's finest restaurants, shopping areas, and visitor attractions. Rowes Wharf and its surrounding neighborhoods offer visitors to the marina a wealth of recreational activities and cultural sites.

Within walking distance of Rowes Wharf is Faneuil Hall Marketplace, the Freedom Trail, the New England Aquarium, the Children's Museum, the Computer Museum, Chinatown, the North End, Beacon Hill and the Boston Common. For fine dining at a downtown Boston restaurant, a family trip to the Children's Museum or a walk through one of Boston's many community neighborhoods, the Marina is at the center of a host of exciting possibilities.

- Boston Tides; High:+0:06, Low: +0:08, Mean 9.6 feet
- $1.75 - $2.00 / ft per day, 38 Berths, 150 LOA
- Fresh Water Elec: 110V, 220V, 50 Amps Max
- Many nearby
- Several in the immediate area
- MasterCard, Visa

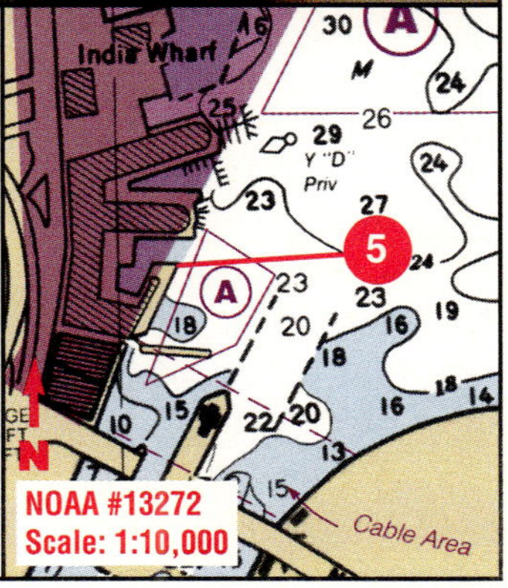

lent facility on this shore is **Boston Yacht Haven**, a conveniently located facility that will allow you to see many sites by foot, such as Quincy Market, the New England Aquarium, and Boston's historic waterfront.

The main channel in the harbor gradually turns north. You'll pass downtown Boston to port and East Boston to starboard. The Coast Guard station in North Boston will loom ahead to the west. The channel branches here—one side continuing north to a row of piers, the other swinging west toward the Charlestown Navy Yard.

Constitution Marina, near the Charlestown Navy Yard, has excellent facilities and offers good protection. It is within walking distance of the historic North End. The Science Museum (617) 723-2500, and the New England Aquarium (617) 973-5200. Farther along the north arm of the channel, you'll find **Shipyard Quarters Marina** in the former Charlestown Navy Yard on piers 6 and 8. This marina offers excellent protection and service for small and large transient boats in a beautiful and historic environment with water shuttle access to Boston. If you choose to continue into the Charles River, you'll have to pass through the Charles River Dam with its series of three locks. Otherwise, you can leave the Inner Harbor and see what else the waters inside the Boston Harbor Islands have to offer.

To reach **Dorchester Bay**, head back toward President Roads. At flashing green 4sec buoy "3," north of Spectacle Island, head southwest toward Dorchester Bay. The flashing green 4sec "5" marks the change from natural to dredged channel. After reaching the flashing red 4sec "12," you can cut south along the marked channel into **Marina Bay**, which has excellent facilities and restaurants. You may also continue beneath the bascule bridge into Dorchester Basin.

The **Weymouth Fore River** and **Back River** are worth a visit if you're in the Quincy Bay area. From Dorchester Bay, you can go beneath the Long Island Bridge if your mast is a vertical clearance of less than 51 feet. Your other choice is to go around Long Island by heading back to President Roads and picking up Nubble Channel. You can duck through West Gut starting at red nun "4" and green can "3." From there, work your way through the channel to red nun "10" where you can pick up the channel into the rivers. If you follow the channel south, you'll find it splits at flashing red 4sec "18." Head east for Back River, southwest for Fore River.

Once through Weymouth Fore River, visit **Quincy**, with Quincy Square and its eclectic gathering of American architecture. Stop by the Adams National Historical Site (617) 773-1177 for a look at one of the large and well-kept colonial homes. **Marina Bay Quincy** is located in North Quincy. You can use the extensive line of facilities at the marina while you enjoy touring the local area or travel to nearby Boston.

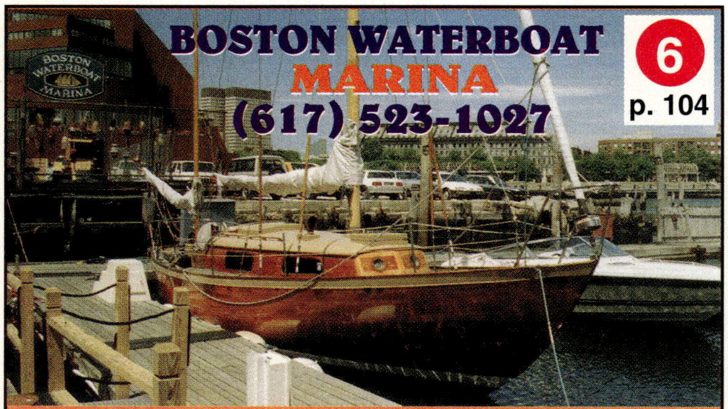

6 p. 104

The Boston Waterboat Marina, located at 66 Long Wharf, is close to Quincy Market and downtown Boston. Within minutes you can be shopping, dining, visiting local museums or the Aquarium, or just enjoying the sights and sounds of beautiful historic Boston.

- See Boston Tides
- Under/Over 50': $1.75 / $2.00; 36 Berths, 140 LOA
- Propeller, Electronics
- Fresh Water
- Christie's Market, Haymarket; Call 617-523-1027 for details
- The Charthouse 617-227-1576 *family*, $18

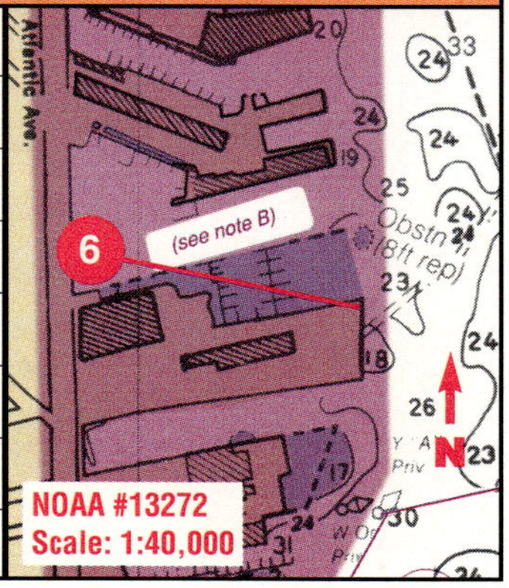

BOSTON YACHT HAVEN

7 p. 104

Boston Yacht Haven is a centrally located marina to Boston's downtown business, historic and cultural interests. So convenient is the location that the suggested mode of transportation is by foot, to such places as Quincy Market, New England Aquarium and Boston's historic North End and Waterfront. Pre-arranged parking is available as well as a friendly staff waiting to accomodate you! Advance reservations are suggested and casual reservations can be accomodated by means of VHF channel 9.

JAMES LANDING

87 Commercial Wharf,
Boston, MA 02110
(617) 523-7352

- Boston Tides; High: +0:04, Low: +0:04, Mean 9.5 feet
- $2.00 / ft per day; 45 Berths, 140 LOA
- Fresh Water
- Boston Merchant & Specialty Shops, Fanueil Hall, North End
- Many Diverse Restaurants
- MasterCard, Visa

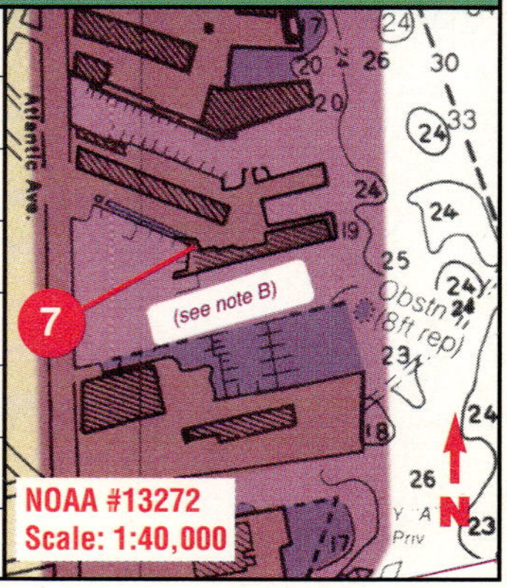

The historic and tourist-oriented areas of **Hull** and **Hingham Bays** are your last stop before leaving the labyrinth of Boston Harbor islands. Though Hull is only a small spit of land and Hingham not much larger, you will find no shortage of things to do. The Hull Lifesaving Station has an observation tower and exhibits commemorating the efforts of Joshua James, a lighthouse keeper who rowed out into the harbor to rescue the crews of stranded ships at the turn of the century. Fort Revere, built during World War I, also has an observation tower which is under restoration. The towers provide a spectacular view of Boston.

Hingham Square is alive with vibrant shops and historic New England sites. The Old Ordinary (617) 749-0013 on the hill above Lincoln Square has fabulous twelve-over-twelve windows and is capped with a wooden, shingled roof. Old Ship's Church on North Main Street was built in 1681 and is the oldest continuously operating house of worship in America.

To get to Hull and Hingham bays, you'll want to take the channel back out of the Weymouth Rivers. For Hingham Harbor, wait until you're at red nun "8" and then pick up flashing green 4sec "1" to starboard in order to begin your entrance into the harbor. Keep the buoy and Bumkin Island Shoal to port as you turn south. Just before the green and red "HH" buoy, the marked channel for the Weir River breaks off to port. The channel into

Hingham Harbor begins at Crow Point and runs for 1.3nm to the inner harbor, which is generally used by smaller boats due to water depths at low tide.

To go through Hull Bay to **Allerton Harbor**, take the main channel back toward Hull Gut. Before reaching the gut, veer off at flashing red 2.5sec "4." Keep well outside of Outer Seal Rock. Once abeam of Outer Seal Rock, head for the gold cupola of the school ahead. Come in close to the beach before cutting north into Allerton Harbor. In the harbor, you'll find anchorages on both sides of the channel. The bottom is muddy and offers good holding for most anchors.

You can make your way back through Hull Gut to Nantasket Roads and then head east to exit the island area. Hull Gut passage gets extremely congested at times. Heed the buoys and beware of the currents, which can run up to 5 knots.

Just a short jump down the coast brings you to **Cohasset Harbor**. You can moor or anchor in the harbor, but transient facilities are usually limited, so call ahead to the harbormaster's office (617) 383-0863. Once you're there, you can enjoy a relaxing stroll to Cohasset Commons, one of the more picturesque commons in coastal New England.

Don't put your navigation skills to rest yet. Cohasset Harbor has three passages through the many rocks and ledges scattered across its mouth: the Western Channel, the Gangway, and the Eastern Channel. The northern approach is by way of the Gangway, the only lighted passage. Leaving Minots Ledge to port, make for the green flashing 6sec bell "1," and then bear south to the town lights of Cohasset and the flashing 2.5sec red "2" tower south-southwest of you. The 5-foot deep channel will lead you to Cohasset Cove. You may want to look for an anchoring spot before the channel in Sandy Cove since so many moorings in the cove are privately owned.

Scituate awaits you, ready to fulfill all your expectations of a quaint Massachusetts seaside town—shingled cape houses slightly weathered by the wind, sand dusting every street, wrap-around porches, and the strong salt smell of the sea. Take all this in on a walk to the Mann Farmhouses and Historical Museum (617) 545-5186, situated on three acres of rolling land, or visit one of many small museums, such as the Cudworth House or the Little Red School House.

Navigating into Scituate is not only a relief after the tougher northern work, but is also quite beautiful. The harbor is best approached from the east between two gorgeous points of land: First Cliff and Cedar Point. Follow the well-marked channel after starting with red nun "2" to starboard and green can "1" to port. You should make arrangements with a marina in the area ahead of time as most space in the harbor is taken by privately maintained moorings.

10 p. 104

- Boston Tides; High: -0:02, Low: +0:03, Mean 9.5 feet
- $1.50 / ft per day, 685 Berths, 200 LOA
- Prop., Sail, Elec., Rigging, Gas Diesel, Wood, Fiberglass
- *Citgo* Gas, Diesel
- Fresh Water, Phone, Cable TV 110/220V, 100 Max Amps
- Groceries, Liquor, Restaurants 847-1800
- (8) 617-847-1800 Casual to formal $5 - 15
- MasterCard, Visa

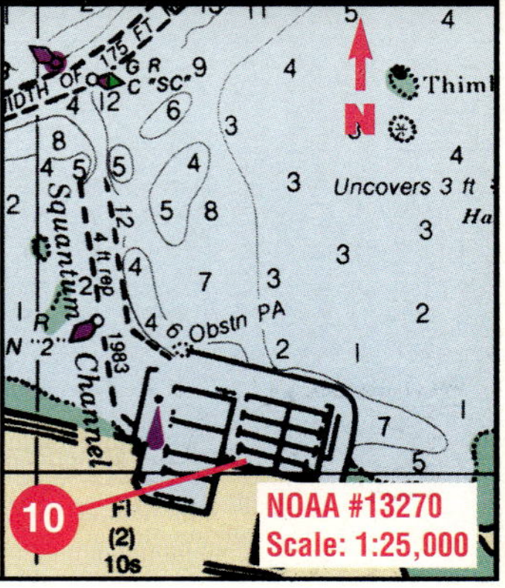

Cape Cod Bay: Green Harbor to Provincetown and the Cape Cod Canal

As you enter **Cape Cod Bay**, you're heading into the nook of the great curled arm with its clenched fist, otherwise known as Cape Cod. In addition to the exciting ports on the bay, you may be lucky enough to spot some of the underwater wonders while you cruise these waters—the whales. Stellwagen Bank, a 110-foot underwater bank located about 12 miles off the coast and extending from Gloucester to Provincetown, is a national marine conservation area. You're likely to see humpbacks and fin whales and Atlantic white-sided dolphin in these waters. Many whalewatching tours are available, but you may want to wander these waters in your own boat in search of the beautiful marine mammals. Be considerate, though! Don't get too close or chase the wondrous mammals.

Green Harbor, located directly west and just slightly north of the fist, introduces you to Cape Cod Bay. Green Harbor is a small fishing community with clapboard or shingle-sided houses set nearly flush to one another along narrow streets. You can take in the local flavor, or take a taxi to Green Harbor proper, three miles away. There you can see the preserved home of American orator Daniel Webster (617) 837-9400 alongside a blacksmith's shop, or visit the historic Winslow House (617) 837-5753.

The entrance to Green Harbor is marked by red nun "4." Coming into the harbor, look for the microwave towers west of its entrance. Leave red nun "8" marking Howland Ledge to starboard as you bank hard to the west. Although moorings and anchorages are hard to come by, there are marinas that provide transient dockage in the harbor.

Both anchorages and moorings are easier to come by in **Duxbury Bay** just around the corner. And though many of the other coastal towns are alluring, Duxbury has its own brand of beauty. The curved shoreline is decorated with mansions; the boat basin is lined with fine shops and a few eateries. You may want to visit the Old Shipbuilders District. The Duxbury Art Complex (617) 934-6634 and Myles Standish's Homestead are also in the area. The Standish Monument on a 116-foot tower rising on Captain's Hill provides a wonderful view of the harbor.

The town of **Plymouth** is a quaint, coastal village with a thriving waterfront, and all the services and amenities you might need. Reminders of our country's past are everywhere. There is, of course, "The Rock." There are other historical buildings in any direction you turn. You can see the sights by hopping aboard the Plymouth Rock Trolley (508) 747-3419, which departs from the Village Landing Marketplace, two blocks west of the *Mayflower II.* If you'd rather bike, there are shops for rentals. But walking will do you fine, too.

Both Duxbury and Plymouth are approached through the opening between the spits of land Duxbury Beach and Plymouth Beach. From Green Harbor, head south toward Plymouth Light, which pulses 3 times every 30 seconds on Gurnet Point. Once there, you'll spot red nun "2" leading you into the channel. The red flashing 5sec horn, known as "Bug Light" is at the entrance to Plymouth and Duxbury. Both channels (south for Plymouth, north for Duxbury) are very well marked. Both areas have anchorages and docking facilities.

From Plymouth or Duxbury you have two choices: head south to the Cape Cod Canal, or head east to **Provincetown** and tour the "Lower Cape." Provincetown, on the "fist" of the Cape, is 25 miles out to sea and forever wrestling with the wind and water. Only "three miles long and two streets wide," P-town is as wild a honky-tonk town as you'll find, brimming with life and the appeal of merriment. The main attraction in town is everybody else in town. Walking, rollerskating, biking down Commercial Street, or just sitting on a stoop, you come here to see and be seen. For a walking map, list of attractions and events in town, call (508) 487-3424.

Provincetown Harbor at the breakwater is 26nm east of Plymouth. If rounding Race Point, be careful of a wreck .85nm south-southwest of the light. Your first mark will be the flashing green 4sec bell south of Wood End. From here, head for the quick flashing green bell "3" east of Long Point, leaving it to port before heading into the harbor. You can anchor just about anywhere in the harbor, as there is plenty of deep water. There are also several facilities available.

Farther down the arm of the cape, your next port of call will be **Wellfleet Harbor**. The foremost attraction in Wellfleet is the water. If you want the salty variety, try the surf beaches on the Atlantic side, or the warmer and calmer bay side at Mayo, Duck Harbor, Power's Landing, and Indian Neck. While out and about, you have the entire Cape Cod National Seashore to explore.

Wellfleet is about 9nm south of Provincetown. Head for the flashing green 4sec bell "1" at the southwest end of Billingsgate Shoal. Coming from the north, you may not want to swing so far west, but there are no other markers between the bell and Billingsgate Island, and the island disappears at half tide. From the bell, head for the lighted green 2.5sec buoy "3" at the southern end of Billingsgate Island, and follow the marked channel north into Wellfleet Harbor. Anchoring space is usually available, as are a number of transient moorings and slips at the town marina.

Another stop en route to the Cape Cod Canal is **Barnstable Harbor**. Barnstable has often been referred to as a microcosm of the entire Cape. The township includes everything from vast, peaceful marshes and historic villages at its north border to the active commercial center of Hyannis and the aristocratic venues of Cotuit, Osterville, and Centerville.

Entering Barnstable Harbor can be tricky. Don't try to enter the harbor in questionable weather. There are shallow areas, and the tidal currents are tricky. Your approach should begin at the red and white vertically striped Morse (A) lighted bell "BH." You then have a straight shot to the flashing red 2.5sec "8" off Beach Point, but watch for strong cross currents of wind and tide. To reach the marinas in Maraspin Creek, follow the marked channel scrupulously. The lighted green 4sec flasher "11" at Horseshoe Shoal marks an obstruction, which nonetheless should be kept very close to port.

At this point in which you cruise, you can head toward the **Cape Cod Canal** and cut from Cape Cod Bay to Buzzards Bay. Miles Standish was the first to suggest that a channel across the Cape could substantially increase the trade between his colony and the other Dutch and English colonies along the Atlantic Coast. The Cape Cod Canal was completed almost three centuries later in 1909. Originally, the canal was owned by an ambitious businessman who charged tolls. In the early 1930s, the U.S. Government bought and enlarged the canal, did away with the tolls, and replaced them with taxes.

Now 200 yards wide, the Cape Cod Canal has plenty of room for everything but supertankers and aircraft carriers. Consult your tide tables for more in-depth information on the 5.5-foot difference and swift currents between the east and west ends of the canal. All boats must observe the no wake, 10-mph speed limit and must complete the transit within 2.5 hours. You must motor through the canal, not sail. Fishing is not allowed, and vessels going with the current have the right of way. For more information or assistance, call the U.S. Army Corps of Engineers (508) 759-4431.

The east (or Cape Cod Bay) entrance to the Cape Cod Canal is easy to make, primarily because it is so well marked and there are no obstacles in the way. The entrance is marked by a green, quick-flashing bell "3" and a quick-flashing red "4." The current at both ends of the canal at peak can reach 6 or 7 knots.

The East Boat Basin is just inside the east end of the canal and a very convenient spot to tie up. The East Boat Basin has a dredged depth of about 9 feet at mean low water. Anchoring and mooring are not allowed, so you should make prior arrangements with the harbormaster's office (508) 833-0808 for an overnight slip. You can visit **Sandwich**, about a mile inland, for a look at what may be the oldest town on Cape Cod. There are several small but fascinating museums, including the Yesteryears Doll Museum (508) 888-1711 sitting between a collection of town offices at River and Main Streets. Around the corner is the Sandwich Glass Museum (508) 888-0251 with brilliantly etched and colored lamps, bowls, and other glasswares, the finest examples from the glass-blowing days of the town.

Buzzards Bay to Nantucket Sound

As you exit the Cape Cod Canal, you have two choices. You can sail along the mainland coast of Buzzards Bay and head towards Rhode Island, or choose the southern route and see the Elizabeth Islands, Martha's Vineyard, Nantucket, or the south shore of Cape Cod—all exceptionally beautiful cruising grounds.

As you exit the canal, you can duck into **Onset Bay** to bask on the white sandy shore, windsurf, or fish. There is only one approach into Onset, from the south. The entrance is marked by the quick flashing green tower "21" off Hog Neck, which you must round because the shoals to the north are unmarked. Staying in the channel, continue toward the town pier that's north-northwest of you. You'll find a safe anchorage after the slight turn at Wickets Island.

Just south of the Cape Cod canal exit in Red Brook Harbor is **Kingman Marine**, a full-service marina that offers 365 slips and moorings, is open year round, and welcomes transients.

Another stop on your way west is **Marion**, located on Sippican Harbor. This quaint shoreside town is the starting point for the Marion-to-Bermuda yacht race and homeport for *Tabor Bay*, a big black schooner operated by the local boarding school. But it's a one-street town with shady trees, picket fences, and a real general store.

Once back in Buzzards Bay, head down the channel, which is very well marked. You'll see the Sand Spit to starboard almost to the end of the channel. As you reach quick green can "1," you'll exit the channel and head for the southeasterly approach to Sippican Harbor. Keep Bird Island to starboard. You'll enter the harbor between the flashing 4sec "2" Centerboard Shoal buoy south of Butler

Moby-Dick
WHARF RESTAURANT

12 p. 104

and Marina

One Bridge Road
Westport, MA 02719
508-636-6500
1-800-417-6256
Monitor VHF 9/68/74

Newport Tides; High: +0:09, Low: +0:33, Mean: 3.0 feet	Fresh Water: 110/220V, 3 Phase, 50 Max Amps
Call for Dockage Rates 15 Berths, 55' Max LOA	Lees Market
Gas & Diesel Fuel, Repairs available at:	Moby-Dick Wharf Restaurant 508-636-6500, *Elegant*, $14-16
FL Tripp & Sons 508-636-4058	MasterCard, VISA, American Express

Recently renovated Moby Dick Wharf Restaurant and Marina offers the finest freshest available fish and lobster. Overlooking historic Westport Harbor you could not ask for a more beautiful spot famous for its magnificent sunsets and view of the commercial docks at the foot of charming Westport Point. Long overlooked and under rated Westport Harbor has escaped the congestion of the Cape and Islands and remains virtually unchanged in its sheer beauty and peacefulness. With miles of riverfront and a fully protected harbor with complete amenities, Westport is a true find and treasure. Come have dinner or lunch at the Moby-Dick Wharf Restaurant and Marina then explore the last undiscovered cruising harbor south of Maine! Westport Harbor has an undeserved reputation for being difficult to approach. Once you've arrived, you will feel you have stepped back in time! Please call for restaurant reservations and slips. Open year round. Come by land or sea!

Point and the flashing green 2.5sec "3" buoy off Converse Point. The channel into town is a little narrow and there are foul spots on either side. Local boaters recommend keeping the green cans close to port. Moorings are available from several of the marine facilities in town.

True to its Indian meaning, there is something very restful and relaxing about **Mattapoisett**, the next stop southwest. You can visit the historic district, watch the line of fishermen angling for bluefish, or spend the evening at Shipyard Park (508) 758-9408 listening to live music from the elegant gazebo.

Pass green can "11" at the Bow Bells to starboard, and head for red nun "2" off Angelica Point. Once at the mouth of Mattapoisett Harbor, it's a straight sail into the mooring area, following the nuns. Keep between the markers to avoid the rocks.

Herman Melville and Moby Dick should come to mind in **New Bedford**, on the Acushnet River. It was from this river that Melville set sail on a journey that inspired his epic novel. By the mid-1800s, New Bedford had a fleet of more than 330 whaleships supported by a battalion of shipwrights, chandleries, sail lofts, and insurance companies. Although New Bedford is heavily industrialized today, don't let it discourage you. There is still a historic part of town which is quite handsome. You can sign up for a walking tour offered by the Rotch-Jones-Duff House (508) 997-1401, or you can stroll on your own.

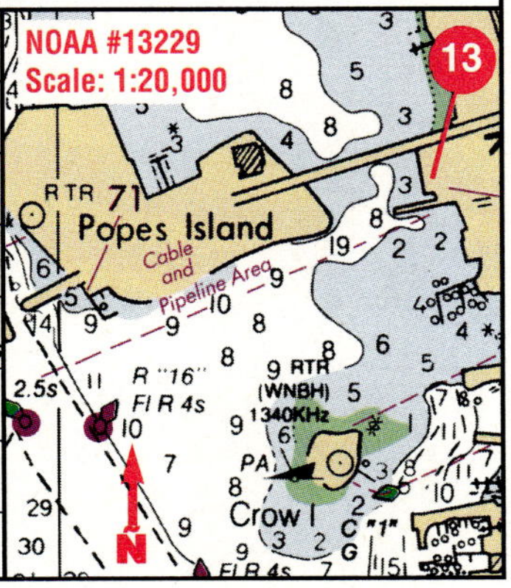

The entrance to New Bedford Harbor at the hurricane gate is about 12.5nm by boat southeast of Mattapoisett Harbor. The main approach to New Bedford is well marked. A deep channel leads into the harbor, starting at Brooklyn Rock, marked by a green 6sec flashing bell, and Henrietta Rock, marked by red nun "6." You can stop at one of the marinas in **Fairhaven** on the east bank, such as **Seaport Inn & Marina**, located just before the bridge. The marina has transient dockage and numerous amenities.

If you're scouring the coast for a small village called **Padanaram**, you won't find it. But it's there. It's known as **South Dartmouth**. Although the name Padanaram became official in 1817, it remains absent from the charts, perhaps as an innocuous way of preventing this unspoiled village from losing its quiet and graceful aura, emphasized by the world-famous Concordia yawls built here.

You'll find Padanaram on Apponagansett Bay. Coming from New Bedford is uncomplicated. A large vacant granite building at Fort Rodman stands at the end of Clarks Point to starboard. The entrance to the bay is marked by the red 4sec flasher on a tower south of Ricketsons Point. Navigating into the town is quite simple, as long as you follow the markers.

As you continue down Buzzards Bay, the last port you'll find in Massachusetts is **Westport**, beautiful and peaceful with clean water. Relaxing on 3.5 miles of public beach at Horseneck Beach (508) 636-8816 or celebrating Westport's annual River Day (508) 636-1000 in June are a few of the choices during your stay at Westport. If you're hungry, we suggest tying up at **Moby Dick Wharft Restaurant and Marina** and sampling their specialty—sea food! Their transient facility will accommodate you while you dine on fresh fish and lobster.

Coming from the east, down Buzzards Bay, you should skirt around Hens and Chickens (the two mounds of rock 0.9nm southeast of Gooseberry Neck) unless you have local knowledge. Head for red nun "6" off Lumber Rock, leaving it 200 yards to starboard. After rounding red nun "6," head for lighted bell "WH" southeast of Two Mile Rock, marking the beginning of the channel. From here, head for green can "3" at the entrance to the channel into Westport Harbor. **F.L. Tripp and Sons** on the south shore is easily approached from the channel and has the best facilities in the harbor.

Back at the southern end of the canal, your alternate choice is to head for the waters south of Cape Cod and see some of the finest cruising grounds in the state—the offshore islands and the south shore of Cape Cod.

For this route, bear toward the south shore as you leave the canal and head to the oceanographic research mecca of **Woods Hole**. You may spot one of the 125-foot schooners operated by the Sea Education Association (508) 540-3954 at the main wharf on Water Street. The town is compact, and everything can be reached on foot.

The western or northern entrance to Woods Hole is marked by the flashing green 4sec bell "13." Follow the channel to the red 4sec flasher "10" off Penzance, where the channel branches eastward. Head for the 34-foot 2.5sec green flasher "5" south of Penzance Point until you pass red nun "6." Turn to the east and stay to the middle of the channel; now head for the yellow quick flasher at the southernmost ferry slip. Continue east to a point about 200 yards beyond red nun "2," where you can line up with **Great Harbor** Range. This range has red daymarks with white center stripes and a quick flashing green light. You can either try for a mooring at the Yacht Club or go to Eel Pond by crossing under the Water Street bascule bridge, which has a 5-foot vertical clearance when closed. Inside the pond, you'll find a marina.

You will undoubtedly be curious about the **Elizabeth Islands**, a string of nine islands that have stretched before you as you've sailed down Buzzards Bay. From **Naushon Island** to **Cuttyhunk**, you can scoot into several of the passages to Vineyard Sound—but caution is required. The islands are almost uninhabited, and without any facilities. But there are many gunkholes for taking a dip and taking in the beauty.

Leaving Woods Hole heading southwest, your first place to stay is on Naushon Island's western end at **Hadley Harbor**. You can go ashore on **Bull Island** for a completely relaxing afternoon with a scenic view.

Begin your approach to Hadley harbor at red nun "6," being sure to watch out for Hadley Rock. You'll find the entrance to the harbor at red nun "2" and green can "1." Take care when entering between the buoys; rocks are just northwest of the nun. The Inner Harbor provides excellent anchorage with good protection and plenty of space. The channel runs between Bull Island and Nonamesset Island.

As you exit the Woods Hole area, you can make your way to a bight south of the Weepecket Islands on the north shore of Naushon Island. There you'll find an unnamed gunkhole, we suggest calling it Mackenzie Harbor, with deep water and good shelter from winds out of the south and east. Just 2.7nm southwest of this spot, still on Naushon's north shore, Kettle Cove is a nice spot for a swim. The cove has deep water but is not well-protected from the weather.

Before leaving the north shore of the Elizabeths, a stop at picturesque **Cuttyhunk Harbor** on the southernmost island is well worth it. During the summer, the 46 moorings and 84 slips fill by early afternoon, so if you're looking for space, go early.

Passing along Nashawena Island, be sure to stay at least 700 yards offshore. You'll see flashing bell "CH" to the north. From here, cut between red nun "2E" and green can "1E" to the harbor entrance.

To enter **Cuttyhunk Pond** from the outer harbor, keep red nun "4" to starboard before rounding red bell "6." The dredged channel opens between green can "9" and flashing red 6sec "8" on the Cuttyhunk North Jetty and is 9 feet deep at a width of 75 feet.

There are three navigable passages through the Elizabeth Islands: Robinsons Hole, Quicks Hole, and the Canapitsit Channel outside Cuttyhunk Pond. Currents run fast through all of these channels. We recommend that you stick to **Quicks Hole,** unless you have local knowledge of the area. Canapitsit Channel at Cuttyhunk is narrow, riddled with rocks, and subject to strong currents. And Sow and Pigs Reef extends a mile and a half southwest of Cuttyhunk at its southern tip. Even the QE2 had trouble here.

While cruising through Quicks Hole, you may want to pull in close to the beach and drop your hook for a while. This is a beautiful, somewhat protected passage and is an excellent place to spend a day of swimming, fishing, or just relaxing. We recommend that you motor through the passage. Keep flashing red 4sec "2" to port when heading southeast.

You may want to stop at **Tarpaulin Cove** on the south shore of Naushon Island. It's a beautiful cove with a soft-sand beach. There are no facilities, but that may very well be the best part of this quiet stop.

From the Elizabeth Islands, you can make a straight shot for **Martha's Vineyard** or Nantucket or head toward the south shore of Cape Cod. We'll go to the islands first.

Menemsha Pond fills the Vineyard's baby boot in the west. To the west of the pond are the Gay Head Cliffs, which tower 150 feet above the sea. It is a perfect jaunt from the Elizabeths.

You'll cut across Vineyard Sound with its currents reaching 4 knots at maximum ebb tide. The water is deep and clear of obstacles. After the green junction buoy at Lucas Shoal, head south to green bell "1" in Menemsha Bight. The holding ground in the bight is an excellent mixture of sand and mud.

While the fishing village of Menemsha is a quiet first stop, **Vineyard Haven** brings you into the center of activity at the island. This is the largest and busiest of the island's harbors. It is where most of the shopping and restaurants are located, and where the ferries dock, meaning that tourists arrive by the hoards. You can rent a bike or a moped here to get around. If you want to get away from the hubbub, you can visit the Felix Neck Wildlife Sanctuary (508) 627-4850 located on the border of Vineyard Haven and Edgartown.

Coming from the west, stay well shoreside of the Middle Ground before rounding West Chop—it's best to head to the east of the flashing red 4sec gong "2" which will keep you clear of the shoals and the point.

In contrast to Vineyard Haven, **Oak Bluffs** is a small, well-protected harbor located on the northeast side of Martha's Vineyard and it is used almost exclusively by pleasure craft. You will witness a tight-knit community that magically comes alive in the summer. In late August, the residents celebrate Illumination Night. They light their gingerbread houses with Japanese lanterns in a century old tradition that symbolizes the closing of the campground season. There's also plenty of good restaurants ranging from traditional candlelit dinners to Japanese sushi.

Round East Chop by staying outside of flashing green 4sec bell "23." Stay about 300 yards offshore continuing toward the flashing red 4sec light "2" on the end of the northern breakwater at Oak Bluffs.

If you decide to pass by Oak Bluffs, **Edgartown** is a 6.2nm away. The town at the end of the island's northern side has become a fashionable summer enclave for upscale city folk. To accommodate them, Edgartown has shop after shop alongside dozens of historical buildings. One of the most intriguing of these is the Gingerbread House on Church Street, inedible although it appears true to its name. The town is well-known among the sailing crowd for the Edgartown Regatta run by the Edgartown Yacht Club (508) 627-4361 in July. There are also numerous beaches in the area. But a walk through this old-fashioned-looking village may just be captivating enough for you.

Coming from the north, look for red flashing 4sec gong "22" off the eastern point of Hedge Fence. Keep this buoy well to starboard. The entrance to Edgartown Harbor is from the north and marked by a red flashing 4sec "6" west of Cape Poge. You'll catch the channel into the harbor as you round red nun "8" hard to starboard. Tucked well into the harbor is **The Harborside Inn**, a waterfront resort with small powerboat, sailboat, and catamaran rentals available.

A wedge of land shaped like the fin of a diving whale, **Nantucket** lies 22 miles south of Cape Cod, weathered and beaten by the sea. For years, the island bore men who spent their lives in search of great sperm whales. Artifacts from this legendary time can be found at the Whaling Museum (508) 228-1736 on Broad Street, one of the more popular spots on the island. But there are other places to visit as well: the Old Gaol [Jail], the Historical Association (508) 228-1894 (with 11 sites portraying many island lifestyles as the town grew from a small agrarian community to a great whaling port), or the Old Mill.

Coming to Nantucket from the west, you'll have to pass along the Cross Rip Channel (otherwise know as Nantucket Main Channel), lying between Cross Rip and Horseshoe Shoals. The water outside of the channel is plenty deep for most cruising vessels, but play it safe and stay within the channel. Also beware of heavy seas. Outside the harbor, you'll see privately owned Muskeget, Tuckernuck, and Esther Islands. The waters there are tricky and should only be explored in a launch. You should stay in the channel until you are ready to make your approach into Nantucket Harbor.

The entrance to the harbor begins at the red and white bell "NB." From here, the channel runs south to Brant Point, marked with range towers. To enter the harbor from the channel, you have to swing east, then west again around Brant Point, marked by a red 4sec 26-foot horn. On the western shore of the harbor, between two bulkheads, you'll find **Nantucket Boat Basin**, which will provide you with all the amenities you need and a convenient location to enjoy this beautiful island.

Should you decide to spend time in ports along the south shore of Cape Cod, you can make the passage from Nantucket Island to Chatham, but there are many shoals and Monomoy Island on that route. Your other choice is to skirt the Cape after your visit to the Elizabeth Islands, which is the route we'll take.

As you hit the Cape, your first stop will be **Falmouth**. One of the attractions of Falmouth is that every amenity the boater will want is within a short walk from the marinas. This tourist center is a stopping place for people on their way to the upper arm of the Cape or to the islands. If you're in town on the Fourth of July, look up in the sky. You're likely to see the colors of hot air balloons as they begin their annual race at a starting point near Falmouth. During the last two weeks of July, you'll see quite a different show in the sky—that's when the Blue Angels and the Thunderbirds perform aerial acrobatics here.

Falmouth Harbor is a narrow haven east of the light on Nobska Point at Woods Hole. When approaching from the west or the Vineyard Sound, you can head for either bell "16" half a mile south of the breakwater or head straight for the breakwater marked by the tower flashing green 2.5sec "1." Approaching Falmouth Inner Harbor, you'll see the sandy bluffs of Falmouth Heights to starboard. The harbor passage is narrow and often crowded.

NANTUCKET

15
p. 104

A Serene Haven in the Heart of Nantucket

Winthrop Hotels & Resorts and the Boat Basin welcome you to Nantucket. We offer all the amenities including water, electricity, ice, telephones, showers, coin-op laundry, ship's chandlery, dock-master's office, a night security guard, and even concierge service. What's more, we're right at the foot of Main Street, close to shops, restaurants, and beaches.

Ashore, our harborside resorts tempt you and your guests with fine accomodations and fine dining. The Wharf Cottages overlook the Boat Basin, while the Habor House, the White Elephant and the Breakers are just a few blocks away. For Hotel reservations please call 1-800-ISLANDS.

For superb food, congenial service and relaxing entertainment, try the Regatta at the White Elephant (228-2500) or the Hearth at the Harbor House (228-1500). Reservations suggested.

Nantucket Boat Basin — Nantucket Island Massachusetts

For information or reserations at the Boat Basin, phone 800-NAN-BOAT VHF 16, 73, 09.

Boston Tides; High: +1:07, Low: +0:52, Mean: 3.0 feet	Freshwater, Phone, Cable TV Elec: 110V, 220V, 100 Max Amps
Call for Rates 243 Berths, 300' LOA	Many Nearby
Sail, Electronics, Rigging, Gas, Diesel, Wood, Fiberglass	
Mobil Gas, Diesel	MasterCard, VISA, American Express

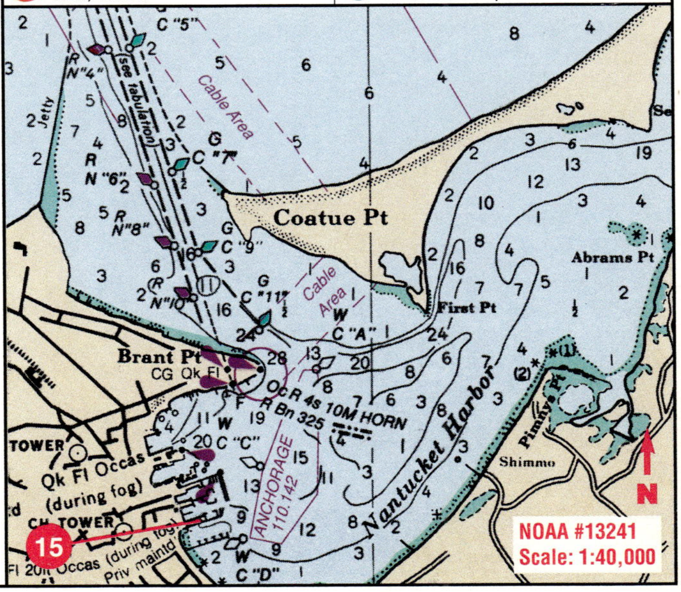

There are no markers to guide you through this deep passage, but you can tell where the channel is by looking at the numerous moorings and other boats on either side of you. You'll find **Fiddler's Cove Marina** nestled in a sheltered cove just northwest of Falmouth Harbor. They have slips, rack storage, and a service department ready to handle any kind of marine repair.

Best known as the home of the Kennedys, **Hyannis** is a tribute to sun, sand, seafood, and saltwater. Billing itself as the "hub" of Cape Cod, the town is packed with restaurants, shops, hotels, and *people!* In mid-June, you'll enjoy the Hyannis Harbor Festival; in mid-August you'll hear the Boston Pops in Pops-by-the-Sea (508) 790-ARTS, and you can watch the art of making potato chips at the Cape Cod Potato Chips factory (508) 775-3358.

Approaching from the southwest, steer clear of Centerville Harbor and its many rocks and shoals. Your best bet is to use the North Channel. You can leave the channel at flashing green 2.5sec bell "5" about 5nm south of Centerville Harbor, or stay closer to shore and head for the red and green flasher (2+1)R at Collier Ledge. From here, pass to the south of Gallatin Rock, and it's a straight run northeast to the breakwater. If you plan an overnight visit on a weekend, make reservations ahead of time; the harbor is busy!

On your way to the end of the Cape, you'll come to **Bass River** with the developed area of **South Yarmouth/West Dennis** on its shores. The river is used mostly by local boaters, but is a nice stop if you're in a small boat that can nose up to shore for a picnic. Farther east is **Allen Harbor,** a completely protected harbor where you'll find the family-owned and operated **Allen Harbor Marine Service** at the head of the harbor (call ahead for reservations). Nearby is the village of Harwich Port and a small shopping center with a pharmacy, laundromat, market, and package store.

Allen Harbor is difficult to see from Nantucket Sound. The best landmark is a white church spire in Harwich Port, just east of Allen Harbor. From the west, your only real navigational hazard is Kill Pond Bar, with depths of 4 to 6 feet mean low water. Be sure to head out to the flashing red 6sec whistle buoy southeast of Kill Pond Bar. From there, approach the first pair of privately maintained buoys "AH1" and "AH2." Stay in the channel in Doanes Creek.

Call it the "elbow," or call it **Chatham**. Either way, it's the last land 'til Portugal. Exposed as it is to the open sea, this land is notorious for its changes—lost islands and channels emerging to split land masses. Visit the Chatham Theatre or the Monomoy Theatre (508) 945-5985, both offering summer-stock performances. For dining, you can get fresh seafood at the Municipal Fish Pier every afternoon, or eat at one of the fine local restaurants.

While the entrance to **Chatham Harbor** is downright dangerous, the approach and entrance to Stage Harbor

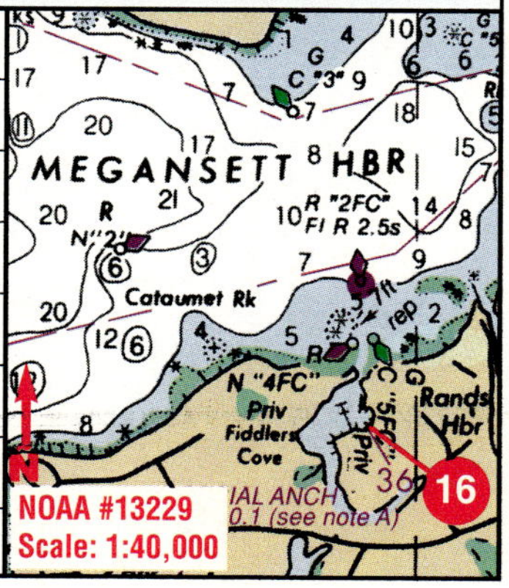

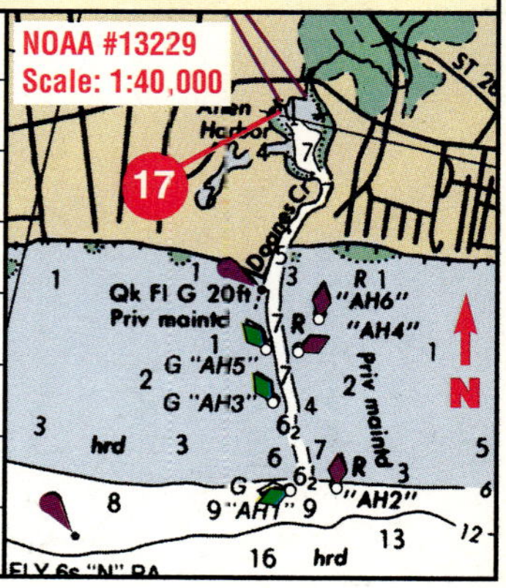

©Embassy

Colorful flowers frame a home on Martha's Vineyard.

(and hence the town of Chatham) is straightforward. Approaching from the west is simple; keep at least a mile offshore to avoid the shallows and fish traps. Head for the red and white Morse code (A) bell "SH" at the entrance to the channel. The old squat lighthouse on Hardings Beach is an excellent landmark. If you follow the channel to the inner harbor north of Stage Island, you'll find transient facilities.

Your trip along the waters south of Cape Cod can take many routes. You can head back along the coast or skip down to Nantucket and Martha's Vineyard. You can come full circle starting at the islands and make your way back by way of the Cape. Any way you choose, by the time you leave Massachusetts, you're sure to agree that you've seen some of the finest cruising grounds in the Northeast.

Shoreside and Emergency Services

Ambulance, Fire, Police: 911

Coast Guard:

Menemsha Station (508) 645-2661
  Emergency (508) 645-2611 or VHF 16 or 22
Woods Hole Station (508) 457-3277
  Emergency (508) 548-5151 or VHF 16 or 22
Cape Cod Canal Station (508) 888-0020
  Emergency (508) 888-0335 or VHF 16 or 22
Brant Point Station (508) 228-0398
  Emergency (508) 228-0388 or VHF 16 or 22
Chatham Station (508) 945-3830
  Emergency (508) 945-0164 or VHF 16 or 22
Provincetown Station (508) 487-0077
  Emergency (508) 487-0070 or VHF 16
Scituate Station (617) 545-3801
  Emergency (617) 545-3800 or VHF 16
Point Allerton Station (617) 925-0165 or VHF 16
Boston Station (617) 223-3225 or VHF 16
Gloucester Station (508) 283-0704 ro VHF 16

Radio Telephone:

VHF 24, 26, 87; I.D.: New Bedford Marine Operator
VHF 28, 84; I.D.: Hyannis Marine Operator
VHF 27, 85, 86; I.D.: Nantucket Marine Operator
VHF 26, 27; I.D.: Boston Marine Operator
VHF 26, 27; I.D.: Quincy Marine Operator
VHF 86; I.D.: Beverly Marine Operator
VHF 25; I.D.: Gloucester Marine Operator

Tow Service:

Marine Rescue Services, Inc., Wickford, RI
  (401) 295-8711 or VHF 16
Sea-Tow Southern Massachusetts, Chilmark, MA
  (508) 645-3618 or VHF 16
Island Marine Service, Newport, RI
  (401) 849-4820 or VHF 16 or 11
Nancy J/Boat U.S. Woods Hole, MA
  (508) 563-3075 or VHF 16 or 9
Menemsha Marine Repair, Menemsha, MA
  (508) 645-2883 or VHF 16
Maciel Marine, Ltd., Vineyard Haven, MA
  (508) 693-4174 or VHF 16
Smallcraft Tow, Bass River, MA
  (508) 362-2874 or VHF 16 or 18
Outermost Harbor Marine, Chatham, MA
  (508) 945-2030 or VHF 16
Systems Marine, Plymouth, MA
  (508) 747-2744 or VHF 16
Cohasset Boat Tow Co., Cohasset, MA
  (617) 383-1343 or VHF 16 or 9
Allerton Boat Works, Hull, MA
  (617) 925-4725 or VHF 16 or 7
Sunset Marine, Hull, MA
  (617) 925-2653 or VHF 16 or 7
Sea-Tow Boston, East Boston MA
  (617) 567-8053 or VHF 16
Wheelers Point Boatyard, Gloucester, MA
  (508) 281-0303 or VHF 16

FOR MORE INFORMATION:
Embassy's Complete Boating Guide & Chartbook
• Rhode Island & Massachusetts (see p. 12)
Embassy's Complete WATERPROOF Charts
• Block Island to Nantucket (see p. 32)
MASS.
RHODE ISLAND
CONN.
Joins page 104
Joins page 132
VHF-FM
WXJ-39
162.40 MHz
Pawtucket
Providence
Warwick
East Greenwich
Bristol
Fall River
USCG
(401) 253-9585
Narragansett Bay
Jamestown
Middletown
Newport
USCG
(401) 846-3676
VHF-FM
KHB-47
162.55 MHz
USCG
(203) 442-4471
Westerly
Point Judith
USCG
(401) 789-0444
Rhode Island Sound
N
ATLANTIC OCEAN
USCG
(401) 466-2462
Block Island
1 inch=7.2 miles
Scale 1:456,000
0
7.2
14.4
21.6
28.8
1
2
3
4
5
6

The Ocean State

Newport is wedged into Rhode Island like the centerpiece of a fine jewelry setting: It's got beauty, draw, and attracts all the attention. What many seem to forget is that it's also surrounded by smaller gems, ports along 400 miles of scenic coastline that offer everything from quiet river scenes to active city life.

Newport is easily accessible directly from the Atlantic. So unlike Massachusetts, where you experience a sort of sailing serendipity by stumbling upon lesser known ports en route, it's easy to overlook the rest of "Little Rhody." Don't do it. Explore Narragansett Bay, your host to most of Rhode Island's sites. Extending 28 miles inland and covering 10 percent of the state's area, Narragansett Bay is actually more an estuary than a bay, characterized by an influx of freshwater from several rivers and streams. Therefore, as you travel north, weaving through the 36 islands scattered throughout, the waters become more brackish, fed mainly by the Taunton, Barrington, and Providence Rivers.

The two largest islands—Conanicut, commonly known as Jamestown, and Rhode Island, also called Aquidneck—split Narragansett Bay into three main channels. West Passage is, as the name implies, the westernmost passage cutting between the mainland and the western shore of Conanicut Island. East Passage separates Conanicut from Aquidneck and is the access route for Newport. And the easternmost route is the Sakonnet River, which is actually not a river at all, but rather a strait beginning at the narrow passage between the north end of Aquidneck and the small section of mainland on the east side of the bay.

Each of these access routes has a distinctly different flavor, an appeal that sets them apart. The vast Sakonnet River is predominantly rural, owing to the narrow passage at its head, which made industrialization here impractical. At the other end of the spectrum, the East Passage is where you're likely to experience the most activity and from which every nook and cranny of the bay, except the Sakonnet River, can be reached. And the West Passage combines the sparse appeal of the Sakonnet with the more urban tastes of the Newport area, giving you a direct route up the mainland coast to Providence.

© Mary Jane Hayes

Visitors shop along the Newport waterfront.

Regardless of the access route, cruising conditions in Narragansett Bay vary little. In general, the tidal range is 3 to 4 feet, increasing a bit as you travel north. Currents in the bay verge on one knot and expectedly increase in narrow passages.

The effects of the bay on the state of Rhode Island are undeniable. The very creation of the colony with its

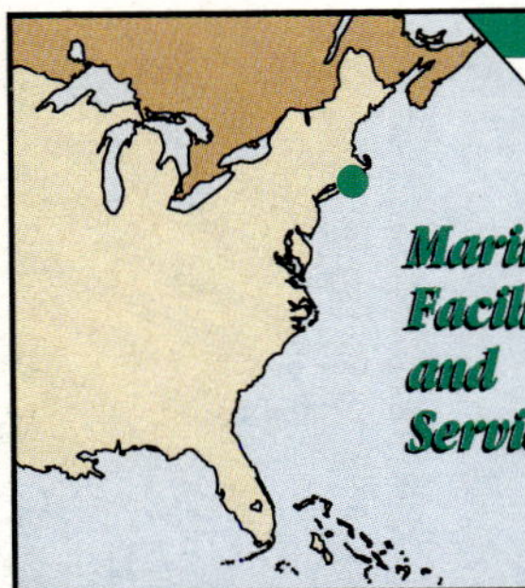

Marine Facilities and Services

		DOCKAGE								SERVICES					SUPPLIES			AMENITIES			COAST GUARD
Marina	Phone	Monitors VHF Channel	Working Channel	Number of Transient Berths	Number of Transient Moorings	Seasonal / Year-round	Maximum Length Over-All	Approach Depth in Feet at MLW	Dockside Depth in Feet at MLW	Hookups: Fresh Water / Phone / Cable TV	110V★ 220V▲ 3 Phase ■ Maximum Amps	Ramp / Dinghy Dock / Launch Service	Railway / Lift / Crane: Capacity (tons)	Repairs: Propeller / Sail / Electronics / Rigging / Gas / Diesel / Wood / Fiberglass	SCUBA: Air / Rentals / Equipment / Charters	Gas / Diesel Fuel	Fuel Brand	Marine / Groceries / Bait / Ice / LPG / CNG	Showers / Laundry / Pumpout / Restrooms	Hotel / Pool / Tennis / Golf	Public Phone / Restaurant / Snack Bar	MasterCard / Visa / Discover / AmEx
1 Goat Island Marina (p. 124)	401-849-5655	9	8	8		S	140	19	15	All	★▲■ 100	L		EGD		GD	Gul	GI	All	HPT	PR	All
2 Newport Yachting Center (p. 125)	800-653-3625	9	8	65		Y	180	17	17	F	★▲■ 100		LC	All		GD	Tex	MI	SLR	P	PR	All
3 Little Harbor Marine (p. 127)	401-683-7100	9	9	50		Y	140	15	15	FP	★▲■ 50		L160 C	All		D		MILC	All		P	MVA
4 Norton's Shipyard & Marina (p. 129)	401-884-8828	9		50	50	Y	250	10	15	F	★■50	DL	L35 C15	All		GD	Tex	MIC	SR		PR	MV
5 Dutch Harbor Boat Yard (p. 128)	401-423-0630	68			15	S	60	15	12	F	★	DL	RC25	PSEGDWF		GD		MI	SPR			MV
6 Norm's Fuel Co. Ltd. (p. 130)	401-783-2084	80	68			Y										D	Tex					MV

See Adjacent Chart for Coast Guard Information

Information in these listings is provided by the facilities themselves.

booming fishing and shipbuilding industries originates from the bay. Contrary to the way Rhode Island's borders blend unnoticed into neighboring states, the waters of Narragansett Bay give Rhode Island a most endearing appeal that's difficult to overlook.

Sakonnet River and Mt. Hope Bay

Many coastal cruisers who frequent Rhode Island and the surrounding area are familiar with Newport—the "Yachting Capital of the World"—and little else. Don't fall into this trap. Newport's nickname is in no way unfounded, but it is hardly the only attraction Rhode Island has to offer. The **Sakonnet River** and points north offer the sea-weary sailor a refreshing change from the long swells of the open Atlantic.

The Sakonnet River is 1.5 miles wide and 10 miles long. And every mile is lined with some of the most beautiful rural scenery on the bay. If you've come in search of constant activity, you've come to the wrong place. But if a peaceful cruise is what you're after, this is it. You'll rarely find more than a handful of other visiting boats, and commercial vessels are few and far between.

Traveling south from Massachusetts, pick up the green and red junction buoy at Elisha Ledge, staying well off the rocks and ledges that pockmark this section of coastline. From there, head for red bell "2" marking Schuyler Ledge, keeping the marker well to starboard to avoid the rocks that extend from **Sakonnet Point**. Begin a long, slow turn to the north, keeping an eye out for the privately maintained pair of 2.5sec yellow flashers that mark fish traps. Maintained until the middle of June, these two flashers should be kept to starboard.

Once above Sakonnet Point, head for the 34-foot 4sec flashing red "2" on the breakwater at **Sakonnet Harbor**. Here you'll find the few marinas on the river below Tiverton. In many ways, this harbor resembles a Maine fishing village. In fact, if you don't hug Sakonnet Point as you approach, you're liable to end up engulfed by the lobster pots and fish traps sprawled across the area—just like Maine.

As you travel north, criss-crossing the river from the eastern shore to Aquidneck and back, the water remains deep right up to the shores. The currents run at less than one knot, and if you heed the charted buoys, you shouldn't have any trouble. As you reach the head of the river at **Tiverton**, you'll find the currents increase significantly, as well as your need to navigate precisely. The Stone Bridge Buttresses, two jetties extending from each side of the river, are all that remains of the Old Stone Bridge, destroyed in the Hurricane of 1938. Green can "15" and red nun "16" mark the entrance to **Tiverton Basin** through these breakwaters. The currents here can reach 5 knots so stick to the center of the channel.

Once inside the basin, which is actually just a narrowing of the river, you'll find that Tiverton marks a change of pace: the rural aspects of the vast Sakonnet begin to fade into marinas and the more urban setting of what lies ahead. The cruising conditions are more or less the same, except in heavy weather when the currents can get tricky. Across the river in Portsmouth you'll find marinas with transient facilities.

As you leave Tiverton, you'll pass under a highway bridge that connects Tiverton to **Portsmouth**. The vertical clearance at the center span is 65 feet. You must keep to one side or the other when passing through here to avoid southbound boats—the first sign that the busier ports of Mount Hope Bay are close by.

Several large and deep shipping channels merge at the northern end of the Sakonnet. Once past the highway bridge, head north leaving red nun "2" to starboard, then turn slightly to the east. The channel will cut through Mount Hope Bay and into the upper reaches of the **Taunton River**.

The town of **Fall River** is the main attraction in **Mount Hope Bay**. A quintessential mill town that inevitably brings you back to days of old, Fall River is where you'll find Battleship Cove, home of the battleship *Massachusetts*. This historic ship weighs 35,000 tons and fought in 33 battles during World War II. Also of interest in the cove are Heritage State Park and the Marine Museum (508) 675-5758.

Mount Hope Bay is also home to several coves, inlets, and estuaries that make the bay a fantastic exploring

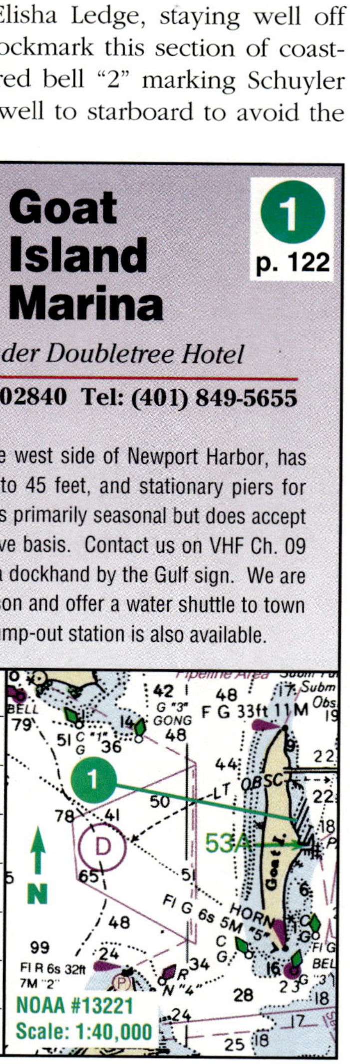

ground. There are very few, if any, major marine facilities available for transients in Mount Hope Bay so we suggest taking advantage of the Lee and Kickamuit Rivers for anchoring. But keep your chart handy, and watch the depth-sounders in these areas—they can get pretty shallow at low tide.

East Passage: Newport to Providence

Rhode Island has long been touted as having an illustrious past, evidenced by the colonial architecture that lines many of its streets. **Newport** is certainly no exception. Along with the stylings of America's formative years, Newport shows another side of history along its shores. The "summer cottages" of America's nouveau riche sprouted up along Newport's south shore at the turn of the century, leaving an indelible mark of high social status.

These "cottages" are actually ornate mansions, and today the guided tours are only a smattering of the many attractions in Newport. The cramped and intimate atmosphere of downtown, the multitude of fine restaurants, the exquisite shops and boutiques, the International Tennis Hall of Fame (401) 846-4567, and the renowned Newport Jazz Festival held late in August are a few examples of the seemingly endless list of things to do. Peace and quiet? No way.

The only difficulty you'll have reaching Newport is the traffic you'll experience along the way. The generous depths and good protection, coupled with all the activities, make Newport one of the most popular ports of call on the Atlantic seaboard. Be sure you have your mooring or dockage reservations confirmed before you set out Newport-bound.

Coming from the east, make your way to quick flashing red whistle "2," and leave the buoy to starboard, giving it a wide berth before turning north. Leave red gong "4" to starboard, then pick up red bell "6" at Butter Ball Rock. Continue north past Castle Hill and the 40-foot flashing red 6sec Ram's Head Light, then turn northeast, and pick up the 32-foot flashing red 6sec horn "2" off of Fort Adams. A straight shot east will put you into the crowds of Newport Harbor.

As you might expect from a popular port, Newport has an abundance of marine facilities, but again, reservations are a necessity throughout the season. **Newport Yachting Center** offers transient slips and facilities along the main harbor in the heart of downtown. If an island visit is more your speed, visit **Goat Island Marina**. Located on the west side of Newport Harbor, they offer all the amenities, including a shuttle to the mainland during the season.

The tiny village of **Jamestown** on **Conanicut Island** is a pleasant respite from the non-stop atmosphere of Newport. Despite its proximity to Newport—located just across the East Passage—Jamestown has somehow been able to retain its quaint, small-town appeal. Why it hasn't become a booming maritime metropolis similar to Newport cannot easily be explained. The only viable explanation is that Jamestown Harbor offers relatively little protection in heavy weather.

Nonetheless, we suggest making a stop at Conanicut Island—you won't be disappointed. Beavertail Point State Park and Fort Wetherill State Park are both adorned with World War II installations and barracks that were, for obvious reasons, never used. But even if the war had made its way to the northeast Atlantic, chances are the enemy would have followed suit, skipped over Jamestown, and gone straight to Newport!

Continuing north in the East Passage, you'll pass several areas that offer no facilities—some are still worth a visit, others are off-limits. **Rose Island**, located just south of the Newport bridge, is difficult to approach in anything save a shoal-draft boat, but since the island is entirely undeveloped, it's worth visiting in a dinghy. The scuba diving is renowned, and hiking is always an option, as long as you respect the nesting gulls. For those who want a first-hand experience of what it's like to be a lighthouse keeper, rooms can be rented in the Rose Island Lighthouse. Built in 1869, the light was deactivated in 1971 and has been restored by the Rose Island Lighthouse Foundation (401) 847-4242 and turned into a museum.

Also worth a visit is **Potter Cove**, located just north of the Newport Bridge on Jamestown (not to be confused with Potter Cove on the northern end of Prudence Island). The Potter Cove on Jamestown is a pleasant, secluded inlet lined with rocks and ledges that invite sunbathers and explorers.

On the opposite side of the passage is a naval base that occupies **Coasters Harbor Island** and **Coddington Cove.** Consequently, both of these areas are off-limits to the public. Also restricted is Gould Island (don't get this confused with Gould Island at the northern end of the Sakonnet River). Located northwest of Coddington Cove, the island is part of the Bay Islands Park System, a nesting ground for birds, and is off-limits from April 1 through August 15. Lack of any facilities or a good anchorage make it inaccessible for large boats, so it may as well be restricted all year round.

The next point of interest in the East Passage is at the center of Narragansett Bay. **Prudence Island** is the largest part of the Narragansett Bay Estuarine Sanctuary, used to study the proper maintenance of many of our natural resources. For this reason, you are allowed to dock only to load and unload, but you can anchor and pull a dinghy up on shore, except where explicitly prohibited. If you go ashore, stick to the roads so as not to disrupt the natural habitat of the sanctuary, and bring your own drinking water.

Across the passage from Prudence Island on Aquidneck is the town of **Melville,** a small section of **Portsmouth.** Melville is fast becoming a haven for recreational boaters, drawn in by the large service facilities located just south of Coggeshall Point. Plans are in the works to expand the facilities at **Little Harbor Marine**. When the expansion is complete, a total of 1,200 slips would be available, making it one of the largest boating centers on the Atlantic seaboard.

The town of Portsmouth, spanning the width of Aquidneck Island, is viewed by some as the birthplace of American democracy. In the mid-17th century, a group of settlers grew tired of the religious constraints they faced in Plymouth, Mass., and decided to go out on their own. The Portsmouth Compact established the village on the principle of separation of church and state—the same principle that later framed the United States Constitution.

Prudence Island and Portsmouth, including Melville, are easily accessible from the south. The water in the East Passage is good and deep, so there shouldn't be any trouble. Starting from Gould Island, make a beeline for the flashing red 2.5sec "24" just northwest of Dyer Island. As long as you keep green can "9" to starboard, a straight shot due east will land you in Melville. If you head northeast and leave red nun "12" to starboard, you'll sail smoothly into Portsmouth. And if you pick up the channel marked by flashing green 4sec "25" and flashing red 4sec "26," Potter Cove on Prudence Island is easy to approach if you watch the chart.

If Block Island is shaped like a pork chop, Conanicut Island like a beaver, then it is certainly no stretch of the imagination to say that **Bristol Harbor** can be likened to a lobster claw. Located just northeast of Prudence Island (which, incidentally, resembles a duck), Bristol is home to some of the most beautiful colonial architecture on the bay and several marine facilities. Of particular interest to boaters is the Herreshoff Marine Museum (401) 253-5000, which commemorates the achievements of one of the most famous yacht-building families in the world. Around the late 19th and early 20th century, Herreshoff designs successfully defended the America's Cup eight times.

And if Bristol Harbor is indeed a lobster claw, **Hog Island** is the tasty morsel about to be clutched. Covering 200 acres and providing Bristol with protection from the south, Hog Island is a pleasant place to visit in a dinghy, as the surrounding waters are rocky and shallow. The cove on the west side of the island is ideal for swimming and shellfishing.

Coming from Potter Cove on Prudence Island, head east, leaving flashing green 4sec "3" to starboard and the group flashing red bell to port. Once around Poppasquash Point, pick up red nun "6," giving green can "3" a wide berth. This will bring you well within the upper reaches of the harbor on the eastern shore where the town of Bristol is easily accessible.

Strong currents and tricky navigation characterize the **Warren** and **Barrington Rivers**, due north of Poppasquash Point. The Warren River turns into the

Barrington River as you head north between the corresponding towns. Don't let this fool you—the rivers may mix together, but the towns are quite different. Warren on the east bank is largely commercial, relying heavily on the harbor, while Barrington, an affluent suburb of Providence, has remained residential. Nonetheless, both offer an interesting change of pace for the cruiser who has not, up to this point, been challenged by the open waters of the East Passage.

From Poppasquash Point, get back to the shipping channel northbound until you pass the flashing red 4sec "8" at the edge of the channel. Then head for green can "3," leaving flashing green 4sec "1" to starboard. Be sure to leave can "3" to port as you approach. From here, keep your eye on the chart, and heed the buoys. Do not be tempted to follow local boaters who cut buoys here and there. At the south end of Little Island, which slices the Warren River down the middle, the Barrington River branches off to the west, then swings north. If you choose to pass the red and green junction buoy at the tip of Little Island, you'll be entering the **Palmer River**.

At the northernmost point of Narragansett Bay lies the capital city of **Providence**, accessible by way of the vast shipping channel that extends from the head of the Providence River to a point between Prudence Island and the town of Melville. Although the industrial appearance of the river's fringe can be intimidating, Providence has so much to offer—art galleries, museums, colonial architecture, historical celebrations, university campuses, dozens of parks, and a zoo. To miss it would be denying yourself an essential part of Rhode Island.

To reach Providence from the south, follow the wide, well-marked shipping channel, a waterway that makes the industry in Providence possible. Although the west bank of the river has several marine facilities and the upper reaches of the river are certainly accessible by boat, we suggest docking in Bullock Cove on the east bank and taking a cab into downtown Providence. Bullock Cove is well protected, and there is so much to see and do in the city that it's necessary to travel by land.

Leave the channel just before flashing red 2.5sec "18," and head north to red nun "2" and flashing green 2.5sec "1." Follow the cans and nuns into the cove, and watch your depth once past Bullock Point.

West Passage: Greenwich Bay to Jamestown

There was a time when **Greenwich Bay** was known mostly for its shellfishing industry. But since then, the town of East Greenwich has grown into one of the most affluent towns in all of Rhode Island. Shellfishing has taken a back seat to tourism. The boats of quahogers have, for the most part, been replaced by the exceedingly frequent visits by transient vessels, and the town is in the process of trying to preserve and upgrade its historic dis-

3
p. 122

LITTLE HARBOR
MARINE

Little Harbor Marine is located on Narragansett Bay, just six miles north of Newport. This well-protected deep water marina features 100 slips with seasonal & transient dockage, indoor & outdoor storage, as well as a fully stocked parts department.

Little Harbor Marine is a full service repair facility recognized worldwide for quality service, attention to detail, emergency repairs, and refits. Visit Little Harbor Marine with forty years of experience in yacht design, engineering, construction, and service in one convenient location.

One Little Harbor Landing, Portsmouth, RI 401-683-7100 FAX 401-683-7118

- Newport Tides; High: +0:08 Low: -0:04, Mean 3.8 feet
- Call for dockage rates 100 Berths, 140' Max LOA
- Prop., Sail, Elec., Rigging, Gas Diesel, Wood, Fiberglass
- Diesel
- Fresh Water, Phone, 220V, 50 Max Amps
- Nearby
- Nearby
- MasterCard, VISA, American Express

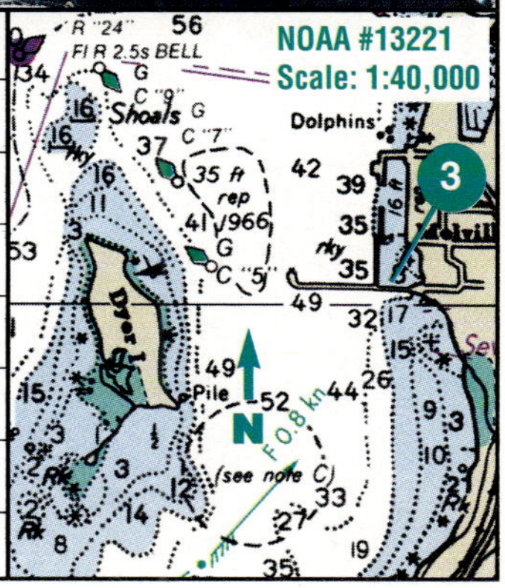

trict. Unfortunately, the smaller, less accessible area in **Warwick Cove** has taken the brunt of the overflow, experiencing mostly commercial and local traffic when once it was considered a haven for transients.

The main attraction in this area is Rocky Point Amusement Park (401) 737-8000 on the east shore of Warwick Neck. The park features 31 rides, including the 144-foot free-fall for those of you who prefer air travel. Across the bay in Greenwich Cove is Goddard Memorial State Park (401) 884-2010, and if you're lucky enough to be in the bay in early September, East Greenwich sponsors its annual Summer Festival with free events and reasonably priced food. The other small coves in this area—Apponaug and Brush Neck—offer little for transients, although Brush Neck has good water for boardsailing and fishing. Greenwich Cove is where to look for marinas. **Norton's Shipyard & Marina** is located just west of Long Point.

From Providence, follow the shipping channel to flashing green 4sec "13." Leaving the buoy to either side, head southwest, setting a course for flashing red 4sec bell "8" off Patience Island. Green can "3" marks the beginning of a shoaling area that covers all of Greenwich Bay. From here, **Warwick Cove** lies just northeast and is the home to several marinas. If you continue into Greenwich Bay, head for green can "5." Green can "7" marks the entrance to Greenwich Cove.

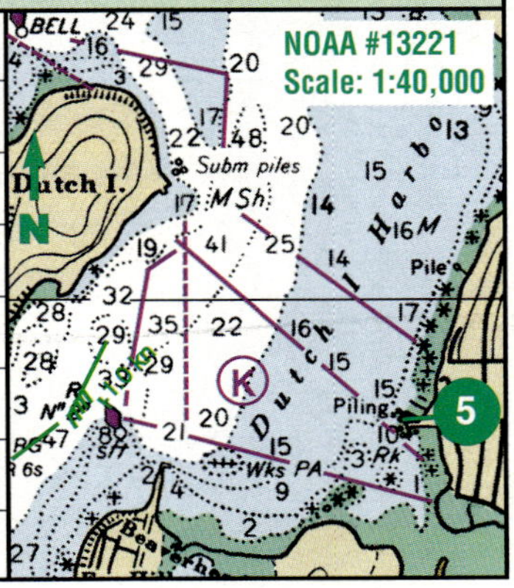

Continuing south along the mainland, you'll pass Allen Harbor, a good port for protection in heavy weather, and Hope Island, part of the Bay Islands Park System which is off-limits from April 1 to September 15. Both of these spots offer a relaxing stopover, but most cruisers opt for the diverse **Wickford Harbor**. Embracing both the industrial age brought about by the 20th century and the old-world village that has survived through it, Wickford is one of the most popular ports on Narragansett Bay. Wickford is known as the birthplace of Gilbert Stuart, the artist responsible for the portrait of George Washington that adorns the dollar bill.

Aside from a visit to the historic buildings that line the streets, such as St. Paul's Church (401) 295-7320—one of the oldest Episcopal churches in the country—Wickford sponsors events throughout the height of the season. The town beach has a series of Wednesday night concerts, and the Wickford Art Festival is held annually on the second weekend in July.

Coming from Greenwich Bay, give a wide berth to flashing green 2.5sec "1" at Round Rock, and head due south. Be sure to avoid the red and green flasher at the south end of Hope Island: it marks a sunken tug which likes to claim a few boats each year. Once past the buoy, head southwest into the harbor where you'll find ample depths and good protection.

As you continue through the southern portion of the West Passage, you'll likely be reminded of the first part of your cruise, up the Sakonnet River. Once past the rumbling traffic on the new Jamestown Bridge, the area becomes conspicuously peaceful, with very few residential areas along the shore all the way to Boston Neck.

Dutch Island, and the harbor of the same name on Jamestown, are worth a visit if relaxation is what you're after. The island was used as a pasture until the early 1900s when fortifications were built. There's little to see there, but you can bring a dinghy up on shore to explore the forts or picnic on the grounds. Most of Dutch Island Harbor is a conservation zone established to maintain the island's rural appeal, so much of the area is undeveloped.

Navigation is simple. Just follow the chart, and keep an eye out for red bell "2" off the north end of the island. Then swing to the east and into the harbor. We suggest docking at **Dutch Harbor Boat Yard, Inc.** in the southeast corner of the harbor. From here you can take a dinghy to the island, or take a trip into the center of Jamestown a few miles east of the harbor. Dutch Island Harbor marks the last port of refuge on Narragansett Bay, and the last harbor of any significance until you reach **Point Judith Harbor of Refuge**. If there's anything remotely resembling foul weather on its way, wait until it blows over before you leave the West Passage.

Point Judith and Block Island

Point Judith is considered the southwestern border of Narragansett Bay and can be more difficult to pass than it seems. Winds out of the southwest can persistently blow toward shore, as there is no protection from land masses, and the submerged rocks extending more than 1,000 feet from the point are easily overlooked. The 80-foot Point Judith Light has a radio beacon and fog signal to provide navigational assistance during foul weather. Just to be on the safe side, stay well off the point as you come around.

The only stopover between Narragansett Bay and Fishers Island Sound is the manmade **Point Judith Harbor of Refuge**, enclosed by three stone breakwaters. Marked by two lighted towers, the East Gap (between the middle and eastern breakwater) is the access route of choice if you're coming from the east. Although the swells and high winds of the open Atlantic can be felt along the way, the Harbor of Refuge is well protected, and the origin of its name becomes immediately apparent. Be careful of crossing traffic as you pass the breakwaters and make your way to the Breachway.

Since Point Judith Harbor of Refuge offers little more than protection and a beach, and since the rest of the Rhode Island coastline west to Connecticut is almost entirely beach, we suggest exploring **Point Judith Pond**, an area that many consider to be a small boating heaven. This sense of heaven is accentuated by the fact that the towns on either side of the pond's entrance are Galilee and Jerusalem. Also, the outside world is difficult to reach while visiting here—and we think that's just fine! The larger islands in the pond are privately owned, but the rest are great for going ashore in a dinghy or shallow-draft boat.

Navigating the pond is actually easier than it would appear to be. From the Harbor of Refuge, leave red nuns "2," "4," and "6" to starboard as you approach the Breachway. This somewhat narrow pass has seen currents ranging from 4 to 5 knots, so it's best to enter with a fair tide. The commercial fishing vessels coming and going from Galilee can also make getting through here tricky. Once past the Breachway, watch the buoys and the chart, and keep an eye out for marsh and rocks. If you need to fuel up before continuing on to Block Island or points west, stop at **Norm's Fuel Co. Ltd.**, located on the west shore near the Jerusalem State Pier.

Perhaps the best-known cruising ground on the eastern seaboard is the pork chop-shaped jewel known as **Block Island**. Roughly 10nm southwest of the breakwaters at Point Judith, Block Island, which rises dramatically out of the sea, has been called "Bermuda of the North"—and with good reason. Throughout the boating season, the island and the waters that surround it are teaming with activity. The town is full of restaurants, shops, and galleries. After all, someone has to cater to the hundreds of tourists that arrive on the ferries each day, not to mention the occupants of nearly 2,000 boats that fill **Great Salt Pond** on any given day. Swimming can be

Norton's Shipyard and Marina, located in a well-protected deep water basin, is a neat, well organized and well maintained facility. From your boat you will have a fine view of the boat basin, Greenwich Cove, and Long Point to your west. In addition to the marina management the shipyard also maintains 160 moorings. The marina management and staff are competent and cordial.

This is a well established shipyard with 50 years of experience in all facets of marine repairs. Norton's is the authorized dealer for Mercury Outboards, Mercruiser, Volvo, Penta, and Yanmar Marine Engines.

Ashore, the marina has a large new building that houses the marina offices, Discount Marine Store and work sheds. The Meritime Restaurant is located on site. This is a family oriented marina. There are Monday night races, summer cruises, and a host of year-round activities.

- East Greenwich Tides; High: + 0:13, Low: + 0:03, Mean 4.0'
- $.75-1.00 / ft per day 50 Berths, 300' LOA
- Prop., Sail, Elec., Rigging, Gas, Diesel, Wood, Fiberglass
- Texaco Gas, Diesel
- Fresh Water Elec: 220V, 50 Max Amps
- Almacs–Thorpe's, (401) 884-6274 or (401) 884-1442
- Meritime (401) 884-7760, Casual Atmosphere
- MasterCard, Visa

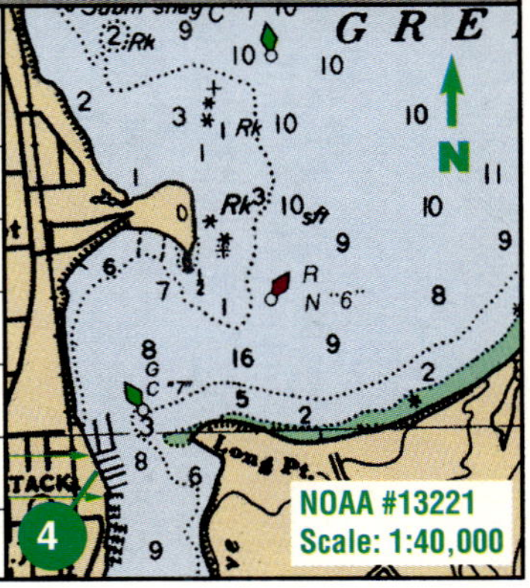

© Mary Jane Hayes

The Great Salt Pond on Block Island fills with boats on summer weekends.

found on the island's east coast, and some of the best dive sites are found at the Pinnacles, a group of enormous boulders piled one on top of the other, just off the southwest end of the island.

Reaching Block Island from any direction is not difficult. From Point Judith, make a beeline southwest for the flashing green 4sec bell "1BI," then head south before picking up red bell "2" at the entrance to Great Salt Pond. Stay to the center of the channel into the pond, marked by the nuns and cans. Once inside, finding room to anchor can be nearly impossible, so we suggest you use one of the many facilities that line the south shore. But make reservations ahead of time—this little gem could be a contender in a popularity contest with Newport!

If you've had enough of the crowds and need a relaxing cruise, head back to the mainland shores, and continue west. This last stretch of Rhode Island is renowned for its Atlantic beaches (17 miles worth) and a little town usually more associated with Connecticut: **Watch Hill**. Watch Hill is vintage Rhode Island—watersports and beach activity abound, the shellfishing is excellent, the surf-casting hot spots are superb, and the town is lined with excellent restaurants and shops.

But Watch Hill and the waters of Little Narragansett Bay can cause some navigational nervousness. A floating island called Sandy Point, a migrating beach named Napatree, and a reputation as the foggiest spot around make this haven seem a bit unapproachable. Simply follow the proper course with care and you shouldn't have any problems.

Head for the Watch Hill Passage between flashing red 6sec "2" and green can "3." Currents in this area tend to be strong, so proceed with care. If the infamous fog has rolled in, the Watch Hill Lighthouse is just north of here and provides an excellent landmark. Continue west to flashing red 4sec bell "6" off Napatree Point before heading north past Bartlett Reef to flashing red 4sec "2." From here, a marked channel leads around the floating island known as Sandy Point and into Little Narragansett Bay. More direct access is prevented by the shifting shoals between Napatree Point and the south end of Sandy Point—they used to be connected.

And here, your cruise in Rhode Island comes to an end. As you leave "Little Rhody" behind and visit other coastal states, you'll become more and more aware of the intimacy of Rhode Island cruising. Other states may offer longer coastlines along which to sail, but none pull you into the state itself quite like Rhode Island's waters.

Shoreside and Emergency Services

Ambulance, Fire, Police: 911

Coast Guard:

Block Island (401) 466-2462 or VHF 16
Point Judith (401) 789-0444
Emergency (401) 783-3021 or VHF 16
Montauk (516) 668-2773 or VHF 16 or CB 9
Castle Hill (401) 846-3675 or VHF 16 or 22
Bristol Station (401) 253-9585

Radio Telephone:

VHF 25, 26, 86; I.D. New London Marine Operator
VHF 28; I.D. KLU 786
VHF 84, 85; I.D. Narragansett Marine Operator
VHF 27, 28; I.D. Providence Marine Operator
VHF 24, 26, 87; I.D. New Bedford Marine Operator

Tow Service:

Island Boat Service, Newport, RI
(401) 846-6920 or VHF 16, 78 or 80
Island Marine Services, Newport, RI
(401) 849-4820 or VHF 16 or 11
Marine Rescue Services, Wickford, RI
(401) 295-8711 or VHF 16
Sea Tow Services of Rhode Island, Wakefield, RI
(401) 294-2360 or VHF 16
Oldport Marine Services, Inc., Newport, RI
(401) 847-9109 or VHF 68
Sea Tow Newport, Newport, RI
(401) 846-6500 or VHF 9 or 16

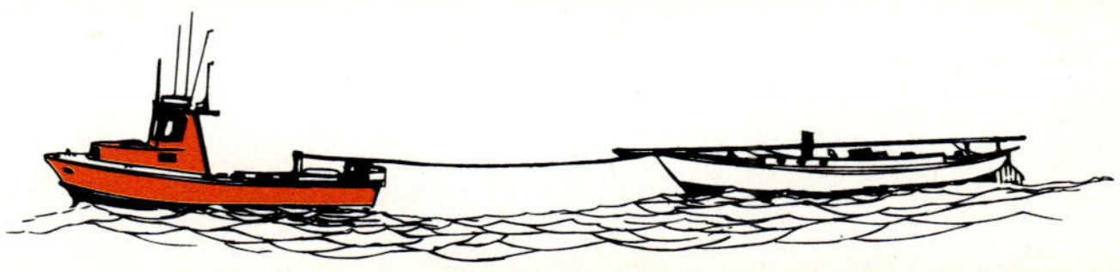

Joins page 122
RHODE ISLAND
CONNECTICUT
FOR MORE INFORMATION:
Embassy's Complete Boating Guide & Chartbook
• Long Island Sound (see p. 12)
Embassy's Complete WATERPROOF Charts
• #1—Long Island Sound (see p. 32)
N
Pt. Judith
U S C G (401) 789-0444
Block I.
U S C G (401) 466-2462
Westerly
Watch Hill Point
Pawcatuck R.
Mystic
Mystic R.
Noank
Fishers I.
Montauk Pt.
U S C G (516) 668-2773
Gardiners I.
Thames R.
Norwich
New London
U S C G (203) 442-4471
THE RACE
VHF-FM KHB-47 162.55 MHz
Niantic R.
Niantic
Plum I.
Orient Pt.
GARDINERS BAY
Shelter I.
Sag Harbor
East Haddam
Connecticut R.
Chester
Essex
Old Saybrook
Westbrook
Clinton
Greenport
Southold
GREAT PECONIC BAY
Mattituck Inlet
U S C G (516) 728-1171
VHF-FM WXM-80 162.475 MHz
Riverhead
Falkner I.
Guilford
Branford
Thimble Islands
U S C G (203) 773-2400
VHF-FM WXJ-42 162.40 MHz
New Haven
LONG ISLAND SOUND
LONG ISLAND (NY)
ATLANTIC OCEAN
Scale 1:475,200
1 inch=7.5 miles
0
7.5
15
22.5
30
Joins page 134
1
2
3
4
5
6
7
8
9
10
11
12
13
14
15

The Constitution State

Old-world charm and America's past cling to the 90 nautical miles of the Connecticut coast. The narrow streets and stoic taverns of Mystic, the old shipbuilding village of Essex on the Connecticut River, the unchanged shores from Guilford to Greenwich, and monuments wherever you turn. Without a doubt, Connecticut brings you back.

The beginnings and preservation of this quaint, old-world heritage may have as much to do with Long Island as Connecticut, though. Because Long Island shrouds nearly the entire coast of Connecticut, it protects the shoreline from the harsh wave activity of the Atlantic Ocean. As a result, the jagged Connecticut coastline with its rocky beaches and rough coastal terrain has remained essentially unchanged for thousands of years.

© Thomas Dimock

Moonrise over Noank.

But don't be fooled. While Long Island may protect the topography of the Connecticut coast, the powerful tidal currents of Long Island Sound can make access to the sites of Connecticut tricky. As tides rise and fall, the Sound acts as a giant bathtub—high tides slosh in from the eastern entrance at The Race raising water levels up to 8 feet. When they pour back out, the current at The Race can run up to 6 knots. Wind opposing current and potential fog can make the passage extremely difficult at best. The upshot: Choose your access route to the Sound carefully. And remember—timing is everything.

Not only time of passage but choosing the proper time of year will add to your enjoyment of boating in

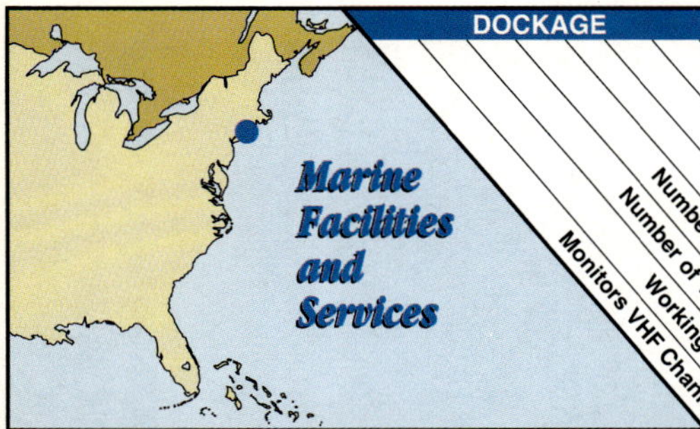

See Adjacent Chart for Coast Guard Information

Marine Facilities and Services	Phone	DOCKAGE: Monitors VHF Channel	Working Channel	Number of Transient Berths	Number of Transient Moorings	Seasonal / Year-round	Maximum Length Over-All	Approach Depth in Feet at MLW	Dockside Depth in Feet at MLW	SERVICES: Hookups: Fresh Water / Phone / Cable TV	110V★ 220V▲ 3 Phase ■ Maximum Amps	Ramp / Dinghy Dock / Launch Service	Railway / Lift / Crane: Capacity (tons)	Repairs: Propeller / Sail / Electronics / Rigging / Gas / Diesel / Wood / Fiberglass	SUPPLIES: SCUBA: Air / Rentals / Equipment / Charters	Gas / Diesel Fuel	Fuel Brand	AMENITIES: Marine / Groceries / Bait / Ice / LPG / CNG	Showers / Laundry / Pumpout / Restrooms	Hotel / Pool / Tennis / Golf	Public Phone / Restaurant / Snack Bar	COAST GUARD: MasterCard / Visa / Discover / AmEx
1 Noank Village Boatyard (p. 136)	203-536-1770	68	68	5	15	Y	45	5	5	All	★▲50	RDL	22	PSRGF					SLR			
2 Spicer's Noank Marinas (p. 137)	203-536-4978		68	15	30	Y	50	7	7	FP	★30	All	L20	All				MGBI	All		All	MVA
3 Mason's Island Marina (p. 138)	203-536-2608	9	6	6	2	Y	45	45	8	F		D	L25	PRGDF				MGI	SPR	P		MVD
4 Fort Rachel Marine Service Inc. (p. 138)	203-536-6647	16	9	3		Y	50	8	7	F	★	D	L30	PGRDWF				MI	SR		P	MV
5 Mystic Harbor Marine (p. 139)	203-536-1210	9	11	20		Y	90		9	All	★▲50			All				MGI	SLR		All	MV
6 Seaport Marine, Inc. (p. 139)	203-536-9681			10		Y	120	17	9	All	★100		R100	PERWF				M	SPR	All	All	
7 Burr's Yacht Haven (p. 140)	203-443-8457	9	68	50	12	S	100	9	9	F	★▲50	DL	L20	All		GD		MGBI	SLR		All	MVD
8 Crocker's Boatyard, Inc. (p. 140)	203-443-6304	16	68	20		Y	90	15	15	FP	▲50		L75	All		GD	GUL	All	SLR		PR	MV
9 Marina at American Wharf (p. 141)	203-886-6363	68	68	100		Y	160	30	25	All	★▲50	All	C25	GDPF		GD	TEX	All	SLR	P	All	MVA
10 Port Niantic (p. 142)	203-739-2155	9	9	10		Y	60	8	8	FC	★▲50	D	L35	All				MIB	All		P	MV
11 Old Lyme Dock (p. 144)	203-434-2267	9		5	2	S	150	16	13	FP	★▲100	D		P		GD	TEX	MI	SLR		PR	All
12 Saybrook Point Inn (p. 144)	203-395-3080	9	10	25		Y	170	15	8	All	■200	D		PSEGDWF	C	GD	MOB	MI	All	HPG	All	All
13 River Landing Marina (p. 145)	203-388-1431	16	9	20		Y	200	23	12	All	★▲100	DL	L60	All		GD	TEX	All	All		All	All
14 The Chandlery at Essex (p. 146)	203-767-8267	9	10	10	20	Y	65	10	10	FP	★▲■50	DL				GD	MOB	GIC	SLR		P	All
15 Cedar Island Marina (p. 147)	203-669-8681	16	68	75		Y	120	8	8	FC	★▲■100		L30	PSRGDWF		GD	GUL	MGBI	All	P	All	All

Information in these listings is provided by the facilities themselves.

FOR MORE INFORMATION:
Embassy's Complete Boating Guide & Chartbook
• Long Island Sound (see p. 12)
Embassy's Complete WATERPROOF Charts
• #1—Long Island Sound (see p. 32)
NEW YORK
CONNECTICUT
LONG ISLAND SOUND
LONG ISLAND (NY)
1 inch=5 miles
Scale 1:316,800
0
5
10
15
20
Joins page 132
Joins page 158
Joins page 158
New Haven
VHF-FM WX-2 162.40 MHz
U S C G (203) 773-2400
U S C G (516) 261-6868
N
Housatonic R.
Saugatuck R.
Shelton
Milford
Stratford
Bridgeport
Westport
Norwalk
Stamford
Greenwich
Mamaroneck
Mt. Sinai
Port Jefferson
Stony Brook
Northport
Huntington
Oyster Bay
16
17
18
19
20
21
22
23

© Embassy

Skiing through Connecticut waters at sunset.

Connecticut. The boating season generally runs from May through October. While late summer conditions can turn the Sound into a lake perfect for those under power, sailing in the fall is a pleasure for those searching for wind.

Timing of your entrance into the Sound is just as crucial. The Race, located between Fisher's Island and Little Gull Island, is the most direct, straightforward route into Long Island Sound. Again, the currents here tend to be strong, but if you start your run at slack ebb, your cruising time will be moderate and your trip comfortable. Keep in mind that taking this route requires backtracking to the northeast in order to visit several harbors, including Mystic, Noank, and Stonington.

Otherwise, accessing the Connecticut coast via Fisher's Island Sound is advisable. The approach is much less direct and requires weaving in and out of various obstructions before finally reaching Stonington Harbor, the easternmost port in Connecticut. But the extra effort is well worth it—several harbors and towns along the eastern shore give tribute to what makes Connecticut cruising an essential entry in any captain's log. Steeped in New England charm and heritage, Stonington Harbor is an appropriate point from which to begin your cruise along the rest of Connecticut's coast.

Eastern Long Island Sound: Fisher's Island Sound to Branford Harbor

Among the five passages into Fisher's Island Sound, Watch Hill Passage is recommended. Coming from Block Island, it is the first accessible passage you'll encounter, and the easiest. Watch Hill Lighthouse and the yellow facade of the Ocean House Hotel lend themselves as visual guides. But keep your eyes open—lobster pots and crossing traffic tend to congest Fisher's Island Sound. Stay between flashing red 6sec bell "2" and can "3" to avoid being swept into Watch Hill Reef by the currents. Once

Marine Facilities and Services

			DOCKAGE							SERVICES						SUPPLIES			AMENITIES			COAST GUARD
		Monitors VHF Channel	Working Channel	Number of Transient Berths	Number of Transient Moorings	Seasonal / Year-round	Maximum Length Over-All	Approach Depth in Feet at MLW	Dockside Depth in Feet at MLW	Hookups: Fresh Water / Phone / Cable TV	110V★ 220V▲ 3 Phase■ Maximum Amps	Ramp / Dinghy Dock / Launch Service	Railway / Lift / Crane: Capacity (tons)	Rigging / Gas / Diesel / Wood / Fiberglass — Repairs: Propeller / Sail / Electronics	SCUBA: Air / Rentals / Equipment / Charters	Gas / Diesel Fuel	Fuel Brand	Marine / Groceries / Bait / Ice / LPG / CNG	Showers / Laundry / Pumpout / Restrooms	Hotel / Pool / Tennis / Golf	Public Phone / Restaurant / Snack Bar	MasterCard / Visa / Discover / AmEx
16	Port Milford (p. 148)	203-877-7802	9	9	10		Y	55	10	8	FP	★▲ 50	DL	L35	All				All	All		All	MVA
17	Milford Boat Works (p. 148)	203-877-1475	68	68	10		Y	50	10	8	F	★50	D	L35	All		GD	TEX	MIC	All		P	MV
18	Stratford Marina, Inc. (p. 149)	203-377-4477	9	10	50		Y	85	13	7	FC	★▲ 50	R	L35	PEGDWF		GD	TEX	MGI	All		All	All
19	Marina at the Dock (p. 150)	203-378-9300			20		Y	110	20	8	All	★▲ 50		L50	All				GI	LPR		All	
20	Norwalk Cove Marina (p. 151)	203-838-2326	9	18	50		Y	150	15	9	F	★▲■ 100		L150	All		GD	GUL	All	All	P	All	MV
21	Yacht Haven Marine Center (p. 153)	203-359-4500	9	14	50		Y	140	12	12	FP	★▲ 100		L60	PERGDWF		GD	TEX	MGI	SLR		All	MVA
22	Harbour Square Marina (p. 152)	203-324-3331	9		20		Y	175	15	8	All	★▲■ 100	All	8	All		GD	TEX	All	All		All	All
23	Greenwich Harbor Inn (p. 154)	203-861-9800	9	78			S	130	9	9	F	★▲ 50										PR	All

See Adjacent Chart for Coast Guard Information

Information in these listings is provided by the facilities themselves.

through the passage, head for flashing red 4sec bell "6" off Napatree Point. Napatree Point Ledge swells on either side of this marker, so stay within 200 yards of the marker. From here you can either continue west along the coast through Fishers Island Sound or stop in at Stonington.

The tiny village of **Stonington** manages to be both cramped and reclusive—its narrow streets and numerous architectural achievements echo old New England. Quaint shops and restaurants as well as the Stonington Lighthouse Museum (203) 535-1440 make a visit here quite rewarding.

To approach **Stonington Harbor**, pass Napatree Point and head for the 46-foot red 4sec flasher known as "Middle Ground." Heading due north, leaving the southwest end of the west breakwater well to port, will bring you into the harbor where the controlling depth is about 11 feet.

Unlike many Connecticut towns that are named after their sister-cities abroad (Greenwich leaps to mind), the village of **Noank** is singular in its namesake as well as its charm. As the only "Noank" in the world, this peninsular community is a testament to the Connecticut Yankee mystique. Some might even say the people of Noank have taken their individuality too far—if you listen closely when they speak, you'll notice the slight tinge of a New England accent that you'll seldom hear elsewhere in Connecticut.

The town centers around a church and a town green and is almost entirely residential. Consequently, your search for activity in Noank will undoubtedly lead you to the water. Gunkholing, fishing, and swimming are among the most popular activities here. Mumford, Palmer, and Beebe coves are great spots for exploration by dinghy, and the area is touted as being a bird watchers dream.

From Napatree Point Ledge, set a course for the 55-foot flashing 6sec bell at Latimer Reef. Stay well south of the reef and once past it, head for the north end of Ram Island, passing Gates Island on either side. The Mystic River channel is just northwest of Ram Island and is the route into Noank.

As a town devoted to the water, marine facilities are plentiful in Noank. Located in West Cove, **Spicer's Marina** is a full-service facility well known for its marine store and fine repairs. While on the east shore, **Noank Village Boatyard** can be found just west of red nun "14" and will cater to all your needs.

Historic **Mystic** is the next stop west, attracting cruisers to its famous Mystic Seaport Museum (203) 572-0711 and Mystic Marinelife Aquarium (203) 536-3323. And downtown Mystic proves that the town has been aptly named. After a short walk through antique shops and taverns, it's difficult not to imagine carriages rumbling over dirt roads and the sails of clipper ships passing up the river—truly mystical!

The channel into **Mystic Harbor** is clearly marked and should pose no problems. As you travel north from

- New London Tides; High: -0:22, Low: -0:08, Mean 2.3 feet
- $1.00/foot per day 10 Berths, 45' LOA
- Prop., Sail, Elec., Rigging, Gas, Diesel, Wood, Fiberglass
- Fresh Water, 110/220V, 50 Max Amps
- Best Bakery, Grocery, and Package Store a short walk away.
- Sea Horse Restaurant (nearby) Casual

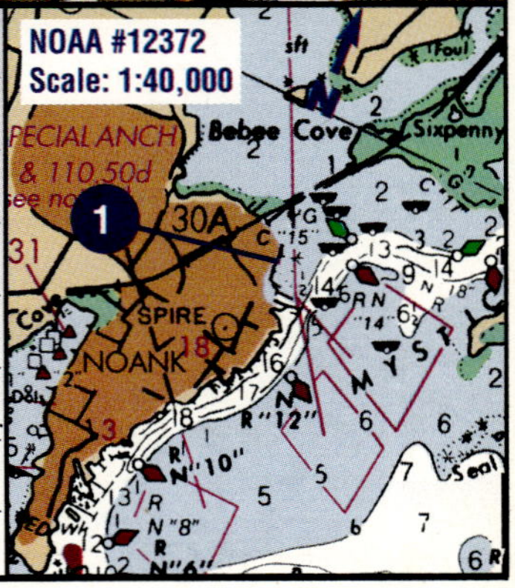

TRANSIENTS WELCOME

★ New Best Bath Facility in Connecticut
★ Concrete Floating Docks
★ Deep Water Moorings ★ Launch Service
★ NEW ENGLAND'S #1 MARINE STORE

2
p. 132

With this Ad 1st Nite Reg, 2nd Nite **FREE!**
Reservations: 203-536-4978 or Ch. 68

TOWING AND SALVAGE PROVIDED BY SOUTHEASTERN MARINE'S
TowBOAT/U.S.
CAPTAIN DON RICH • CH 9, 16, 68 • 24 HRS (203) 445-8381

BOAT U.S
Boat Owners Association of The United States

EMERGENCY ENGINE & HULL REPAIRS
DIESEL ★ GAS ★ FIBERGLASS
FASTEST TURN-AROUND

Spicer's Noank Marina

The place to come when you need something fixed.

93 Marsh Road • Noank, CT 06340 • 203-536-4978

7' MLW easy access day marked channel just west of Mouse Island through our mooring area and past our new 680' breakwater (not on chart yet—call for details).

We're the place you've heard about! Our marine store competes with the catalogs. And if Murphy's Law should occur while you're in Long Island or Block Island Sound, we'll get you back in action fast with the job done right—we've been doing it for over 60 years.

Noank charm
3 restaurants within walking distance
Tennis and beach close by
5 minutes to Mystic
No rivers • No bridges • No hassle

- New London Tides; High: -0:22, Low: -0:08, Mean: 2.3 feet
- $1.50 per foot; $25 moorings
  340 Berths, 150 moorings, 50 Max LOA
- Prop., Sail, Electronics, Rigging, Gas, Diesel, Wood, Fiberglass
- Ice, Bait, Tackle
  Soda, Snacks
- Freshwater, Phone
  Elec: 110V, 30 Max Amps
- Universal Food Store
  203-536-3767
- Fisherman 536-1717 / Abbott's 536-7719
  Sea Horse 536-1670 Casual, $10+
- MasterCard, VISA, American Express

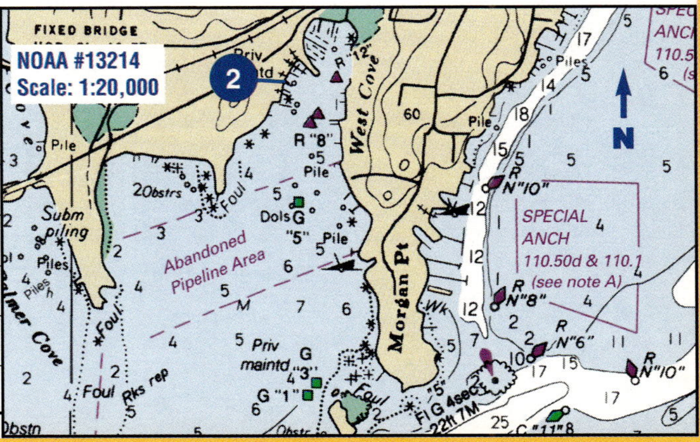

WE ACCEPT BOATER CARDS!

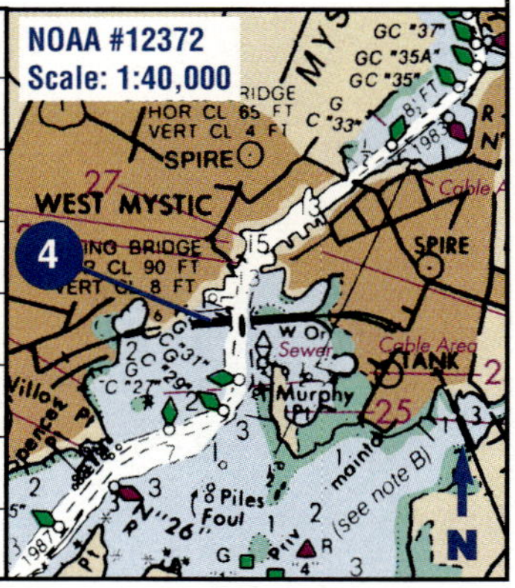

Noank, you'll pass the east side of **Mason Island**, a mostly residential piece of land with some great anchorages. If peace and quiet away from the crowds is what you're after, stop in at **Masons Island Marina** on the north side of the island. Call ahead to reserve your slip, especially if you plan to visit on a weekend.

Several marinas in and around Mystic Harbor offer complete facilities, including transient slips and repairs. **Fort Rachel Marine Service Inc.**, **Mystic Harbor Marine**, and **Seaport Marine, Inc.**, will accommodate your needs in the harbor.

From Noank, the route into West Harbor at **Fisher's Island** is direct. Head for green can "1," leaving it well to starboard, then head south, setting a course for flashing red 4sec bell "8," passing between red nun "2" and Pulpit Rock along the way.

The people of Fisher's Island are generally wary of transient vessels because of past experiences with loud parties. Respect their privacy and a visit to the island will be satisfying. Although there are no tourist attractions on the island, there is a grocery store and a few small shops—a short walk or bike ride is the way to go. *(See New York chapter.)*

If you're in search of more activity, you'll find it along the **Thames River** in the busy stretch from **New London Harbor** to **Norwich**. From flashing red 2.5sec bell "2" off of North Hill on Fisher's Island, make a

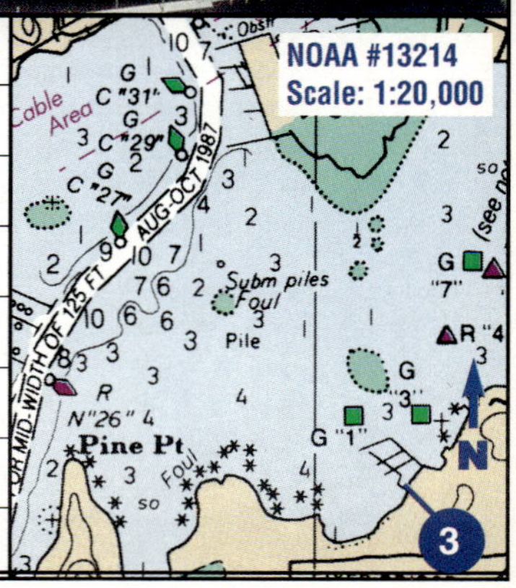

© Mary Jane Hayes

Yachts rest along the Mystic waterfront.

MHM — Mystic Harbor Marine

5 p. 132

40 Water St.,
Mystic CT
(203) 536-1210

Located in the heart of friendly and historic Mystic, Mystic Harbor Marine offers supreme protection from storms. Shops, restaurants and museums are just steps away from your dock. From Mystic Pizza to riverfront gourmet, you're sure to find a restaurant to suit your taste. Whether Mexican, seafood, Australian, al fresco or Nouveau Cuisine, Mystic has it! There's also plenty of places to play in Mystic. Shops and stores galore to suit even the most insatiable of shoppers, and the famous Foxwoods Casino is only a $17 taxi ride away. Car rentals are available, and a bus & train depot is nearby.

If you're in need of repairs, there are mechanics nearby who can fix every problem imaginable. Relax! Just pull into Mystic Harbor Marine and ask for a referral.

- New London Tides; High: -0:22 Low: -0:08, Mean 2.3'
- $1.50 / ft per day; 42 Berths, 90' LOA
- Nearby facilities for all repairs
- (fuel)
- Fresh Water, Phone, Cable TV; 220V, 50 Max Amps.
- Groceries & shops within walking distance
- Several within walking distance
- MasterCard, Visa

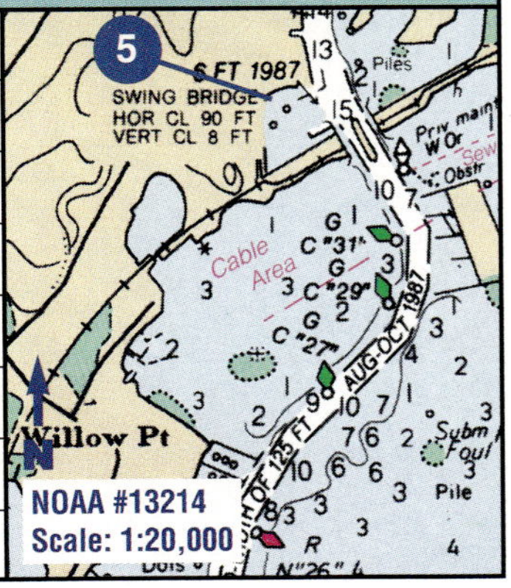

SEAPORT MARINE, INC.

6 p. 132

1 Washington St., Mystic, CT 06355
(203)536-9681 Fax (203)572-0076
Since 1959

Seaport Marine, located in charming, historic Mystic offers complete yacht yard facilities, including a full service yard, 60-ton travelift, 100-ton railway, and yacht brokerage. While docked at the Seaport Marine one can enjoy the many shops, boutiques, galleries and restaurants all packed into a small area, just west of the bascule bridge.

Seaport Marine is the closest big-slip marina to Foxwoods Casino in Ledyard. From storage to slip rental, Seaport Marine offers all the amenities!

- New London Tides; High: -0:22 Low: -0:08, Mean 2.7 feet
- $1.25 / ft per day; 10 Berths, 120' LOA
- Prop, Elec, Rigging, Wood, Fiberglass
- (fuel)
- Fresh Water, Phone, Cable TV; 110V, 220V, 100 Max Amps
- Many shops within walking distance
- Several restaurants only one block from the marina
- (credit cards)

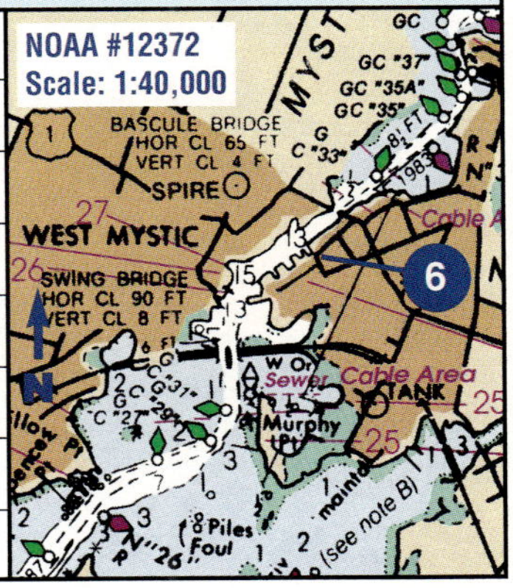

beeline for the pair of 2.5sec flashers that mark the beginning of the channel.

Roughly four miles from the west end of Fisher's Island, New London Harbor will show you another side of Connecticut's history. Although the old-world characteristics of Mystic and Stonington can be found in the inner reaches of New London (on the west bank of the harbor) and Groton (on the east), the fringes of the harbor disguise it well. The rough-hewn industrial features of this urban area suggest a modernized version of an old commercial port—and that's exactly what it is.

Aside from the many attractions in the harbor area—Fort Griswold and the U.S. Naval Submarine Base in Groton and the U.S. Coast Guard Academy in New London—many cruisers choose to dock in New London or Groton in order to use public transportation services. They include Amtrak service to New York and Boston, Groton-New London Airport, and ferry service to Block Island and Long Island.

Burr's Yacht Haven in Greens Harbor and **Crocker's Boatyard** in Shaw Cove offer docking facilities for those who wish to use New London as a starting point for shoreside excursions. Even if a brief stopover is what you're after, both marinas maintain slips and moorings and welcome transients.

Just above downtown New London, a bascule railroad bridge separates the harbor from the upper Thames River. When venturing up the river, there are a few things to keep in mind. The Coast Guard strictly enforces a 5-mph speed limit throughout the harbor, as well as sections of the upper reaches of the river. And as you pass through the harbor, be mindful of "Security Zones" marked on the chart. If you have photography buffs aboard, you should steer clear of these areas, particularly the Coast Guard Academy and the sub base—the Navy and Coast Guard are not fond of pictures!

Stay alert and a cruise to the head of the river to **Norwich** will be pleasant. The prevailing southwesterly winds of summer and generous depths make for easy cruising, giving you the freedom to fully appreciate the sights and scenery. While you're there, a visit to **Marina at American Wharf** is essential. They'll provide not only all the amenities of a full-service marina, but there are also several activities ranging from music and arts and crafts shows to children's events.

After you've absorbed the hustle and bustle of New London Harbor and points north, continue west to **Niantic Bay** for a more leisurely visit. There are two recommended approaches. If you choose the route via Twotree Island Channel north of Bartlett Reef, head south of the red and green buoy "R" at Rapid Rock and green can "3" at Little Goshen Reef. This should give you ample time and room to pick up the flashing red 4sec bell at the outskirts of the bay.

One American Wharf
Norwich, Connecticut 06360
203-886-6363 **VHF 68**

THE MARINA
AT AMERICAN
WHARF

A Full-Service
200 Slip Marina

9
p. 132

New London Tides; High: +0:13, Low: +0:25, Mean: 3.0 feet	Fresh Water, Cable TV 110V, 220V, 100 Max Amps
Wkend $1.50/ ft Wkday $1.25/ ft 100 Berths, 160' LOA	Courtesy Van to local stores, casino
Propeller, Wood, Fiberglass	Americus - Nautical 203-887-8555 - $ Avg.
Texaco Starport Gas, Diesel	MasterCard, Visa, American Express

With the advent of Foxwoods Casino (the only casino in New England) and the Marina at American Wharf, Norwich is fast becoming one of the most popular destinations in eastern Connecticut. Located at the head of the historic Thames River, this all-new marina offers solicitous service to visiting yachtsmen. They can arrange transportation to the casino, casino tickets, continental breakfast and more. If you want to stay put for the day, enjoy the pool, spa, restaurant and open-air café.

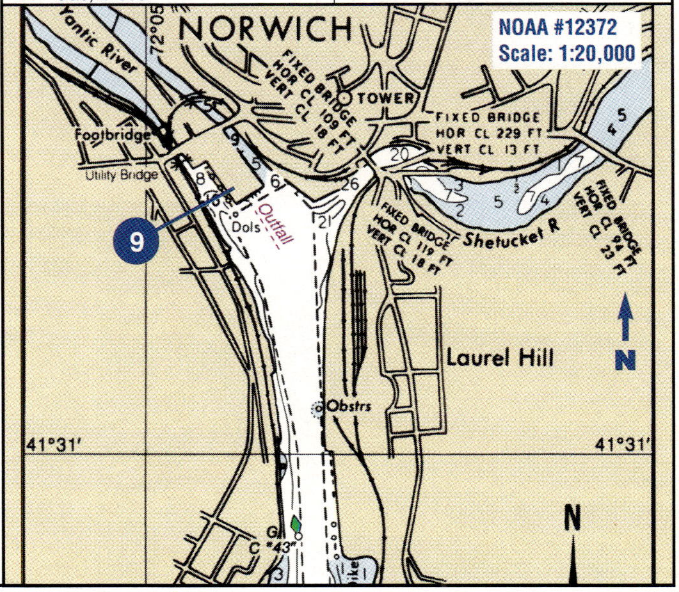

Authentic Italian Cuisine
Americus
ON THE WHARF

Dine "al Fresco" or in the charm of our open air cafe.
Daily Dinner Specials

* *Free docking (for restaurant customers)*
* *Free water taxi from city dock*
* *Tues - Sun 11-12am*

© Embassy

A lone sailboat breaks from the Connecticut River to approach the solice of Hamburg Cove.

You may also consider heading south past Bartlett Reef, marked by a 35-foot flashing white light, a fog signal, and a red and white diamond on a skeleton tower. Because of the strong currents, it's advisable to stay close to the reef as you come around, but not too close—there are lots of lobster pots in this area.

Entering Niantic Bay itself, give a wide berth to White Rock and Little Rock off the eastern shore, marked by flashing red 4sec "4." The width of the bay will occasionally make for rough sailing if the winds are heavy. The narrow channel into the Niantic River, at the eastern tip of a jutting land mass known as "The Bar," has a controlling depth of 6 feet, and the sequential order of the buoys should be followed—several sharp turns could make for tricky negotiation. It's best approached at slack water or against the current.

Inside the west end of The Bar, you'll find **Port Niantic**, a full-service marine facility with transient dockage that's only a five-minute walk from the supermarket, pharmacy, bakery, bookstore, and spirit shop in quaint downtown Niantic.

The **Connecticut River** flows through the heart of Connecticut and is a must-see for any coastal cruiser who hopes to ever truly understand Connecticut. Beginning from small springs just south of Canada, it winds through New Hampshire, Vermont, and Massachusetts before cutting Connecticut nearly in half and emptying into the Long Island Sound. The conspicuous absence of a city at

Saybrook Point Inn, Marina and Spa

Transients and Cruising Clubs Are Welcome

- State-of-the-art Floating and Fixed docks
- 700' Pier with Breakwater
- Indoor & Outdoor Heated Pools
- Steam Bath & Whirlpool
- Marine Store
- Mechanical & Diving Services
- Free cable TV

- 30, 50, 100 & 200 amp. Electric
- Monitoring VHF CH 9
- Fax, Copy & Mail Service
- Miniature Golf
- Bike Rental
- 6 Movie Theaters
- Summer Stock Theater
- Casino Nearby

This brand new multi-million dollar facility features a complete health and fitness spa, heated indoor and outdoor pools, and a state-of-the-art fitness center. Sauna, steam and whirlpool baths and therapeutic massage will revitalize your body and mind.

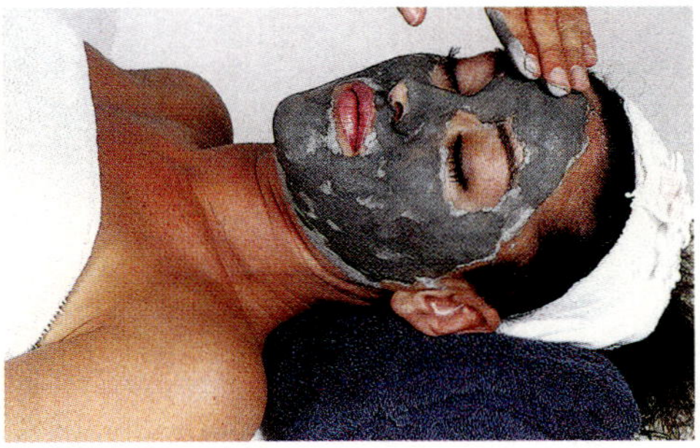

By land

From I-95 take Exit 67 if traveling North, or Exit 68 if traveling South. Follow signs to Route 154 South, approximately 2 miles.

Dine al fresco on our Patio or enjoy fine dining or Sunday Brunch in our waterfront restaurant.

The summer trolley offers hourly service for shopping, antiquing, movies, provisions and points of interest.

SAYBROOK POINT INN
Marina • Conference • Spa

2 Bridge Street
Old Saybrook, CT
06475

Outside Connecticut 800-243-0212

Marina 203-395-3080
Hotel 203-395-2000
Fax 203-388-1504

© Embassy

The Connecticut River Museum at Steamboat Dock in Essex.

the mouth of the river testifies to the dominant rural existence you'll encounter as you cruise northward.

You can approach the mouth of the river through the Saybrook breakwater, but getting there from the Niantic Bay is indirect. Round red nun "6," leaving the marker to starboard, and head for flashing red 4sec bell "8." The tip of the western breakwater is marked by a 58-foot green flashing light. Stay between this light and red nun "2." Shifting shoals east of here make it hazardous to access the river any other way.

Old Saybrook on the west shore and **Old Lyme** on the east can be seen as demographic aberrations. It seems as if early settlers opted to pass by Saybrook and Lyme in order to explore and develop the interior of the river valley. But for whatever reason, the streets and shops of both Saybrook and Lyme are much less densely situated than, for example, Essex, about five miles upriver.

For this reason, Old Saybrook lacks a certain feeling of antiquity that is so prevalent in other areas of the coast. The advantage: the streets are less crowded, which can make for a more relaxing visit. There are a number of good marinas in and around Old Saybrook. **Saybrook Point Inn**, located at Saybrook Point is a favorite because of its proximity to the open waters of the Sound. Stop in and visit with us at **River Landing Marina**, or try the new Boater's Restaurant & Lounge. Across the river, you'll find the **Old Lyme Dock Co.**

Old Lyme Dock Co.

11 p. 132

323 Ferry Rd.
Old Lyme
Connecticut
06371
(203) 434-2267

A yellow main building, a big Texaco sign, and an oversized American flag welcome visitors to Old Lyme Dock Company, which offers some of the lowest rates on the river. Transients are always welcome. Large, easy-access fuel dock. Old Lyme Dock Company also boasts one of the longest seasons, opening early in the spring and remaining open late into the fall. Highly service-oriented, Old Lyme Dock will make reservations for you at local restaurants and arrange for transportation. Marine supplies, water, ice, extra-clean showers and a laundry are all provided. Call for reservations on VHF Channel 09 or by phone at (203) 434-2267.

- New London Tides; High:+1:25 Low: +1:10, Mean 3.1'
- $.95 / ft per day 25 Berths, 150' LOA
- Propeller
- *Texaco* Gas, Diesel
- Fresh Water, Phone 220V, 100 Max Amps.
- A&P groceries - transportation provided
- The Hideaway - Casual Bee & Thistle / Old Lyme Inn - formal
- MasterCard, Visa, Discover, American Express, Texaco

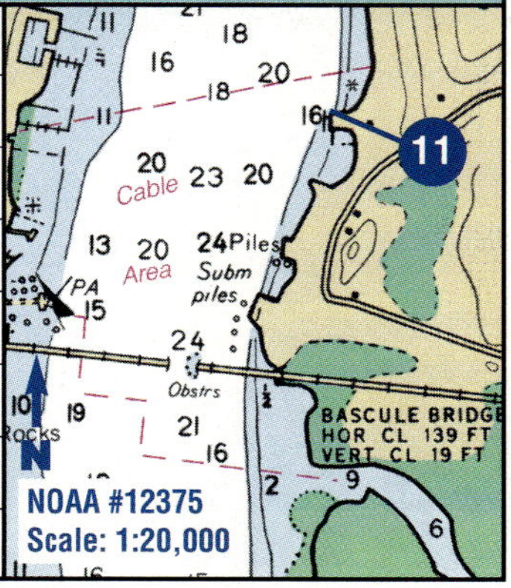

Saybrook Point Inn
Marina • Conference • Spa

12 p. 132

2 Bridge Street
Old Saybrook, CT 06475
Outside Connecticut 800-243-0212
Marina: 203-395-3080 Fax: 203-388-1504

If you're looking for the finest marina on the Atlantic Coast, your search is over! This brand new multi-million dollar facility boasts several fine features, including:

- Floating and Fixed docks
- 700' Pier with Breakwater
- Indoor & Outdoor Heated Pools
- Steam Bath & Whirlpool
- Marine Store
- Mechanical & Diving Services
- Free Cable TV
- Fax, Copy & Mail Service
- Miniature Golf
- Bike Rental
- 6 Movie Theaters
- Summer Stock Theater
- Casino Nearby
- Trolley Service to Village

- New London Tides; High:+1:11 Low: +0:53, Mean 3.2 feet
- $2.00 / ft per day 120 Berths, 250' max LOA
- Prop, Electronic, Gas, Diesel,
- Gas, Diesel *Mobil*
- Fresh Water, Phone, Cable TV 220V, 200 Max Amp.
- Ships Store, 24hr Stop & Shop, 6 Cinemas, Trolley to Village
- Saybrook Point Inn Restaurant Smart Casual Dress, $13-25
- MasterCard, VISA, Discover, American Express

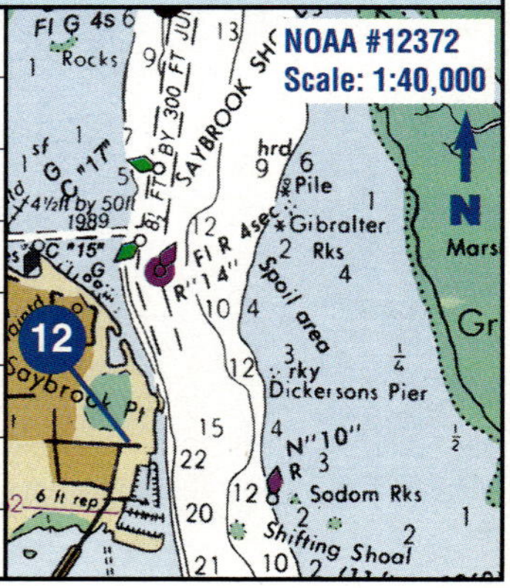

TEXACO | MerCruiser STERN DRIVES & INBOARDS | AWLGRIP | MARINE AIR | DETROIT DIESEL

New London Tides; High: +01:11, Low: +00:45, Mean: 3.5 feet	Fresh Water, Phone 220V, 100 Max Amps
$1.50 / ft per day 150 Berths, 120' LOA	Deli & Ship's Store 203-388-1431
Prop., Sail, Elec., Rigging, Gas, Diesel, Wood, Fiberglass	The Boaters Restaurant & Lounge 203-388-3100 Entrée: $15.95
Texaco Gas, Diesel	MasterCard, VISA, Discover

Located at the edge of historic Old Saybrook, Connecticut, River Landing is ideal for those who need access to both the upper Connecticut River Valley and Long Island Sound. Since we're just a few miles off exits 67 or 69 on I-95, you can be on your boat and in the Sound in no time. From there, Block Island is only 38 miles away. Montauk Point is just 32 miles distant.

Its not just the location that makes River Landing Marina so desirable. We offer complete facilities including a full-size swimming pool and a veranda. In the off season or year-round, we're equipped to professionally haul and store your craft. In our inside rack storage facilities we can hold up to a 40' performance boat.

The advantages of docking at River Landing Marina don't stop at the dock. Old Saybrook and the surrounding area has all the charm, history and beauty of a great New England sea town. Here you will find renowned country inns, superb restaurants, numerous antique and craft shops, wonderful colonial history, and the work of talented, well-known artisans.

Just a few steps from your dock is *The Boaters Restaurant & Lounge*, and *The Yacht Club* banquet facility. We can organize parties for yacht clubs and boating groups up to 300 people. Simple or fancy, the possibilities are as endless as your imagination.

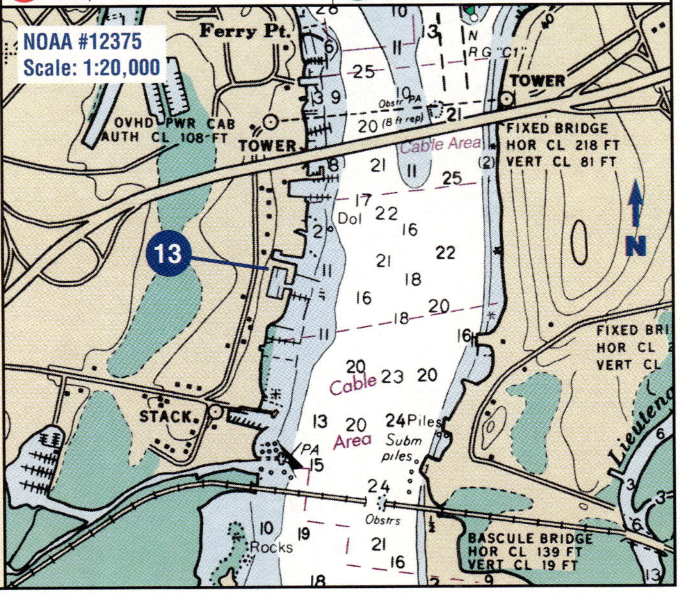

The charming village of **Essex** is a favorite among Long Island Sound yachtsmen, comparable in all respects except size to Newport, R.I. Famous for its shipbuilding industry, of particular importance during the War of 1812, Essex has built a reputation for its stylish physical features—exquisite boutiques, one-way streets, and even an old-fashioned barber shop.

Because it's located on a peninsula of the west bank, Essex can be approached several different ways. When coming from the south, we suggest following the Coast Guard buoys to Haydens Point, leaving the 30-foot flashing green 4sec "25" to port. Then, your only problem is choosing a marina from several fine ones that border on Essex. We recommend **The Chandlery at Essex**, located just north of Thatchbed Island.

A few miles north of Essex lie the sister villages of **Deep River** and **Chester**. In the mid to late 19th century, businesses in this area began to realize the potential of harnessing the energy of the river for industrial purposes. Consequently, Deep River and Chester became well-known for their river-assisted production of brushes, combs, knitting needles, ink wells, lace, and other goods. This section of the river is well-marked and easy to navigate. From the south, you may approach Deep River from either side of Eustasia Island.

The town of **East Haddam**, north of Gillette Castle on the east bank, resides atop a cliff—unlike the low banks of Essex, Deep River, and Chester. The town is the sight of the Goodspeed Opera House (203) 873-8668, one of the many landmarks in the area built by 19th century entrepreneur William Goodspeed. The opera house, accommodating 400 theater-goers and dedicated to the American musical, is singular in its architectural design, and all of East Haddam is equally beautiful.

Much of the river valley from East Haddam north to Hartford is as picturesque as it has been thus far in your cruise. A trip up to Middletown and beyond can be particularly refreshing because much of this area borders on beautiful state parks.

Once back out of the Connecticut River, continue west to **Westbrook** and **Duck Island**. Docking in the Patchogue River will allow you to visit downtown Westbrook, where you'll find more than enough supplies. There are more marine facilities crammed into this small port than at many of the large harbors on the Sound. And if exploring is what you're after, Westbrook will supply that as well. We suggest taking a dinghy to Duck Island, or into the inner reaches of the Patchogue and Menunketesuck Rivers where you're likely to see swans, geese, tern, and heron.

The major obstacle in this area is Long Sand Shoal. If you're leaving the channel from the Connecticut River, you'll be well north of the shoal and should not have a problem. Once past Saybrook Point Light, the 58-foot flashing green at the end of the breakwater, head for red 4sec bell "2," staying south of Cornfield Point Shoal, Hens and Chickens, and Crane Reef. Pass between Menunketesuck Island and Duck Island (keeping at least 400 yards off Menunketesuck), and leave flashing red "2" to starboard to enter the channel.

If you're coming from flashing red 4sec bell "8" at the east end of Long Sand Shoal, you have a choice to make. Tides tend to be slightly stronger south of the shoal, so when heading against the tide, stay north and follow the navigation already described. When going with the tide, stay south and head for flashing red 6sec horn "W," leaving flashing red bell "8a" to starboard.

The channel into Westbrook is actually the mouth of both the Menunketesuck and the Patchogue Rivers. Although it doesn't appear on the chart, there is a channel into the Menunketesuck River to the west which has been recently dredged to 11 feet. Otherwise, the Patchogue River is navigable and has several marinas.

Moving west along the coast, **Clinton Harbor** is a popular stopover, largely because of the ease of entry, the protection provided by the harbor, and the beach and facilities at Hammonasset State Park (203) 245-2785 west of Clinton. The straight coastline of Hammonasset Beach makes it the most popular spot on Long Island Sound for swimming, and the camping facilities attract transients in search of an inexpensive place to spend the night ashore.

Approach Clinton Harbor from the flashing 2.5sec 33-foot **Kelsey Point Breakwater light. Set a course for** flashing green 4sec "3," which marks the entrance to the channel, making sure you leave the white and orange nun "C" that you pass along the way to starboard. If you're coming from Westbrook, follow the same route between Menunketesuck Island and Duck Island. After rounding red nun "6," head for Kelsey Point. Once in the channel, which is continually dredged to 8 feet, pay strict attention to its boundaries—just outside the channel can be as shallow as one foot.

While there may not be any major tourist attractions in Clinton, the marinas you'll find are attraction enough, offering all the amenities. We recommend well-protected **Cedar Island Marina**, behind Cedar Island. The marina has everything you'll need—a swimming pool, recreation areas, saunas, whirlpool, picnic ground, and a shopping arcade.

Hammonasset Beach also attracts people to **Madison** and **Guilford**. Much like Westbrook, Guilford Harbor is where the waters of two rivers meet before emptying into the Sound. As a result, it's a port highly frequented by cruisers in search of exploration. But the rivers—called East and West River—are only half of the attraction. The town green in Guilford is one of the most beautiful and spacious in New England. The wildlife sanctuary and 94-foot lighthouse of **Faulkner Island**, located just a few miles off the Guilford shore, provide peaceful getaways. And Hammonasset Beach is always a favorite for swimming or sunning.

Coming from Clinton, the safest route to Guilford Harbor is south of Madison Reef and red nun "14" at Charles Reef. Give a wide berth to red flashing bell "4," and turn north toward the harbor, leaving The Tailings green can "5" to port before turning into the dredged channel. The channels into each river are marked and have been dredged. Navigating them, you should leave the green cans to port and the red nuns to starboard.

Sachem's Head, as the name implies, has an unusual history, which makes it worth a visit. Given its name for the head of a Pequot sachem, decapitated by a rival chief and placed in the nook of an oak tree, it should come as no surprise that Sachem's Head is not actually a town—it's the head of the larger body of Guilford!

From Guilford Harbor, start from red nun "16" just southwest of Indian Reef, and head for red nun "20," staying well south of the marker before turning northwest toward Sachem's Head Harbor.

There isn't much to see or do in Sachem's Head, and aside from their yacht club, there aren't any marinas. But don't lose your head over it—the village of Stony Creek is just around the corner.

The largest group of islands in the Sound, **The Thimbles** will certainly provide boaters with navigating practice and picturesque coves, but little else. The largest of the 33 main islands is only 10 acres, so aside from some fantastic anchorages that you can use to observe several species of waterfowl, you won't find too many activities here. The mainland town of Stony Creek boasts of impressive architectural feats, has hourly cruises throughout the islands, and a few marinas that can accommodate shallow-draft boats.

To approach Stony Creek and The Thimbles from Sachem's Head, start from flashing red 4sec bell "22" and head for red nun "4" into the well-marked channel. Keep in mind: If you choose to weave in and out of the islands, they are made of granite bedrock, not sand. And if you're seeking shelter, as Captain Kidd purportedly did in The Thimbles, we suggest continuing west to Branford.

The whole of **Branford Harbor**, including the marinas, will provide good protection, except from the south. Because of the deep-water anchorages and shoreside facilities, it's a popular port but seldom crowded. Neighboring Short Beach Bay usually takes some of the overflow when Branford gets busy.

From Outer Island, the westernmost of The Thimbles, make your way cautiously to flashing red 4sec "28," staying north of both green can "3" and red nun "2NW." Head due west, then due north between Cow and Calf rocks, marked by flashing red 2.5sec bell "34," and Five Foot Rock. Stay west of flashing red "2" at Blyn Rock. The route into the channel will pass between Big Mermaid and Little Mermaid, the two large boulders that bracket the entrance to Branford River. But if you draw more than 5 feet, we suggest keeping to the outer harbor where depths range from 10 to 14 feet.

While Branford offers excellent marine facilities, getting to a restaurant or shopping center will require a cab. And while you're in town, be sure to visit the Shore Line Trolley Museum (203) 467-6927 in nearby East Haven for a taste of urban history.

Western Long Island Sound: New Haven Harbor to Greenwich

Strikingly different from eastern Long Island Sound, both land and sea of the Western Sound are much busier. From New Haven to Greenwich is almost entirely urban and, consequently, harbors here tend to experience considerably heavier traffic.

For a first-hand experience of such urban life, take the time to explore **New Haven**, the second largest commercial port in Connecticut. Home of Yale University (203) 432-2300 and recognized largely by the skyline of its Gothic and Georgian architecture, New Haven stands as one of the oldest cities in the country, founded in 1638.

Perhaps because of its intimidating size, or perhaps because of its harsh industrial outskirts, many recreational boaters overlook **New Haven Harbor**. Don't be discouraged by the large commercial vessels entering and exiting. Just give them plenty of room and continue into the heart of the city—it's well worth it!

Dining and entertainment alone make New Haven worth visiting. Italian cuisine dominates the inner city and its suburbs, but you'll also find French, Jamaican, Indian, and the finest burgers around. After dinner, take in a show at one of several theaters, which include the Shubert (203) 562-5666, Yale Repertory (203) 432-1234, Long Wharf (203) 787-4282, and the Palace (203) 789-2120.

Approaching New Haven from the east is rather easy, but not even the largest of harbors on Long Island Sound can be taken for granted (or granite for that matter)! Leave red nun "36" to starboard and come in at Morgan Point at the eastern end of the east breakwater. Pass between the breakwater and red nun "4," then head for red nun "6" off Lighthouse Point. The channel is about 50 yards wide, and as you would expect from such a large harbor, it carries plenty of water. Marine facilities, however, are fairly scarce relative to the size of the harbor.

Perhaps all the restaurant service in New Haven will inspire you to fend for yourself during your next stop—and **Milford Harbor** is the perfect place to start. Oystering here has been popular and plentiful for more than a century. The only restriction is that you use hand tools and limit your take to a half bushel per person each day. To accommodate the traditional fisherman on board, bunkers and blues are also in abundance in theGulf between Charles Island and the beach.

To reach Milford, start from flashing green 4sec "9" at the edge of the channel in New Haven, and head for flashing red 4sec "2" at the west end of the west break-

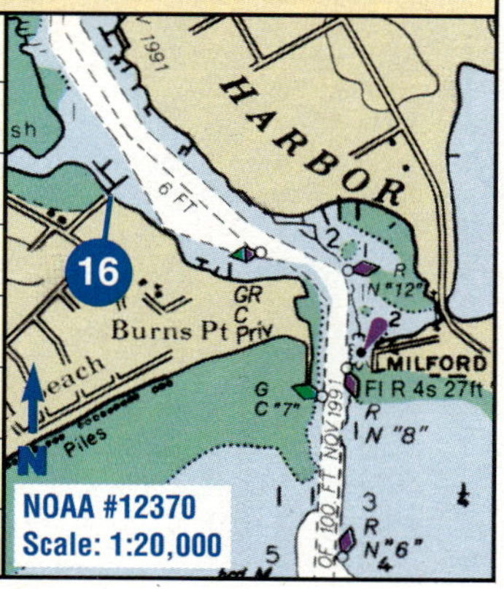

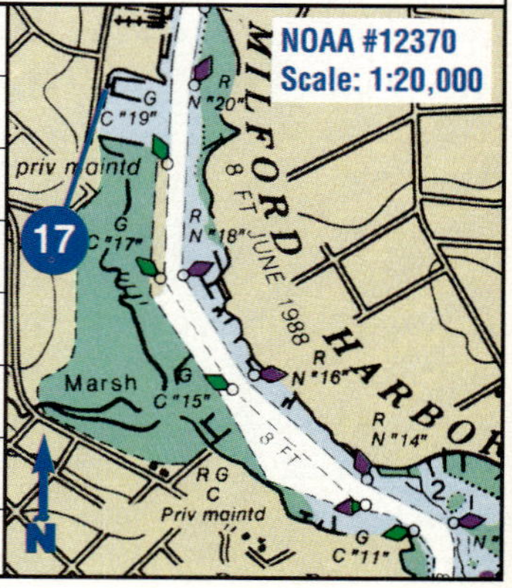

water. Leave red nun "2" to starboard as you round Welches Point into The Gulf, then head for the flashing red 4sec "2A" at the beginning of the channel. Once you're inside the harbor, stay well within the channel—the waters outside tend to be very shallow.

Marinas in Milford are plentiful. **Port Milford**, on the west side of the harbor, offers full-service facilities and is close to downtown Milford where you can catch a cab to the Jai-Alai fronton (203) 877-4211. Located further into the harbor on the west side is **Milford Boat Works**, also a full-service facility, which is one block from downtown Milford.

Visitors flock to Captain's Cove Seaport in Black Rock.

Courtesy of Captain's Cove Seaport

If theater is more your speed, continue west to **The Housatonic River**. It shouldn't surprise you that the town of Stratford, named after Stratford-on-Avon, is also home to the American Shakespeare Theater, modeled after the famous Globe Theatre in London. Built right on the river in 1955, the theater thrived until very recently and is now owned by the state and operated as a park.

Visiting the theater or downtown Stratford is easily accomplished from any one of the many marinas found here. We recommend **Stratford Marina, Inc.**, about one and a half miles from the mouth of the river, or **The Marina at the Dock**, situated between two bascule bridges farther upriver. Both offer floating slips and complete marine repairs.

From Milford, the approach to the Housatonic is simple and direct. At flashing red 4sec bell "16" just south of Charles Island, set a course for flashing green 4sec "1," which marks the beginning of the channel. The Stratford Point Light, flashing every 20 seconds and marked by a brown band, is the second most powerful beacon on the Sound and an excellent landmark when heading for the river.

For those of you who haven't gotten your fill of urban cruising ports, **Bridgeport** is around the corner, just over five miles from the Housatonic. Although not as well known as Hartford or New Haven, Bridgeport is the largest city in Connecticut and one of the three deep-water commercial harbors in the state. If you're lucky enough to be in Bridgeport in July, don't miss the Barnum Festival (203) 367-8495, held in honor of P.T. Barnum, the city's most celebrated citizen. The festival features a parade, fireworks, concerts, and other events.

From Stratford Point Light, pick up flashing red 2.5sec bell "10" at the entrance to the channel, which maintains depths of 35 feet all the way to the head of navigation at the Interstate 95 bridges.

Bridgeport Harbor is well protected but highly commercial—the overconfident skipper can easily find trouble by not taking seriously the wakes of large freighters and tankers. So as long as such vessels are given the respect they deserve, ambitious cruisers can explore the three tributaries of the harbor: Johnsons Creek, Yellow Mill Channel, and Pequonnock River. All three offer marine facilities from which the city can be reached.

Black Rock Harbor is considered by many to be part of Bridgeport, probably because the city incinerator and other signs of urban blight are apparent from the water. But any yachtsman familiar with Long Island Sound can tell you that no matter how close they are to each other, these two ports are vastly different. Black Rock Harbor is predominantly a sailor's port (with Wednesday night yacht races), whereas Bridgeport is much more commercial.

Captain's Cove Seaport (203) 335-1433 is the main attraction in Black Rock Harbor, providing entertainment throughout the day. Home of the *H.M.S. Rose*, the largest operational wooden sailing ship in the world, Captain's Cove offers an amusement center, a restaurant, Sunday afternoon band concerts on the boardwalk, and full-service transient facilities.

From green can "13" at the edge of the channel in Bridgeport, head for the 43-foot flashing red 4sec "2A," which marks the entrance to Black Rock Harbor. Boats drawing more than 4.5 feet should honor the nuns and cans that mark the edges of the channel.

Southport Harbor, a few miles west of Black Rock, offers a pleasant contrast to the urban sprawl of the Bridgeport area. Because of its size and limited facilities, Southport does tend to be crowded during the height of the season. Cruisers familiar with Connecticut will know Southport as the access port for the town of Fairfield, home of the Audubon Society of Connecticut, as well as some of the finest restaurants and shops on the coast. It will be immediately apparent why the area from here west along the shore has been known as "the gold coast"—this is one of the wealthiest areas of the state.

Unfortunately for transient vessels, this area offers very little in the way of marinas. There are two yacht clubs on the west bank of the harbor which offer fuel and a few transient moorings, but overnight docking is scarce. If you plan to visit the harbor for an extended period of time, we recommend calling ahead.

The approach to Southport is unobstructed. From the red and white Morse (A) whistle "BH" southeast of Black Rock, set a course for flashing red 4sec bell "22." From here, the entrance markers for Southport to the northwest should be in view. The channel is very narrow—50 to 75 feet wide—and after passing red nun "10," it's advisable to stay west of center for the deepest water.

For a continued look at the rich history of Connecticut's coast, head for the mouth of the **Saugatuck River** and the town of **Westport**. Here you'll find an outcropping of Manhattan's elite, the Wall Street execs who so frequently seek refuge in the southwest corner of the state. Consequently, the homes in this area are exquisite and make for picturesque sailing.

From green can "1" and red nun "2" at the end of the channel in Southport, head southwest, giving the rocks off Frost Point plenty of room. Then pick up flashing green 4sec "3" at the entrance to the river. The river is shallow, and the channel generally hugs the western shore—night sailing here is not suggested.

An abundance of facilities and fun can be found in **Norwalk Harbor**. The Maritime Center at Norwalk (203) 852-0700 is a popular stopover for the transient with some spare time—you can easily spend an entire day here. The center has a harbor seal show, IMAX movies on a giant screen, boatbuilding demonstrations, and a great aquarium. If you prefer to spend more time in your boat, Norwalk Harbor is also famous for its oyster crop. And if you happen to be in the harbor the weekend after Labor Day, don't miss the Norwalk Oyster Festival.

Getting to Norwalk is somewhat indirect. The harbor itself is only about two miles from the Saugatuck River, but all the reefs and rocks of the Norwalk Islands make a direct approach difficult. Your best bet is to start from the green flasher "3" at the mouth of the Saugatuck River and head southeast giving a good berth to the east end of Cockenoe Reef. After leaving the two flashing red 4sec bells, "24" and "26" to starboard, head for red nun "28." Navigate your way carefully around the south end of Sheffield Island until you reach flashing red "2" marking the entrance to the channel.

Located in sheltered Norwalk Cove at the mouth of Charles Creek, Norwalk Cove Marina offers the finest of facilities from which to enjoy all Norwalk has to offer. Norwalk Cove Marina is Southern CT's leading yachting center, offering complete repair, 3 Travel-lifts for up to 150 tons and refueling services as well as friendly yacht brokers.

Enjoy shopping at the area's factory outlet stores, or dine at one of the many fine restaurants in the quaint South Norwalk district. Bring the kids and enjoy our beautiful beaches, spend the day at the famous Maritime Center, or visit the 50-room Lockwood-Mathews Mansion.

Come discover Norwalk Cove Marina and find out for yourself why we are Southern Connecticut's best! New mega yacht slips located adjacent to the main channel and dredged to 15' MLW.

Bridgeport Tides; High: +0:09, Low: +0:15, Mean 7.1 feet

$1.50 / ft
50 Berths, 150' LOA

Propeller, Sail, Elec., Rigging, Gas, Diesel, Wood, Fiberglass

Gulf
Gas, Diesel

Fresh Water, Phone, Cable TV
Elec: 30/50/100A, 100 Max Amps

Skippers 203-838-2211, Casual, $16.95

MasterCard, Visa

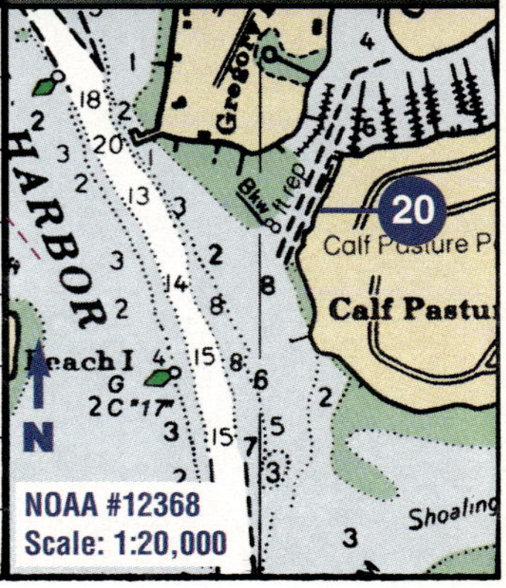

NOAA #12368
Scale: 1:20,000

Norwalk Harbor is divided into three channels, all offering good protection largely because of the **Norwalk Islands**. The first channel is Charles Creek, between Gregory Point and Calf Pasture Point on the east side of the main harbor. Here you'll find **Norwalk Cove Marina**, with excellent facilities for transients. The second of the three channels, also on the east side, is the East Norwalk Channel, and the third is the main channel created by the Norwalk River.

If Norwalk Harbor doesn't suit your fancy, perhaps the Norwalk Islands will. These 16 islands have attracted boaters for their beauty and potential for exploring, but the bird life here is of particular interest. Sheffield and Chimon islands, the two largest, have been central parts of the Stewart B. McKinney National Wildlife Refuge. Chimon Island, in fact, is one of the most important heron nesting grounds and one of the three largest wading-bird colonies in the Northeast. So if you plan to visit the islands, be mindful of where you walk and don't leave any trash behind. Hiking is prohibited during the nesting season (April 15 to August 15).

After exploring the Norwalk Islands, **Darien** and the **Five Mile River** are perfect for relaxing. A peaceful day in this area will be much appreciated when you reach the more urban ports farther west. About a mile from Sheffield Island, the area offers pleasant gunkholes and friendly marinas but little else. The mostly residential surroundings house a few shops, restaurants and delis, but you'll be hard-pressed to find tourist attractions—and that's exactly what makes it appealing.

From the west end of Sheffield Island, head for red nun "2," leaving the buoy to starboard to avoid Ballast Reef. The entrance to the channel into Five Mile River is marked by flashing green 4sec "3" and red nun "4." The channel is narrow and carries about 6 feet of water. You'll find a number of good facilities in and around Five Mile River, but many of them are not accessible in larger boats.

If more activity is what you're after, **Stamford Harbor** is the place to go. One of the major yachting centers on the coast, Stamford has flourished as a commercial center and boasts of being home to many corporate headquarters. Much like Westport, this city is replete with Manhattan commuters and the money usually associated with them—the average cost of a house here exceeds $300,000.

The advantages of such money to visitors are obvious. The Stamford Museum and Nature Center (203) 322-1646, the Whitney Museum of American Art (203) 358-7652, the Stamford Center for the Arts (203) 358-2305, and the Palace Theater (203) 325-4466 are just a few examples of what you'll find in downtown Stamford, a 5-minute cab ride from the water. **Yacht Haven Marine Center**, and **Harbor Square Marina** will accommodate you while you venture inland.

From Five Mile River, pick up flashing red 6sec bell

YACHT HAVEN MARINE CENTER

21 p. 134

Foot of Washington Boulevard • Exit 7, I-95 Conn. Tpke • P.O. Box 1435, Stamford, CT 06904
203-359-4500 • Monitor VHF Channel 9/14

Foreground: Yacht Haven West • Background: Yacht Haven East

Bridgeport Tides; High: +3 min, Low: +8 min, Mean: 7.2 feet	Freshwater, Phone Elec: 110V, 220V, 100 Max Amps
$1.50 / ft per day 600 Berths, 140 LOA	Dockside Deli 203-325-3595 Harbor Deli 203-348-5769
Propeller, Electronics, Rigging, Gas, Diesel, Wood, Fiberglass	The Rusty Scupper Casual 203-964-1235
Texaco Gas, Diesel	MasterCard, VISA, American Express

Stamford's YACHT HAVEN MARINE CENTER is conveniently located in the well-protected "inner" Stamford Harbor boasting direct access and deep water to the Long Island Sound. Yacht Haven has long been a favorite of day sailors and mega yachts up to 140 feet. Yacht Haven's two locations, YACHT HAVEN WEST and YACHT HAVEN EAST offer large professionally-run marinas close to Metro-North, I-95 and the cultural and culinary offerings of downtown Stamford.

YACHT HAVEN WEST, the "working yard" of the Marine Center is a full-service marina complete with land storage for over 500 pleasure yachts. If while cruising you need repair, the staff of mechanics, painters and expert carpenters offer a full range of services to get you back on your cruise. Yacht Haven's "Texaco Starport" is open seven days a week for your convenience.

YACHT HAVEN EAST is easily recognized from the Stamford Breakwater by the large brick buildings surrounding the docks. The Yacht Haven East Marina is the home of the Rusty Scupper Restaurant, Harbor Deli, a mini mall, picnic areas and jogging paths. Just a short walk to the Stamford Yacht Club, markets, public beaches and tennis courts. Yacht Haven East's suburban setting has been a favorite destination to the visiting yachtsman for over forty years.

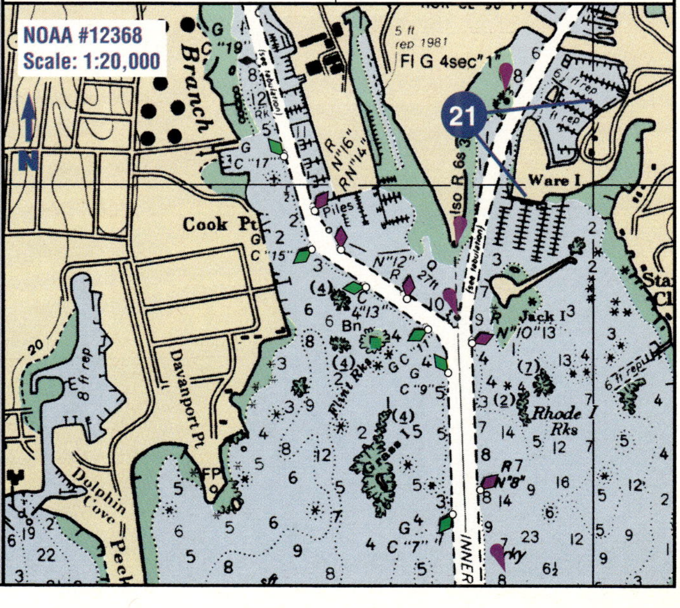

GREENWICH HARBOR INN

23 p. 134

500 STEAMBOAT ROAD, GREENWICH, CT 06830
(203) 661-9800

ATLANTIS Restaurant

Located at the head of our beautiful harbor, the 88-room Greenwich Harbor Inn offers the luxury and gracious amenities that sophisticated yachtsmen seek. You and your guests aboard will find this the perfect spot to come alongside, whether it's for a few hours or a few days.

Savor the gourmet pleasures offered by our acclaimed Atlantis Restaurant while dining harborside...or request Atlantis cater aboard your yacht or in private dining rooms with the same fine gourmet touch and service. Then browse on picturesque avenues where elegant boutiques abound.

We offer the perfect spot to make connections to major cities. New York is just 45 minutes via Metro North trains two blocks away...nationwide Amtrak service 12 minutes away. Major New York area airports are served by dockside shuttle service. Private limousines and taxis are on call day and night.

You are always assured a warm welcome and service dockside. We monitor Channel 9 (work 78) from 11:00am daily May to October. **Beginning May 1, 1994 we will offer Twin 30 AMP/120V and Twin 50 AMP/220V power.** Reservations a necessity. Please call our Dockmaster at 203-861-0000, the Inn 203-661-9800 or Atlantis Restaurant 203-861-1111 for further information.

Bridgeport Tides; High: +0:01, Low: +0:01, Mean: 7.4 feet	Fresh Water 50Amp/220V, 30Amp/120V
$.50/ ft,day-4hrs – $2/ft, overnight 500' floating dock	Delis, Markets, within a mile Food Emporium, Grand Union
	Atlantis 203-861-1111 Casual Atmosphere - 3 Stars
	MasterCard, VISA, American Express, Diners Club

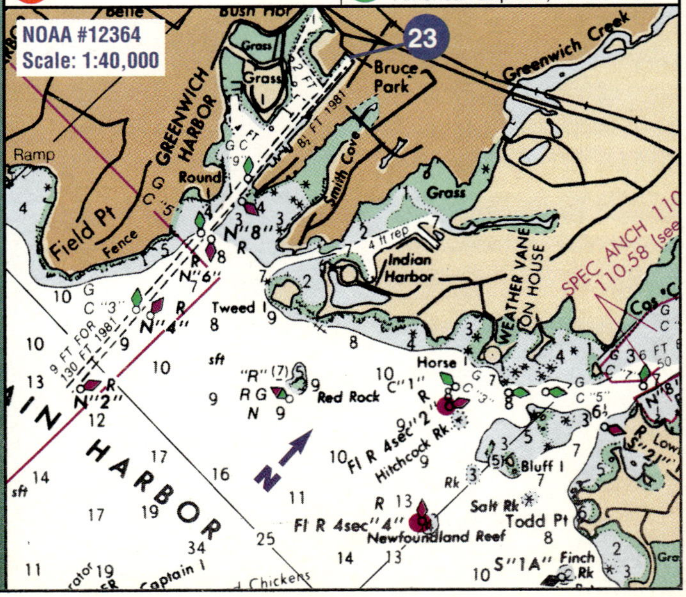

"32" south of Shippan Point. Stay south of the marker to avoid running into The Cows, then turn to starboard and head for red nun "2." Once inside the main harbor, a 38-foot red 6sec flasher should be visible on the headland. This divides the channel into East and West branches.

The last stretch of the Long Island Sound along the Connecticut shore is **Greenwich**, which consists of five harbors: Captain Harbor, Cos Cob Harbor, Greenwich Harbor, Indian Harbor, and Byram Harbor.

Captain Harbor is actually the body of water between the **Captain Islands** and the mainland and is the first harbor you'll encounter. If you're coming from the east, stay south of the islands and pass between bell "1" and red nun "2," leaving green can "1A" to port. Once in the harbor, you have two choices—either explore the islands, (Calf Island has a beach and campsites), or continue into one of the other four harbors.

Regardless of which harbor you choose, amenities are scarce in Greenwich. The conspicuous lack of industrial scenery and the gracious depths of the harbors make it a favorite layover port for those who wish to anchor. If you're in need of a marina, we suggest pulling into **Greenwich Harbor Inn**, but reservations are necessary due to its proximity to many fine restaurants, boutiques, parks, and a museum. A town fuel dock provides pumpout facilities.

It's rather fitting that the last harbor you'll encounter in Connecticut is Greenwich. The name itself, taken from Greenwich, England, substantiates the New England heritage so deeply established in Connecticut. The five harbors that make up the Greenwich coast are evidence of the diversity you've experienced in Connecticut cruising. And the sheer number of boats that you'll find pulling into Greenwich tells the tale—those traveling west are in no hurry to leave, and those traveling east are eager to get a taste of Connecticut.

Shoreside and Emergency Services

Ambulance, Fire, Police: 911

Coast Guard:

New London (203) 442-4471 or VHF 16
New Haven (203) 773-2400 or VHF 16
Eatons Neck, NY (516) 261-6868 or VHF 16

Radio Telephone:

VHF 24, 27; I.D. Bridgeport Marine Operator
VHF 25, 26, 86; I.D. New London Marine Operator
VHF 28; I.D. Riverhead Marine Operator
VHF 85; I.D. Bay Shore Marine Operator

Tow Service:

Harbor Marine Services, Mamaroneck, NY (914) 698-3373 or VHF 16 or 9
Marine Rescue Services, Wickford, RI (401) 295-8711 or VHF 16
Saybrook Towing and Salvage, Old Saybrook (203) 388-4065 or VHF 16
Seatow Central L.I. Sound, Branford (203) 488-3442 or VHF 16
Seatow New London, New London (212) 884-8936 or VHF 16
Seatow Southern CT, Bridgeport (203) 331-0410 or VHF 16 or CB 9
Seatow Western L.I. Sound, City Island, NY (718) 885-0101 or VHF 16
Sound Tow Co., Bronx, NY (718) 885-3420 or VHF 16
Southeastern Marine Towing, Groton (203) 445-8381 or VHF 16

✦

Local Services

Wilcox Marine Supply
Est. 1879

- Marine Hardware and Boat Supplies
- Nylon-Dacron-Manila Rope
- Foul Weather Gear
- Complete Rigging Services
- Netting

OLD STONINGTON RD., MYSTIC (1.5 Miles East of Mystic on Rte. 1)
Monday-Friday 8:30 am-5pm / Saturday 8:30 am-3pm
TEL : (203) 536-4206 / FAX : (203) 536-8326

Sound inflatables

- BOATS
- ENGINES
- LIFE RAFTS
- ACCESSORIES
- SALES
- REPAIR
- REPACKING
- WARRANTY CENTER

304 BOSTON POST RD • OLD SAYBROOK, CT 06475 • 203-388-6634

METAL FABRICATOR

Announcing... Northeast Metal Works

New England's fastest growing metal fabricator!! We build anything in Stainless Steel or Aluminum specializing in: T-Tops, Hard Tops, Gin Poles, Radar Mounts, Rocket Launchers. Mobile service is available. Call us today for a quote on your custom fabrication.

Toll Free 1-800-235-2628
In CT (203) 669-5608
Fax (203) 669-6982

Northeast Metal Works is a division of MARCON International Ltd.

COMPASS PROBLEM?

CONNECTICUT COMPASS SERVICE

Complete magnetic compass repair and adjusting
All major brands
Fast turn-around
1 year warranty
Compass and navigation accessories
Compass adjusting with Gyro Compass
13 years of experience
Serving the entire North East

3 ANCHORAGE LANE
OLD SAYBROOK, CT 06475
203-388-2019

SURFSET

The Little Flag Store
"On The Corner"

Personalized glasses, napkins, tee-shirts, sweatshirts, robes, jackets
Custom flagmakers

2 ESSEX SQUARE
ESSEX, CT
(203) 767-FLAG

WOODWORKERS STORE AT SOUTH NORWALK

Where Boaters Come For Great Woodworking!

LUMBER, PLYWOOD AND MOLDINGS
Large selection of mahogany, teak and marine plywood.

HARDWARE & SUPPLIES
Brass cabinet hardware, non-corrosive fasteners; WEST system; abrasives and marine adhesives.

CUSTOM MILL & CABINET SHOP
On-board repairs & fabrication, custom boat cabinetry; one-off replications.

HAND & POWER TOOLS, SHOP MACHINERY

3 Blocks So. of Maritime Center

85 WATER STREET
SO. NORWALK, CT
TEL 800-422-2783
FAX 203-857-4223
Visitor Docking Available

ATLANTIC

Detroit Diesel Allison

- Generators
- Northern Lights Generator
- Luggar
- Kohler
- Onan
- AC
- Stewart Warner
- Racor
- VDO

DETROIT DIESEL

Perkins Engines

- Allison
- E.M.D.
- Mercruiser
- Twin Disc
- Air-sep

Mobile Dockside Service
Factory Trained Technicians
Full Line of Parts & Inventory
Year Round Service

Noank Shipyard
Pearl Street
Noank, Connecticut 06340

Fax (203) 572-8854
Phone (203) 536-6726
Corporate (203) 632-0218

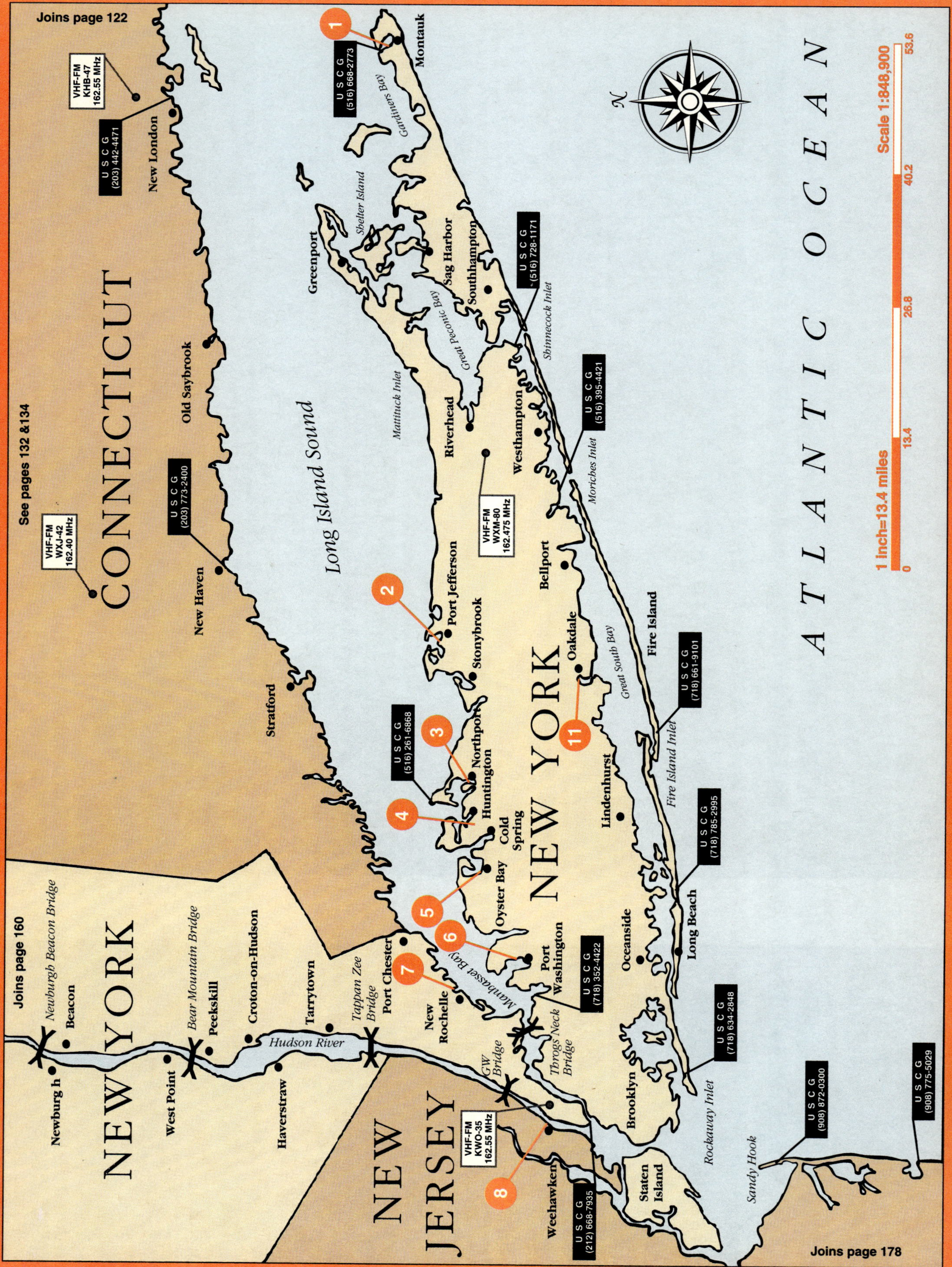
Joins page 122
See pages 132 &134
Joins page 160
Joins page 178
CONNECTICUT
NEW YORK
NEW JERSEY
ATLANTIC OCEAN
Long Island Sound
VHF-FM KHB-47 162.55 MHz
VHF-FM WXJ-42 162.40 MHz
VHF-FM WXM-80 162.475 MHz
VHF-FM KWO-35 162.55 MHz
U S C G (203) 442-4471
U S C G (203) 773-2400
U S C G (516) 668-2773
U S C G (516) 728-1171
U S C G (516) 395-4421
U S C G (516) 261-6868
U S C G (718) 661-9101
U S C G (718) 785-2995
U S C G (718) 352-4422
U S C G (718) 634-2848
U S C G (908) 872-0300
U S C G (908) 775-5029
U S C G (212) 668-7935
New London
Old Saybrook
New Haven
Stratford
Montauk
Gardiners Bay
Shelter Island
Greenport
Sag Harbor
Southampton
Great Peconic Bay
Shinnecock Inlet
Mattituck Inlet
Riverhead
Westhampton
Moriches Inlet
Port Jefferson
Stonybrook
Bellport
Oakdale
Great South Bay
Fire Island
Fire Island Inlet
Northport
Huntington
Cold Spring
Oyster Bay
Lindenhurst
Port Washington
Manhasset Bay
Oceanside
Long Beach
Throgs Neck Bridge
Brooklyn
Rockaway Inlet
Sandy Hook
Staten Island
Weehawken
GW Bridge
New Rochelle
Port Chester
Tappan Zee Bridge
Tarrytown
Croton-on-Hudson
Peekskill
Bear Mountain Bridge
Beacon
Newburgh Beacon Bridge
Newburg h
West Point
Haverstraw
Hudson River
1
2
3
4
5
6
7
8
11
Scale 1:848,900
1 inch=13.4 miles
0
13.4
26.8
40.2
53.6
N

The Empire State

Start with a panorama of windswept dunes. Add towering manmade wonders. Mix in the awesome splendor of rocky cliffs. And finally, blend in sites where the course of history has changed. Such an eclectic recipe could yield only one place—New York. Since the early 1500s, when Giovanni da Verrazano and his crew explored the uncharted coast of America, the shores of New York have provided awesome scenery, practical support for daily living, and resources for vital economic growth.

Verrazano sailed away never to return, but 85 years later, an English captain sailing under the Dutch flag came to New York. His name was Henry Hudson. Hudson and his crew were the first to sample the native foods of the Algonquin Indians of the great Hudson River valley, and the first to catch the huge sturgeon that filled its waters. He failed in his quest for a new route to China for his employer, the Dutch East India Company, but he claimed the fertile valley lands for the company. From the settlement of New Amsterdam, where Manhattan now glitters, to the heights of Fort Orange, now Albany, the Dutch left the legacy of their lifestyles.

The Dutch East India Company was a harsh landlord, however, and in 1664, the Dutch willingly allowed the English commander James, Duke of York, to lay claim to New Amsterdam and its neighboring lands for his brother, King Charles II of England. The Duke arrived with five ships and a contingent of troops, and promptly renamed New Amsterdam for himself: "New York." The settlers kept their lands, and the taxes imposed by the Dutch East India Company were abolished.

© Mary Jane Hayes

The Manhattan skyline glows at sunset.

Meanwhile, sandy, 96-mile Long Island held little interest for the first European captains or the settlers. It was not until the 1700s, when whaling and regular shipping commerce were established, that the villages of the eastern end of the island came to be. Later, the shoreline spots with a good view, cooling breezes, and a place for the family yacht were claimed by the Vanderbilts and families of like stature for summer homes. Even today, with access by train, plane, and car, the port towns of Long Island remain unique in their proximity to New

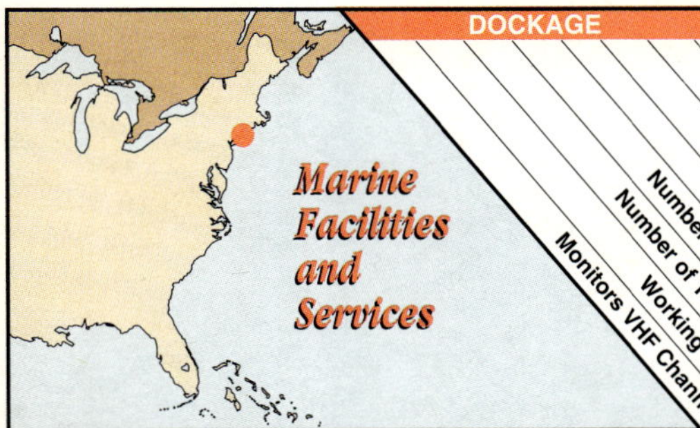

See Adjacent Chart for Coast Guard Information

Facility	Phone	DOCKAGE								SERVICES					SUPPLIES			AMENITIES		COAST GUARD
		Monitors VHF Channel	Working Channel	Number of Transient Berths	Number of Transient Moorings	Seasonal / Year-round	Maximum Length Over-All	Approach Depth in Feet at MLW	Dockside Depth in Feet at MLW	Hookups: Fresh Water / Phone / Cable TV	110V★ 220V▲ 3 Phase■ Maximum Amps	Ramp / Dinghy Dock / Launch Service	Railway / Lift / Crane: Capacity (tons)	Repairs: Propeller / Sail / Electronics / Rigging / Gas / Diesel / Wood / Fiberglass	SCUBA: Air / Rentals / Equipment / Charters	Gas / Diesel Fuel	Fuel Brand	Marine / Groceries / Bait / Ice / LPG / CNG	Showers / Laundry / Pumpout / Restrooms	Hotel / Pool / Tennis / Golf	Public Phone / Restaurant / Snack Bar	MasterCard / Visa / Discover / AmEx
1 Montauk Yacht Club (p. 163)	516-668-3100	9	9	150		Y	250	11	10	All	★▲■ 50					D	Tex	All	All	All	All	All
2 Danfords Inn, Marina & Conference Ctr. (p. 164)	516-928-5200	9	9	75		S	250	12	12	All	All	All		PSERGDF		GD	Cit	BI	All	All	All	All
3 Britannia Yacht & Racquet Club (p. 165)	516-261-5600	9		40		Y	107	5	6		★▲ 100			GDPH		GD	Tex	MGI	SP	HP	R	MV
4 West Shore Marina (p. 165)	516-427-3444	9	73	50		Y	170	8	18	FP	★▲■	D	LC30	PSERWF				MGI	All	P	P	MV
5 Oyster Bay Marine Center (p. 166)	516-922-6331	71	71	2	10	S	150	13	18	F	★▲	LS	10	All		GD	Cit	MI	SR		PS	MV
6 Manhasset Bay Marina (p. 167)	516-883-8411	9	72	10		Y	110	9	6	All	★▲■ 50	L	L80	All		GD	Tex	All	SLR		PS	All
7 Wright Island Marina (p. 169)	914-235-8013			10		Y	65	15	7	All	★▲50		L50	All		GD	BP	MI	SLR	P	All	MVA
8 Port Imperial Marina (p. 171)	201-902-8787	09	88	50		Y	150	8	6	F	★▲■ 100		L35	PEGD		GD	Mob	BI	All	G	All	MVA
11 Oakdale Yacht Service (p. 175)	516-589-1087	9	11	50		Y	80	7	6	F	★▲■ 50	RD	L50	PSRGDF		GD		All	All	P	All	MVD

Information in these listings is provided by the facilities themselves.

Troy
Albany
10
Castleton-on-Hudson Bridge
Scale 1:913,750
1 inch=14.4 miles
0
14.4
28.8
43.2
57.6
N
MASS
(See page 104)
Hudson
9
Rip Van Winkle Bridge
Catskill
Hudson River
Kingston
NEW YORK
Mid Hudson Bridge
Poughkeepsie
CONN
Joins pages 132 & 134
Newburgh Beacon Bridge
Newburgh
Beacon
West Point
Bear Mountain Bridge
Peekskill
Haverstraw
Croton-on-Hudson
Tarrytown
Tappan Zee Bridge
Long Island Sound
2
Port Chester
USCG (516) 261-6868
5
4
3
NEW JERSEY
Yonkers
New Rochelle
7
Port Jefferson
Northport
Huntington
Oyster Bay
George Washington Bridge
Port Washington
Long Island
8
Throgs Neck Bridge
6
VHF-FM KWO-35 162.55 MHz
USCG (718) 352-4422
USCG (212) 668-7935
Great South Bay
Jamaica Bay
Staten Island
USCG (718) 785-2995
USCG (718) 661-9101
Joins page 178
USCG (718) 634-2848
ATLANTIC OCEAN

York City while still maintaining their isolation from urban pressures.

Cruising New York waters, you'll find every type of port imaginable, from small fishing villages to high-style resort facilities. You'll also find every obstacle imaginable, from heavy commercial traffic and floating debris to strong currents and lobster pots. The boating season is nominally from Memorial Day to Columbus Day. Most marinas and facilities provide full dock service, launches, and dockside amenities during these months. However, more year-round facilities have been springing up in all coastal areas, so inquire about services that are available at either end of the season. Of course, many New York State harbors serve commercial interests, and dockage, repairs, and access to towns are virtually always nearby for the mariner in need.

Eastern New York: Fisher's Island to Long Island's Forks

Until recently, **Fisher's Island** remained isolated due to rumors of its inhospitable attitude toward transients. But if you pack out what you pack in and respect the peaceful atmosphere of the island, the locals will be more than happy to accommodate you. Originally the summer retreat of a few very wealthy families, there are now a few marinas that offer dockage and moorings for transients. At the very least, Fisher's Island offers a convenient and pleasant overnight anchorage en route to other destinations. The island is at the northern edge of Block Island Sound and marks one of the entrances into Long Island Sound. The marinas are located in West Harbor on the north side of the island. A well-marked channel leads you through the harbor entrance beginning at flashing red 4sec "8."

Montauk Harbor, 4nm west of Montauk Point on the north shore of Lake Montauk, is the first port of refuge for mariners on Long Island. Montauk Downs State Park (516) 668-5000 and Montauk Point State Park (516) 668-2461, site of the Montauk Point Lighthouse, provide peaceful diversions for sea-weary travelers. A museum dedicated to the lighthouse is operated by the Montauk Historical Society (516) 668-5340 and is located at the point. Montauk is touted as the sportfishing capital of the Northeast, evidenced by the many elaborate fishing vessels in the harbor.

Red and white bell "M" is the outer buoy for Montauk Harbor. The entrance is between flashing green 4sec "1" and flashing red 5sec horn "2." As you near Lake Montauk, Star Island will be straight ahead. The Montauk Coast Guard Station is located here, as well as the **Montauk Yacht Club**. Under new ownership, the club is being extensively renovated and already has two restaurants, three pools, tennis courts, and a health club on the premises. Golf courses and horse stables are nearby.

Many who have traveled the "Hamptons" find **East Hampton** at **Three Mile Harbor** to be the most beautiful. The main sight to see here is the Home Sweet Home Museum (516) 324-0713, once the home of actor-poet John Howard Payne who wrote the famous folk melody and lyrics in the early 1800s. The village of East Hampton is about a mile from any marina in the harbor, but is well worth the trip. Entrance to Three Mile Harbor begins at the red and white buoy "TM" about 0.6nm offshore. Look for red nun "2" and green can "1," and follow the channel buoys into the harbor. Do not cut any of the buoys before the red and green "A" just beyond Wings Point (erroneously labeled Maidstone Point on NOAA charts), as the flats are very close to the channel here and are bare at

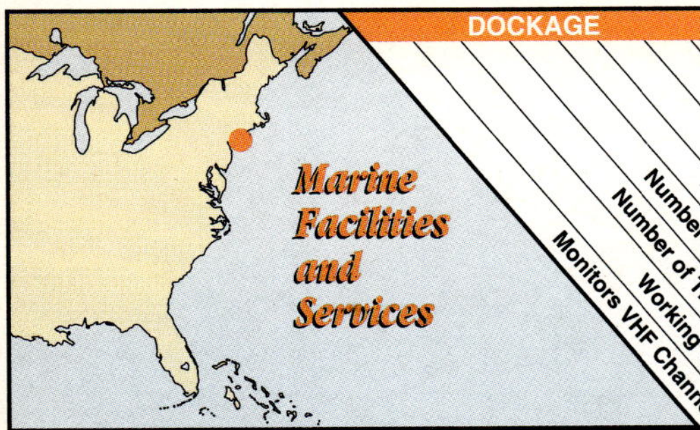

		DOCKAGE								SERVICES					SUPPLIES				AMENITIES				COAST GUARD
		Monitors VHF Channel	Working Channel	Number of Transient Berths	Number of Transient Moorings	Seasonal / Year-round	Maximum Length Over-All	Approach Depth in Feet at MLW	Dockside Depth in Feet at MLW	Hookups: Fresh Water / Phone / Cable TV	110V★ 220V▲ 3 Phase ■ Maximum Amps	Ramp / Dinghy Dock / Launch Service	Railway / Lift / Crane: Capacity (tons)	Repairs: Propeller / Sail / Electronics / Rigging / Gas / Diesel / Wood / Fiberglass	SCUBA: Air / Rentals / Equipment / Charters	Gas / Diesel Fuel	Fuel Brand	Marine / Groceries / Bait / Ice / LPG / CNG	Showers / Laundry / Pumpout / Restrooms	Hotel / Pool / Tennis / Golf	Public Phone / Restaurant / Snack Bar	MasterCard / Visa / Discover / AmEx	See adjacent chart for Coast Guard information
2 Danfords Inn, Marina & Conference Ctr. (p. 164)	516-928-5200	9	9	75		S	250	12	12	All	All	All		PSERGDF		GD	Cit	BI	All	All	All	All	
3 Britannia Yacht & Racquet Club (p. 165)	516-261-5600	9		40		Y	107	5	6		★▲ 100			GDPH		GD	Tex	MGI	SP	HP	R	MV	
4 West Shore Marina (p. 165)	516-427-3444	9	73	50		Y	170	8	18	FP	★▲■	D	LC30	PSERWF				MGI	All	P	P	MV	
5 Oyster Bay Marine Center (p. 166)	516-922-6331	71	71	2	10	S	150	13	18	F	★▲	LS	10	All		GD	Cit	MI	SR		PS	MV	
6 Manhasset Bay Marina (p. 167)	516-883-8411	9	72	10		Y	110	9	6	All	★▲■ 50	L	L80	All		GD	Tex	All	SLR		PS	All	
7 Wright Island Marina (p. 169)	914-235-8013			10		Y	65	15	7	All	★▲50		L50	All		GD	BP	MI	SLR	P	All	MVA	
8 Port Imperial Marina (p. 171)	201-902-8787	09	88	50		Y	150	8	6	F	★▲■ 100		L35	PEGD		GD	Mob	BI	All	G	All	MVA	
9 Catskill Marina (p. 173)	518-943-4170	9	68	25		S	110	8	15	All	★▲■ 50					GD	Mob	MGIL	SLR	P	P	MVA	
10 Troy Town Dock and Marina, Inc. (p. 174)	518-272-5341	16		30+		S	200	35	21	All	★▲■ 100					GD	Coa	MGIL	All		All	MV	

Information in these listings is provided by the facilities themselves.

© Mary Jane Hayes

The fishing fleet at Montauk.

low water. Proceed to the south end of the harbor to a channel leading to the inner harbor and marine facilities.

Sag Harbor, one of the earliest federal shipping ports for both the West Indies and Europe, should be your next stop. Sag Harbor was once a bustling whaling port that rivaled Nantucket and New Bedford, and since the dissolution of that industry, the town's population has declined by one third. Visit the Custom House (516) 725-0250 and the Sag Harbor Whaling and Historical Museum (516) 725-0770. If you're up for a walking tour, the whole village is a National Historic District. The Chamber of Commerce (516) 725-0011 has a visitor's center located in a recreated windmill at Long Wharf.

Heading for Sag Harbor from the east, leave red nun "8" and the 10-foot flashing red 4sec "10A" (south of Mashomack Point on Shelter Island) to starboard. They mark a sand spit and rocks. Also stay north of green can "9," and mind the channel markings for the surest passage. There are several marinas with transient berths and moorings available.

Shelter Island is the aptly named, well-protected land that "floats" between Long Island's two forks. Each of the harbors on Shelter Island has a different attraction, although all are fairly quiet. **West Neck Harbor** should be your next stop if you are approaching from the South Fork and want to drop anchor. A well-protected anchorage inside Shell Beach is a great place to spend a few days on the hook. Be sure to observe the few channel markers when entering and leaving this area—the depths near the red buoys are not very generous.

Dering Harbor, across from Greenport, is perhaps the most developed harbor on the island. Several marinas offer dockage and moorings. The ferry to Greenport, which is on the North Fork, runs frequently and is a short walk from any of the marine facilities in Dering Harbor. Greenport is where you will have to go for most of your shopping and dining needs. **Coecles Harbor** on the northeast side of the island is another delightful stop for anchoring. The waters are clear and deep, although you may find the narrow entrance channel a bit daunting. If you draw more than 6 feet, enter on a rising tide. Be sure to keep green can "5" (just inside the entrance) to port before turning into the anchorage. Although the broad expanses of water are used by locals for water sports, the anchorages are very pleasant and calm.

Ever since the Yankee Clipper *Flying Cloud* set a world record in its passage from New York to San Francisco with a crew from **Greenport**, the town has been making maritime history. Many old houses in Greenport are home to art galleries and antique shops. There are usually at least two grand old schooners offering hour-long sails from Preston's Wharf. The town is in the heart of Long Island's wine country, making it a good port to restock your "hold." Call the Long Island Grape Growers Association (516) 475-5492 for information. You can find plenty of dockage at Greenport, either in the commercial main harbor or at Stirling Basin. Call the Greenport Village harbormaster for moorings in Stirling Basin. Anchoring is possible just inside the Greenport breakwater, marked by a flashing red 4sec light, or east of the breakwater at Gull Pond.

Little Peconic Bay and **Great Peconic Bay** offer the yachtsman miles of clear water for sailing or fishing. Although surrounded by land, the relatively shallow bays do experience quite a chop, especially in Great Peconic Bay. The perimeters of these bays are lined with shallow spots, so don't venture too close to shore. There are several cozy anchorages in both bays, and there are lots of good exploring grounds accessible by dinghy.

The main harbor for **Southampton** is **Sebonac Creek** at the southeast side of Great Peconic Bay and the northern entrance to the Shinnecock Canal. Sebonac Creek gives you easy access to the upscale summer resort village of Southampton: a good stop if you need supplies. There's an anchorage and a town dock at West Neck where you can tie up while shopping for supplies.

The North Shore: Plum Gut to Manhasset Bay

Leaving the shelter of **Gardiners Bay** to continue on to Long Island Sound, plan carefully for your passage through **Plum Gut**, between Orient Point and Plum Island. This narrow channel is deep but is subject to tidal currents in excess of 5 knots. Wind opposing current can

MONTAUK YACHT CLUB

RESORT MARINA

1 p. 158

P.O. Box 5048
Montauk, New York 11954
516-668-3100

New London Tides; High: -0:24, Low: -0:16, Mean: 1.9 feet	Fresh Water, Phone, Cable TV 110/220V, 50 Max Amps
$2.00-2.50/ day 150 Berths, 250' Max LOA	Markets within walking distance
	Two Restaurants on Premises *Informal & Formal*
Texaco Diesel	MasterCard, VISA, American Express

Pull ashore this season at the Montauk Yacht Club on one of our three docks, located on Star Island in Montauk, on the Eastern end of Long Island – ports of call from Nantucket to Shelter Island are just a jaunt away. We offer 230 slips, drafts of 5-14 feet, 30, 50, and 100 amp electrical service, deep sea fishing, golf and stables nearby, local airport service, cable TV, and we monitor Channel 16. Our Dockmaster and attendants are present seven days a week to assist you.

Dine at breakfast, lunch and dinner in one of our two restaurants, swim in any of our three pools (one indoor), enjoy daytime or evening tennis on one of our nine courts, exercise in our health club, relax in our sauna, and participate in a host of other waterside activities.

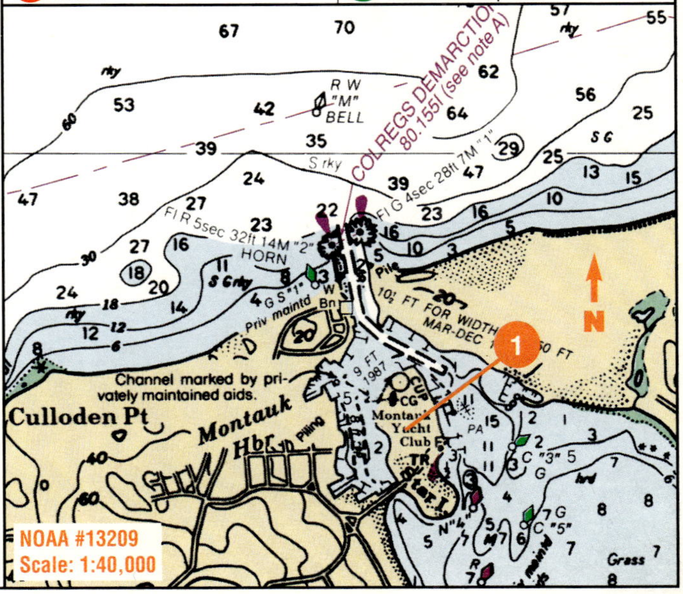

create choppy seas, so it's best to time your passage with a fair current or at slack water.

As you set your course along the North Shore of Long Island, watch for the green coastal buoys, which should be kept to port. At **Mattituck Inlet,** 6.7nm southwest of Horton Point Light, you'll find the only good sheltered port between Greenport and Port Jefferson. The jetties are low, and the westernmost jetty is marked by 25-foot flashing white 4sec "MI." The channel is narrow at low tide and turns sharply to port soon after entering. An anchorage is located at the head of the inlet, which is a peaceful place for transient vessels to drop a hook.

Port Jefferson is the first major harbor you come to as you work your way west. Along with the numerous shops and eateries, the town is full of turn-of-the-century architecture, and the streets are alive with people enjoying its nautical atmosphere. The town has a full schedule of special events, so call the Port Jefferson Chamber of Commerce (516) 473-1414. Walking tours are available from the Historical Society of Greater Port Jefferson (516) 473-2665.

Approaching Port Jefferson from the east, stay outside of green can "11." Head for flashing green 2.5sec "1," and leave the buoy to port. **Danford's Inn & Conference Center** is at the head of Port Jefferson Harbor. With extensive facilities, it can host both social and business affairs as well as cater to yachtsmen of all tastes. Its marina offers floating docks with hookups for every need.

Just west of Port Jefferson, **Smithtown Bay** is a wide open body of water that offers little protection for anchoring. But **Stony Brook Harbor** and the town of **Stony Brook** offer a picturesque backdrop for boaters who want to venture through the narrow channel leading from flashing green 4sec "1." Stony Brook's claim to fame is the State University of New York at Stony Brook. Its contemporary medical school can be seen for miles around. The Museums at Stony Brook (516) 751-0066 include the History Museum, the Carriage Museum, and four historic buildings on a nine-acre site just off Main Street. Walking tours are available from the Stony Brook Community Fund (516) 751-2244 and the Three Village Historical Society (516) 751-3730. Enter Stony Brook Harbor about three hours before high tide. There's tremendous shoaling and strong currents in the channel. The entrance is marked by a daymarker, and the buoys are right on the edge of the channel, so stay close to avoid the shoals on the starboard side. If there's a northerly wind, the water may be lower than the charts indicate. Once inside the harbor, there's a marina and a few yacht clubs.

The next major port west of Port Jefferson is actually the confluence of several harbors off **Huntington Bay**. The Huntington Bay area and the town of **Huntington** offer many sightseeing opportunities, whether you prefer museums and historic homes or recreational hiking trails. From your berth in Huntington Harbor, a visit to Walt

Whitman's Birthplace (516) 427-5240, or the Heckscher Museum (516) 351-3250, will give you some insights into local history. For observing the native wildlife, visit the Target Rock National Wildlife Refuge (516) 271-2409 on the eastern end of Lloyd Neck. When entering Huntington Bay from the east, keep to the north of green can "13" off Eaton's Neck Point. The Huntington Harbor entrance is marked by the Huntington Light. Leave the light to port, and pass red "2" and flashing red 4sec "4." Be wary of the strong current at Wincoma Point, and don't cut the flashing green 4sec "7" south of the harbor entrance. In Huntington Harbor, **West Shore Marina** is an outstanding facility that welcomes transient boaters. A new pool and picnic area, recently renovated docks, shoreside conveniences, and major repairs are all available at the full-service facility. It's at the head of the harbor, providing shelter from any weather and easy, quick access to Long Island Sound.

Northport Bay and the town of **Northport** are to the southeast of Huntington Harbor. The Northport Museum (516) 757-9859 has interesting exhibits on the history of the area and its shipbuilding industry. If you're moored on the Northport side of the harbor, you'll want to take a short cab ride to the Vanderbilt Museum (516) 262-7888 which is on the peninsula known as Little Neck. Once the home of William K. Vanderbilt II, it now houses a museum and planetarium that reflect his interest in maritime

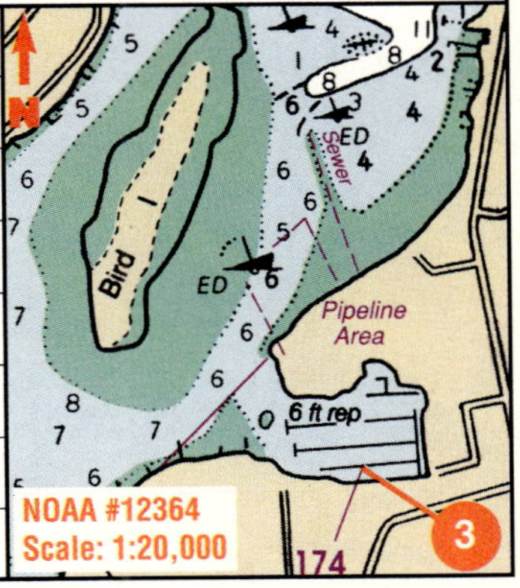

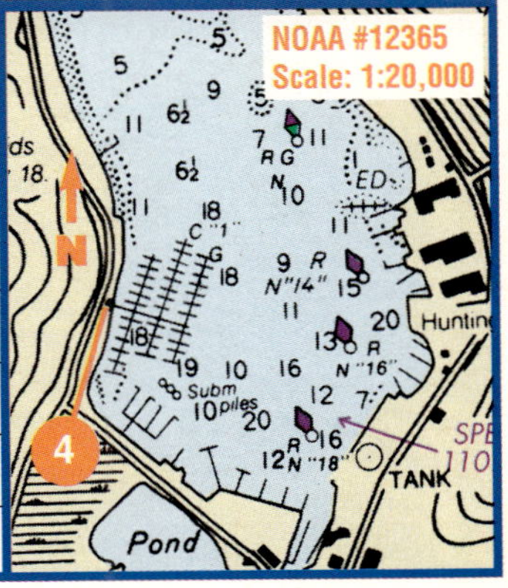

research and science. Northport Bay and Northport Harbor are wide and easily navigated from Huntington Bay. Enter the bay between flashing green 4sec "1" and red "2." Deep-draft boats should watch the privately maintained channel markers. Tucked well inside the Northport Harbor, east of Bird Island, you'll find the new facilities of the **Britannia Yacht & Racquet Club** at **Northport Marine**, a full-service marina with ample transient berths.

Anchoring possibilities abound in the Northport area, but you'll usually find lots of company. The protected waters of **Eaton's Neck Basin** provide for a quiet overnight anchorage unless a thick fog rolls in. The signal from the lighthouse can be alarming! Anchoring in Northport Bay behind Asharocken Beach is another good spot. **Duck Harbor** is fine for shallow-draft boats, and depending on the wind and weather, pulling into one of the many coves is possible.

Oyster Bay is the last town on Long Island that still has the feel of a suburb, as opposed to the city atmosphere of the towns farther west. Shops abound, and there are occasional festivals and street fairs that attract boaters to the area. Call the Oyster Bay Chamber of Commerce (516) 922-6464 for details. Sagamore Hill (516) 922-4447, once the summer White House and home of Theodore Roosevelt, and Raynham Hall (516) 922-6808 are especially interesting local landmarks. When approaching Oyster Bay from the east, keep flashing green 4sec gong "15" off Lloyd Point and green gong "1" off Whitewood Point to port. Follow the channel into Oyster Bay Harbor between nun "2A" and flashing green 4sec "3." The town of Oyster Bay and most of its marine facilities are on the south shore as you follow the channel around Centre Island. **Oyster Bay Marine Center,** on the south shore in the bend past red "4," is located close to the historic town, but it is also close to the railroad, thereby providing easy crew changes and access to New York City.

West Harbor, on the west side of Centre Island, is an ideal place to drop a hook, dinghy ashore, and walk across the sand spit to Ferry Beach for a swim in the Sound. This spot has a reputation as the area's "hurricane hole," and though it can get crowded on weekends, there is usually plenty of space. After nun "4," pick up private channel markings that lead you into West Harbor. These will keep you off the rocks that extend from Brickyard Point.

Hempstead Harbor was home to the great yachts of J. Pierpont Morgan in the 19th century. One can still see signs of a bygone era, although the harbor does have plenty of 20th-century commercialism to tarnish the nostalgia. The harbor itself offers excellent anchorages with good holding ground and protection from the prevailing southerlies. In the fall, though, be sure to be cautious of the wind shifts.

In the town of **Glen Cove**, Garvies Point Museum and Preserve (516) 571-8010 is devoted to preserving the story of the Long Island Indians, as well as the area's geology. Rocky beach, rugged cliffs, and sedimentary soil sculpted by the great glaciers are all explained in the exhibits and trails at this facility. The Webb Institute of Naval Architecture (516) 671-2213 is also in Glen Cove, and next door is the 200-acre Welwyn Preserve that offers some beautiful hiking trails. On the south side of Glen Cove Creek is the village of **Sea Cliff**, established as a Methodist camp meeting site in the 19th century. Many of the gingerbread cottages that replaced the original tents are well-maintained today, and are interspersed with antique shops.

Approaching **Hempstead Harbor** from the east, stay outside green 4sec gong "21" off Matinecock Point, then turn southwest past Weeks Point. The Glen Cove Landing breakwater light (flashing green 4sec "5") marks the beginning of two anchorages offering good overnight protection: one to the north of Glen Cove Creek, the other to the south.

Manhasset Bay was once known as the Gold Coast, home to the wealthy who built castles and mansions there. The Sands Point Preserve (516) 883-1612 contains the restored Guggenheim family homes, both of which offer several unique exhibits and tours. There is a dinosaur exhibit is in a huge stone castle, and aeronautical memorabilia in an elegant manor house. These tours are by escort only, so make sure you call ahead for times and reservations.

Manorhaven and **Port Washington** are busy towns that offer services and resources for transient boaters. Plenty of restaurants and delis can be found in this area, as well as fine shopping. A cab ride will take you to the "Miracle Mile," a strip of expensive shops at the foot of the harbor that will separate you from your money in no time at all!

Leaving Hempstead Harbor for Manhasset Bay, be sure to avoid the rocky coastline along Manhasset Neck—the charts are well marked, so keep them handy. Head northwest to flashing green 4sec gong "23," then bear to port and pick up flashing green 2.5sec "25." Leave flashing green 6sec "27A" and gong "27" at Gangway Rock to port before heading south into Manhasset Bay. Port Washington has several fine marinas. **Manhasset Bay Marina** is home to the Manhasset Bay Cup Races. The full-service facility provides many amenities including a waterside restaurant. Many fine restaurants, convenience stores, and movie theaters can be found closeby.

Leaving Manhasset Bay you'll be heading toward the **East River** and **New York Harbor**. Be sure to keep the flashing green 4sec "1" off Plum Point to starboard as you exit the harbor. Near Hewlett Point, keep flashing green 4sec "29" to port. Before leading you into the city, we will guide you along the other shore of Long Island Sound, from the Connecticut state line, across Westchester County to the Throgs Neck Bridge.

Mainland Shore: Port Chester to the East River

The shoreline area of New York from its boundary with Connecticut to the Throgs Neck Bridge is part of highly developed Westchester County. While you'll find the best resources here in the way of provisions, services, transportation and communication, little of the region's rich history remains near the waterfront. But these ports are a perfect berth from which to reach New York City for some sightseeing.

Port Chester Harbor, just 1.7nm west of Greenwich Harbor, is at the mouth of the Byram River, the natural boundary between New York and Connecticut. The river supposedly got its name from the days of Prohibition, when smugglers frequented its shores and it was a good place to "buy rum." Today, large homes line the river's banks, both in New York and Connecticut. The Byram River is a busy commercial waterway, but the area farther north affords pleasant scenery if you have some time to take your dinghy. The town of **Port Chester** offers all the urban amenities you could want, and lots of interesting ethnic delis, seafood, and pizza restaurants.

From the north and east, the safest course to Port Chester Harbor is south of red nun "2" near Great Captain Rocks. Local captains enter between Great Captain Rocks and the unmarked Channel Rock, but this is tricky if winds and current are running strong. Keeping red nun "2" to starboard and the charted wreck to port, head for flashing red 2.5sec "2A" on the end of the breakwater, and you'll have fair water to the main channel. Several charted but unmarked obstructions are in the approach to Port Chester Harbor. It is best to enter on a rising tide.

More than 1700 vessels call **Mamaroneck** home port. **Mamaroneck Harbor** is well protected and divides into two channels. The busy town of Mamaroneck is a few steps away from water's edge. Mamaroneck Avenue provides markets, delis, drugstores, and laundromats. Restaurants abound to suit everyone's palate and purse.

Mamaroneck Harbor is 4.4nm southwest of Port Chester Harbor. The buoys for Rye and Milton Harbors (where there's little room for transients) are nearby, so note your approach carefully as you will be leaving several green buoys to starboard! For Mamaroneck Harbor, it's best to keep flashing red 2.5sec bell "42" to starboard, as this will give you a clear passage to the channel entrance. This is known locally as the 42nd Street buoy, and many skippers approaching or leaving to eastward will cut it, but leave this route to local captains. Marinas for transients are found in both the West and East Basins. The channel into the West Basin is unmarked but straight.

One of the most important yacht racing centers on the Atlantic Coast is **Larchmont Harbor**. This small cove becomes the focus of hundreds of competing yachts during its Larchmont Race Week in July. On any summer day, the waters may be fluttering with colorful spinnakers and white sails competing in the regattas. Larchmont Harbor abounds with rocks and shoals, so we recommend a daytime approach. The preferred approach to the east of Hens and Chickens Ledge is made between the breakwater's flashing red 4sec light "2" and the green can "3" marking Dauntless Rock to the west. While the harbor itself is crowded, guest moorings may be available at Larchmont Yacht Club for smaller yachts. Larger vessels can find space just south of the mooring field in the southern part of the harbor, north of Umbrella Rock.

History buffs will enjoy a stop at **New Rochelle Harbor**. New Rochelle was once the home of Thomas Paine, Revolutionary War era author of *Common Sense.* The Huguenot/Thomas Paine Historical Association (914) 632-5376 maintains the Paine Monument and the Paine Cottage. **New Rochelle** has supermarkets, delis, and eateries of all kinds. Downtown there's an enormous shopping mall; its location 10 miles from New York City makes a Metro North train ride to see the sights a convenient diversion.

Several islands lie off New Rochelle, many of which have been used as military sites, amusement parks, and secluded living for the wealthy. **Glen Island** was the home of a famous 19th-century amusement resort, replete with a zoo, aviary, aquarium, and an exquisite replica of a German castle.

New Rochelle Harbor and **Echo Bay** are filled with rocks and obstructions. To enter New Rochelle Harbor, the best approach is from flashing red 4sec "2" just north of Huckleberry Island. Keep this to starboard to avoid the rocks to the north, especially at low water. Proceed west to nun "4," and then look for green can "7" and red nun "8." Take care to avoid Spindle Rock in the middle of the area between the Davenport Neck and Davids Island. Head towards green can "9," leaving it to port, then follow the buoyed channel into the harbor.

Wright Island Marina is a full-service facility behind Davenport Neck. Here you can have a relaxing swim in the pool, dine at the convenient restaurant, or have your vessel serviced or repaired. The marina is located just two blocks from the town. From the east, your approach to **Echo Bay** should begin at flashing red 4sec "2" about 1nm east of Middle Ground. The landmarks to look for are the gold-domed clock tower of the New Rochelle Rowing Club on Beaufort Point and the long red clubhouse of the Echo Bay Yacht Club on Echo Island. A municipal marina is located in Echo Bay.

Off New Rochelle, you'll spot one of the area's most infamous lighthouses. **Execution Rocks Lighthouse**, built in 1867, marks a treacherous spot in Long Island Sound where many vessels and lives have been lost.

The biggest surprise of a cruise in New York is **City Island**, a yachting center at the apex of the Metropolitan area! Connected to the mainland by a single bridge, City Island is a one-and-a-half mile oasis for boatsmen, packed with yacht clubs and marinas, marine suppliers, marine services, and restaurants catering to every crew's tastes. If you're reluctant to try just any place, ask a local "clam digger," as the natives are known—no two will agree on a recommendation! Visit the North Wind Nautical Museum (718) 885-0701 with its fine collection of whale exhibits and nautical art. Also on "the avenue" is the City Island Public Library (718) 885-1703. With more than 2,000 volumes covering the past, present, and future of sailing and the sea, it has the largest collection of maritime literature in New York City.

When navigating the water west of City Island toward **Eastchester Bay**, the most notorious features to watch for are "Big Tom," a rock that bares itself at low water, and Cuban Ledge. The well-charted approach from Long Island Sound brings you to the deepwater City Island Harbor between the east shores of City Island and Hart Island. Known locally as Hart Island Roads, the harbor serves as a harbor of refuge and is well protected from both easterly and westerly winds. Transient berths are located on this side of City Island, as well as on the western side.

You can now choose to cross over to the North Shore ports of Long Island, or journey through New York City via the East River and connect to the Hudson River for points north or south. Whatever your next port of call, take time to cruise by the Throgs Neck Bridge for a dramatic view of man and nature coming together.

Throgs Neck is where prudent captains can spend some time while waiting for fair currents through the infamous Hell Gate to New York City. Here is where Long

Island Sound technically ends and the East River begins, at the breathtaking 138-foot Throgs Neck Bridge that gracefully swoops from shore to shore across the river. While the area doesn't offer too many attractions accessible by boat, there are plenty of landmarks to view. On Throgs Neck you can see the impressive Fort Schuyler, which in 1938 became the State University of New York Maritime College (718) 409-7276. Built to protect New York City from Spanish raids, construction of the fort was not completed until 1856—long after the danger had passed. Often the huge training ship of the college is tied up alongside the docks west of the bridge.

Coming from Eastchester Bay, a popular anchorage is located in the enclosed "bay" formed by the north span of the bridge as it crosses between Locust Point and the north side of Throgs Neck. Although somewhat open to winds from the east, it is well protected from the prevailing southwesterlies, and offers some relief from the heavy traffic in the area. The center span of this section of the bridge has a vertical clearance of 60 feet. You can use almost any of the channel ways under the bridge, except the first five sections on each end where shoals extend from the shores.

Little Bay is across the river from Throgs Neck and west of Willets Point. The western side of the bay is a good location for anchoring, but stay clear of the approach ramp for the bridge, which marks the beginning of shallow water. **Little Neck Bay**, just east of Willets Point from Little Bay, offers some excellent anchorages. The U.S. Merchant Marine Academy at Kings Point (516) 773-5527 is on the eastern shore of this bay and the prominent granite walls on its western shore are of the closed Fort Totten, which now houses the Fort Totten Coast Guard station. The grounds of the Merchant Marine Academy and its National Maritime Hall of Fame are open Saturdays and weekdays, except during the month of July. Facilities for transient boats are limited because of the many private landings here and the extensive Merchant Marine fleet. A small-craft facility is located on the bay's west side. Midway along the east side of the Bay, a cove provides 2.5 to 7 foot depths at mlw.

The East River and New York Harbor

The **East River** makes its way along the eastern side of Manhattan, connecting Long Island Sound with the Upper Bay of New York Harbor at the Battery. A tidal strait, the river has strong currents that can cause choppy or rolly conditions when influenced by heavy winds. The effect is heightened because the river currents run opposite the currents in the Sound: The East River current sets eastward at flood and westward at ebb, while Long Island Sound currents set westward at flood and eastward at ebb. Although the river is 14 miles from its entrance to the Battery, the chaotic mix of waters is most evident at Hell Gate, which is just northeast of Roosevelt Island and southwest of Wards Island, where the East River meets the Harlem River. Here the tide can swirl itself into two-foot waves, with crosscurrents and whirlpools ready to set a small craft spinning.

Navigating the East River must be carefully planned with a current chart. If you are heading westward from Long Island Sound, time your arrival at Hell Gate with the end of the Long Island Sound flood (slack current). When traveling eastward, it is best to plan your passage through Hell Gate on the river's flood current, catching the eastward ebb of the Sound. Although you can pass to either side of Roosevelt Island, the western channel is the wider and generally preferred, if only for its views!

On the East River there are many landmarks. Seven bridges cross the river at towering heights. The famous Brooklyn Bridge is the last of these as you approach the Battery. Just after the Brooklyn Bridge, the South Street Seaport (212) 669-9424 makes the Manhattan waterfront a wonder to behold from the river. At any time there may be a mix of unique vessels of different vintages, some open for tours, some being restored at the Seaport, and some just visiting. Transient vessels may tie up at the Seaport's Pier 15 when activities at the Seaport allow. Reservations are a must.

Before entering **New York Harbor**, check your charts and familiarize yourself with the names and directions of the channels, bays, and various landmarks. New York Harbor is deep, but its water can be misleading to the unwary. Strong currents follow the tidal pathways, and eddies are often created by the winds that channel between buildings. New York Harbor is divided into three sections: the Lower Bay, The Narrows, and the Upper Bay. **The Lower Bay** is the water that extends from Sandy Hook, New Jersey, west to Raritan Bay, and north to The Narrows. **The Narrows** is the aptly-named narrow portion of the bay crossed by the Verrazano Bridge. On the charts, it is the area between Fort Wadsworth on Staten Island and Fort Hamilton in Brooklyn and ends where the anchorage channel widens to meet the Upper Bay. **The Upper Bay** is the water between The Narrows and the Battery and is the location of the Statue of Liberty, Ellis Island, and Governor's Island, famous for its military bases and now the primary Coast Guard Station for the Northeast.

Coming from the **East River,** you'll enter the Upper Bay from the east. The Upper Bay actually leads into the Hudson River in the shadow of Manhattan. This busy junction has perhaps the most breathtaking sights. On one side, you have the towers of the World Trade Center and the other skyscrapers of the city; on the other, a unique perspective of the Statue of Liberty and Ellis Island. Don't get too carried away with the sights, however, as this waterway is also filled with ferries, tugs, barges and other vessels of such size that they demand your full attention. Remember that commercial traffic has the right of way, and these vessels use VHF Channel 13, so monitor it yourself for updates on traffic. The famous Staten Island Ferry traverses a route from the tip of Manhattan, the area known as the Battery, to the southernmost of the five New York City boroughs, and there always seems to be one crossing the channel at a rapid pace. Facilities for smaller craft are very limited in New York Harbor, but as you journey up the Hudson River, you will find the dockage and services you need.

Hudson River: The Battery to the Troy Locks

The **Hudson River** begins high in the Adirondack Mountains of the north, and flows almost 275 miles to the Battery at the southernmost tip of Manhattan in New York City. Navigation of the Hudson is less daunting than you might imagine; it is straightforward with hazards marked on large-scale charts. Just keep a lookout for floating debris, especially after high tides and heavy rains, and watch for fish traps, particularly in the spring.

Most facilities for coastal cruisers in the Manhattan area are located on the western shore in New Jersey. We recommend **Port Imperial Marina**, in Weehawken, N.J. Just as you pass the Empire State Building and directly across from the aircraft carrier *Intrepid,* you'll see the marina on the New Jersey shoreline. Port Imperial Marina is a full-service facility with 24-hour security, the marina also offers a restaurant, golf center, café, and ferry service to and from Manhattan (a four-minute crossing and the only civilized way to approach New York).

Of course, the list of things to do once you're in Manhattan is endless, but be sure to include a visit to the observation deck at the 107th story of the World Trade Center (212) 435-7377. For a circumnavigation, take a guided cruise aboard a Circle Line (212) 563-3200. Access to Miss Liberty and Ellis Island are by ferry from the Battery in Manhattan or Liberty State Park in New Jersey (201) 915-3400. For more information on these and other events, call the New York Convention & Visitor's Bureau (212) 397-8222.

Proceeding north on the Hudson River, the heights of the Palisades on the western shore will begin to compete with the majesty of the George Washington Bridge (the current at the bridge can be strong). The rock cliffs range between 200 and 540 feet, and run about 15 miles north to just below the Tappan Zee Bridge. Be sure to have your camera ready when passing Englewood Creek. This stream ends in a surprise tumble down the rocky side of the Palisades—an unexpected view in the middle of urban America!

You'll also find some marine facilities on the western shore in New Jersey. Dockage can be found at **Tubby Hook** on the western shore just past the bridge, which will allow you to visit the Cloisters at Fort Tryon Park (212) 923-3700. This is a fascinating restoration of medieval monasteries that have been brought to America piece by piece. A branch of the Metropolitan Museum, its medieval collections of art and artifacts are unequaled and well worth a visit.

Spuyten Duyvil Creek on the eastern shore leads to the **Harlem River** and ultimately to the East River. Continuing north on the Hudson, you'll see densely populated **Yonkers** spreading over the eastern shoreline, but marine facilities are limited.

Upriver, the village of **Hastings-on-Hudson** offers more for transient boaters with shops, markets, and restaurants, all within walking distance, albeit a long walk. Here, the urban sprawl begins to lessen and hence, more historical sites remain in place. The first landmark brought to mind is a picturesque old castle, now a private school, perched atop the mountain overlooking the river.

On the west side of the river three miles north, the village of **Piermont** is nestled at the base of the steep hillside with marinas and restaurants on its shores. From here, you can explore picturesque Sparkhill Creek by dinghy. When approaching Piermont, be sure to stay in the channel on the north side of the long pier. The south side of the pier has many submerged obstructions remaining from the old ferry dock and other pilings. Watch your depths on both sides of the pier, as this is not a deep area of the river.

From Piermont, the Hudson widens to almost two and a half miles for a distance of more than 12 miles. Henry Hudson's men first named this "Zuyder Zee" thinking the width of this body of water resembled their own small sea. Today, it is called Tappan Zee ("tappan" is an Indian word for "cold spring"; "zee" is Dutch for "sea") and is crossed by one of the longest bridges in the United States, boasting a length of almost three miles. Few vessels have to be concerned about vertical clearances, though, as the center span has a height of 139 feet, and the east and west spans are 123 feet.

Tarrytown, on the east shore, provides the first developed waterfront area above the Tappan Zee Bridge. This is the region of magical hills and valleys that inspired Washington Irving to write *Rip Van Winkle* and *The Legend of Sleepy Hollow*. Irving's grave is in the Old Dutch Church burial ground on the way to **North Tarrytown,** which you can reach by car. Within sight of the Old Dutch Church is the site of the second of the Sleepy Hollow Restorations, Philipsburg Manor, Upper Mills (914) 631-3992. The restoration is maintained as an authentic replica of pre-1750 Hudson Valley life featuring costumed guides who lead you through its exhibits. You can also tour Tarrytown Lighthouse. It was first lit in 1883, but has been inoperative since 1965. Navigating into Tarrytown is easy. Barges are anchored off Tarrytown as protection from the surges of river traffic, and two well-marked channels can accommodate boats with up to a 6-foot draft. The southern channel begins at the eastern span of the Tappan Zee Bridge at green "1" and red "2" and is marked by a lighted set of range marks. The northern channel starts at flashing red 4sec "2a."

Cruising north of Tarrytown, the majestic heights of the west bank come right to the water's edge. The Riceland Lake Light (flashing 6 sec) warns you off submerged ruins on the west shore. On the east shore, you'll pass the infamous Sing-Sing Prison, marked by its two water towers. As you follow the river northward, the main channel is well marked.

Haverstraw Bay is the name given to the 15 miles of the Hudson River extending north of Croton Point to Stony Point. The river is at its widest here, stretching to 3 miles across. The main channel in the bay favors the western shore, while depths of 6 to 9 feet over a soft mud bottom are to the east. The largest number of marinas are on the western shore and are located both below and above Grassy Point.

Stony Point, the high rocky headland that towers over the river at the top of the bay, was a strategic spot during the Revolutionary War. General "Mad Anthony" Wayne fought a major battle from the fort that was located here. The Palisades Interstate Parks Commission maintains the Stony Point Battlefield Museum (914) 786-2521 that commemorates this 1779 event where a reenactment is staged each year. In the park, the Stony Point Lighthouse, the oldest lighthouse on the Hudson River, can be visited during park hours. To navigate Stony Point Bay, stay between the privately maintained poles until near the shore and deeper water.

Bannerman's Castle on the Hudson.

As it narrows to less than a third-of-a-mile across, in the area known as the Hudson Highlands, we see yet another side of the Hudson River—a more intimate side. This begins at the 1,000-foot Dunderberg Mountain, the legendary home of a troublesome Dutch goblin who is responsible for the fierce summer storms that thunder around the river valley. Just before **Peekskill** on the east shore, the Indian Point Atomic Power Plant is a prominent landmark. Take extra caution as you approach the sharp turns in the river here, especially Jones Point at the base of Dunderberg Mountain on the west shore. Services can be obtained at **Peekskill Bay** by following the U-shaped channel to the marinas. The bay is shallow, so stay within the channel.

After rounding Jones Point, Bear Mountain State Park (914) 786-2701 lines the shores with its centerpiece, the 1,305-foot-high Bear Mountain. The park offers a museum and zoo, pool, hiking trails, picnic areas, and other recreational activities. Wharves are available at Day Line Park for transient vessels to tie up, but space is limited. Anchoring for park visitors is allowed between the Day Line excursion boat dock and Iona Island.

Traveling north, you will soon find yourself in a narrow section of the river known as "The Race." As you pass under the 155-foot Bear Mountain Bridge, the river turns to starboard, and its beautiful wooded west shore is the beginning of the famed West Point Military Reservation (914) 938-2638.

The view of the United States Military Academy at **West Point** (914) 938-4011 is probably the most spectacular you'll see on the river. You can appreciate the strategic location of the facility, built in 1802 on the spot known as the Gibraltar of America, as you make your way along the river. The Academy's grounds are filled with monuments, and the museum has extensive collections of military interest. The West Point Yacht Club may have dockage or mooring space available, and is under the authority of the Military Harbormaster who monitors VHF Channel 16. This spot is exposed to river traffic, though, and is a bit bouncy. Several marinas south of the Academy are a long walk, or short ride, away.

Connected by the twin spans of the Newburgh-Beacon Bridge, the cities of Newburgh on the west and Beacon on the east flank the river. General George Washington made his headquarters in **Newburgh** at the Jonathan Hasbrouck home (914) 562-1195. Newburgh was actually a whaling port in the early 1800s, and many of the Hudson steam vessels were built at the Marvell Shipyard here. **Beacon** got its name for the signal fires on its promontory, which warned Revolutionary soldiers of the advance of British troops. At Beacon, you can go to the top of the north peak of Mount Beacon, an elevation of about 1,530 feet, by way of a unique railway designed and built by the Otis Elevator Company. History buffs can visit the Van Wyck Museum (914) 896-9560, a 1732 homestead used by the Revolutionary Army.

Traveling north from Cold Spring, you'll see a medieval-style castle dominating **Pollepel Island** (or Bannerman's Island, as it's known locally) off the east shore. Across from Pollepel Island is Cornwall-on-Hudson (marine facilities are limited here). From Cornwall to Chelsea, the river is known as Newburgh Bay, and widens to about 1.5 miles. Marine facilities can be found on both shores along Newburgh Bay, although the southern end of Newburgh's shoreline is mostly commercial.

Poughkeepsie has been called the "Queen City" of the Hudson. Here also begins a 16-mile stretch filled with the former (and present) estates of the rich and famous, as well as many elite colleges and private schools. In Poughkeepsie itself, a restored area has restaurants and shops amid gas lights and brick walkways. Vassar College is in Poughkeepsie, and just north of the city is the famed Culinary Institute of America (914) 471-6608, where you can make reservations for lunch or dinner prepared by tomorrow's four-star chefs. Your sightseeing itinerary in this area can include Springwood, the home of Franklin D. Roosevelt (914) 229-2501, and nearby is Val-Kill (914) 229-9115, the retreat of Eleanor Roosevelt, which later became her permanent home. Here, too, is the Vanderbilt Mansion (914) 229-7770, a palace with stunning views of the Hudson and the Catskill Mountains. A stop at any number of shore facilities on the east shore just north of Poughkeepsie itself will give you easy access to these sites, either by taxi, rental car, or even by foot.

Poughkeepsie is located approximately 11nm after the Newburgh-Beacon Bridge at the next bridge on the river, the Mid-Hudson Bridge. This is one of two bridges off Poughkeepsie, both with ample clearance for recreational craft. The city is located on the eastern shore.

Kingston was the first capital of New York State, the site where the state constitution was drafted and ratified, and where the New York Senate first convened in 1777. Although the fortification is long gone, the historic district called the Stockade still has a few historic stone houses remaining. The waterfront has been largely restored, and the centerpiece of it is the Hudson River Maritime Museum (914) 338-0071. A floating dock with utilities is available for museum members to tie up to, even for an overnight stay. The nominal membership may be purchased on the spot, allowing instant access to the museum's facilities.

Kingston lines the northern banks of **Rondout Creek**, a tributary that empties into the Hudson at a large, natural harbor. The entrance to the tributary is well marked by flashing green 4sec "1" to port and flashing red 6sec red horn to starboard. A fixed bridge with a 56-foot clearance crosses the creek about one mile from its entrance. The creek has a controlling depth of about 13 feet to Eddyville and is somewhat more shallow from there on. Several small craft facilities offer transient services along the creek, and you will find this to be a pleasant, quiet area in which to spend some time.

Beyond Kingston, the west shore of the Hudson gives you a taste of the lovely Catskill Mountains, which have attracted explorers, settlers, artists and vacationers for years. As you continue to this region of the river, "The Flats," a large area of shoals, begins to form in the middle of the river and continues northward. Although the river runs deep on either side of The Flats, the channel to the west is usually favored, except around Cruger Island, where the east shore offers better passage.

Esopus Creek enters the Hudson at the west shore just above quick flashing green "39." On one of the two dikes protecting the entrance, the restored Saugerties Lighthouse stands. Esopus Creek is a quiet place to stop, mostly to relax. The town of **Saugerties** is a short walk from the few marinas on the creek and is filled with old homes and churches.

Located 1nm before the Rip Van Winkle Bridge and 9nm north of Esopus Creek, you'll come to **Catskill Creek**, which leads to the quaint the town of **Catskill**. This is an excellent area to stop and enjoy the peaceful and picturesque setting of the region. Plan a stay at **Catskill Marina**. The marina offers a park-like setting, as well as a pool and full services. It is a short walk, a mere two blocks, to the town of Catskill and its restaurants. Entrance to the creek on the west shore is at flashing green 4sec "67," and the marina is only 0.7nm away.

The capital region encompasses the cities of Albany, Troy, and Saratoga as well as the smaller cities on both shores. This area is rich in history dating back to the Dutch settlers, and alive with cultural activities. Most sights are easily reached by public transportation, taxi, or rental car. In addition to the attractions that follow, stop in at the visitor's center in each city.

The first area you'll come to on the river is **Albany**, where you'll find no shortage of things to do. Be sure to include a visit to the Empire State Plaza, the hub of both state government and the city itself. Guided tours take you through the New York State Capitol Building (518) 474-2418 and include the State Assembly and Senate Chambers, the Governor's Office, Hall of Governors, and a military museum. Albany is located just after the Dunn Memorial Bridge, which is a fixed bridge with a clearance of 60 feet.

Just upriver, **Troy**'s Urban Cultural Park also offers self-guided and guided walking tours. Don't miss the Troy Savings Bank Music Hall (518) 273-0038, acclaimed as one of the finest acoustical theaters in America. Nearby Saratoga Springs has remained a summer center for the "Sport of Kings." If the horses are not racing during your visit, the National Museum of Racing & Thoroughbred Hall of Fame (518) 584-0400 offers decades of history and memorabilia related to the sport. The Walworth Museum/Casino & Saratoga Historical Society (518) 584-6920 highlights the opulence of the 19th-century gambling casino and mineral spring resort that made the area famous, and the Saratoga Performing Arts Center (518) 587-3330 offers performances of today's world-class performing artists.

The ideal place for transient vessels to find dockage in the capital region is at **Troy Town Dock & Marina,** located at the newly restored bulkhead in Troy. This facility is just below the Troy Federal Lock in the New York State Barge Canal System. It provides a facility to step or unstep your mast and purchase traditional straw fenders to protect your hull if you're heading north on the canal system. A stay here will put you within convenient walking distance of the historic district while your boat is equipped for your upcoming trip, whether it be farther inland or back down the beautiful Hudson River.

Long Island's South Shore: Shinnecock Bay to Coney Island

Protected by barrier beaches, the bays along the South Shore of Long Island are generally shallow. You should have the largest scale charts possible and take advantage of local knowledge whenever possible. While following the Inland Waterway and the State Boat Channel, remember that the buoys are set with Inland Waterway markings, so if you are traveling west, the red buoys will be left to port. The red-right-returning rule will apply as usual when entering inlets or harbors. Your best bet—always keep your charts handy.

TROY TOWN DOCK AND MARINA, INC.

10 p. 160

427 River St., Troy, NY

518-272-5341

at Troy seawall
1/4 mi. S. of Federal Lock

At Troy Town Dock you'll find the ideal Hudson River dockage to prepare for transit through the New York State Barge Canal System or as a home port to see the sights in the Capital-Saratoga region. Our 1400 feet of new floating docks on Troy's restored waterfront are outfitted with features to make your visit to the area pleasant.

Our specialty is readying vessels for canal passage or cruising on the Hudson. Masts up to 55 feet can be unstepped or stepped and our ships' store stocks straw fenders, charts and other necessities. Complete repair services, diving, towing and salvage are available.

Provisions, shopping and dining are conveniently located within walking distance in the historic waterfront area. You and your crew will enjoy your stay here!

- Albany Tides; High: + 0:08, Low: + 0:10, Mean 4.7'
- $1.00 / ft per day ($30 min) 1400' linear ft.,200' LOA
- Prop, Sail, Elec, Rigging, Gas, Diesel, Wood, Fiberglass
- *Coastal* Gas, Diesel
- Fresh Water, Phone, Cable TV 110V, 220V, 100 Max Amps.
- Many within walking distance
- Castaway 518-273-2244
- Mastercard, Visa,

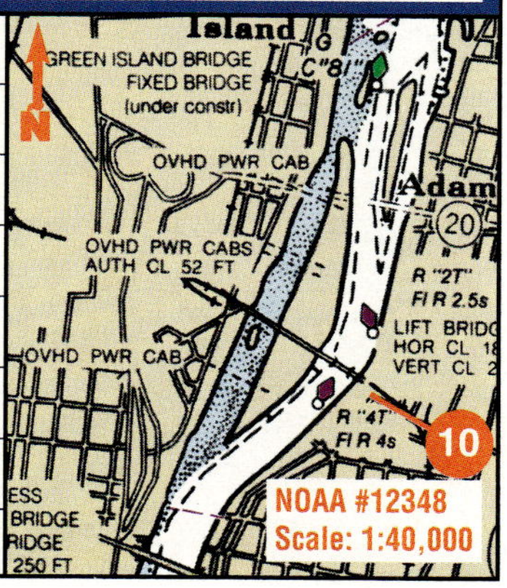

Southampton was settled in 1640 by a group of disgruntled pilgrims from Lynn, Mass. The Halsey Homestead (516) 283-3527, dating from 1648, is a rare example of colonial Southampton, and the Southampton Historical Museum (516) 283-2494 gives a good picture of early life in the village. But today, posh shops cater to the inhabitants of grand mansions. The primary yacht harbor for Southampton is actually on the Great Peconic Bay side, but many choose to reach the town by way of **Shinnecock Bay**, known for its sportfishing.

Access Shinnecock Bay via **Shinnecock Inlet**, the first inlet from the Atlantic to the inside route on Long Island's South Shore. The inlet shoals each time a strong storm blows through and is subject to extreme tidal currents. It is imperative to have local knowledge if you choose to use the inlet, and favoring the west side is advised. From the 36-foot flashing green 2.5sec light on the breakwater, stay close to the west shore where Shinnecock Light is located. After clearing the point, head west and pick up the channel buoys at flashing red 4sec "8." At flashing red 4sec "24," bear northeast toward Southampton. There are several marinas on the bay that will provide services and transient dockage.

Along the north shore of **Moriches Bay**, boaters will find many creeks, coves, and inlets to explore by dinghy or shallow-draft boat. The Long Island Game Farm & Zoo (516) 878-6644, with rides and a collection of exotic and domestic animals, is a short taxi ride from any of these ports. Heading to **Moriches Bay**, which offers the best transient facilities in the area, pick up the Quantuck Canal at the southwest corner of Quantuck Bay, and follow the channel markings. There are two bascule bridges that open on demand between 8:00 a.m. and 10:00 p.m., May through September.

If traveling the outside route, **Moriches Inlet** is located 44nm from Montauk, south of the shoals that protect **Tuthill Cove**. Every year, storms change the entrance channels, so they are unbuoyed and very risky to navigate. If you must use this inlet for passage, you should put in a call to the East Moriches Coast Guard Station (516) 395-4400 for an update on conditions.

The south shores of Moriches Bay and Great South Bay are formed by the 32-mile **Fire Island National Seashore**. You can wander along the many paths beneath a canopy of holly trees and enjoy the variety of preserved wildlife. Smith Point West, accessible by bridge, offers boardwalk nature trails. The Fire Island Lighthouse on the western end of the park has a visitors center, museum, and nature trail.

Many of the marinas are part of private communities on the island. There is a marina with transient dockage at **Watch Hill**, 12nm west of Moriches Inlet. This is part of the National Seashore park, and you'll also find a campground and restaurants. Enter the channel to Watch Hill at flashing green 4sec "1." Depths are between 5 and 6 feet depending on the tide.

Patchogue is the largest town on the north shore of the **Great South Bay**, the biggest of the South Shore bays. The town is filled with shops and is the major shopping spot for most of the area. You will immediately realize that this is not your sleepy resort town! Patchogue is home to the headquarters of the Fire Island National Seashore (516) 289-4810, which can give you detailed information about the barrier islands. Patchogue has two waterfronts, one on the bay, and one on the **Patchogue River**, but most choose the river. The entrance to the channel is almost directly across from the Watch Hill channel at flashing green 4sec "1." A privately maintained 21-foot 2.5sec green flasher marks the entrance to the river, but keep in mind that depths here are charted as low as 5 feet.

Fire Island Inlet, across from **Babylon**, is the passage between the Atlantic Ocean and Great South Bay. The Coast Guard advises that this is an unsafe inlet due to strong currents.

Oakdale is where many of the great turn-of-the-century country estates are located. You can catch a glimpse of the past at Bayard Cutting Arboretum (516) 581-1002, a superbly designed mansion and park built by a railroad magnate. The millionaires who spent summers in Oakdale enjoyed good sport, and the grounds of their shooting preserve are now the Connetquot River State Park (516) 581-1005. The park and grounds have extensive walking trails and a variety of botany and wildlife programs, but to tour the 1880 shingle-style clubhouse or the operational 17th-century grist mill requires an appointment. **Oakdale Yacht Club** is the premier facility for transient vessels in Great South Bay. Located in the northwestern part of the bay on the northern shore of the Connetquot River, Oakdale caters to your every need with all the amenities of a first-class resort. It's convenient to Fire Island's sandy beaches and shopping.

The town of **Bay Shore** on **Great Cove** provides a secure basin protected by a bulkhead spit. Fishing is the main activity in this harbor, and several creeks provide marine facilities. Transients can usually find dockage or services along West Creek (the local name for the combined Watchogue and Pentaquit Creeks).

At **Captree Island**, the water outside of the channel becomes extremely shallow, so the best course is to stay in the State Boat Channel, which runs along the barrier beaches. The channel will take you through **South Oyster Bay** and past **Zachs Bay** to Jones Beach State Park (516) 785-1600. Jones Beach is one of the state's largest and one of the most famous beaches in the Northeast. There are facilities in **Middle Bay** and **Hempstead Bay**, but the channels are constantly changing, and tidal currents are strong. For access to the park and beach, the eastern part of Zachs Bay offers a sheltered, secure holding ground.

Jones Inlet is the point where many cruisers will take their vessels outside on their journey west and south. The inlet is not recommended for smaller vessels or inexperienced captains, but it does avert the tedious passage through **Reynolds Channel** to **East Rockaway Inlet.** Vessels needing more than 14 feet of vertical clearance that continue into the State Boat Channel will have to turn north at Zachs Bay to bypass the fixed bridge at Jones Beach Causeway. Follow the channel around Green Island, east of Snipe Island, and rejoin the State Boat Channel at green can "27."

At the confluence of the State Boat Channel and Jones Inlet, you can turn north at red nun "12" to approach **Freeport**, considered one of the finest harbors in Nassau County and offering more than 40 marinas. Guy Lombardo (the late band leader) was a town resident and has given to the community in many ways, including his name to the avenue along which most of the marinas are found. Once a fishing village, suburbia has taken over this area, but some of the charm of its origins remain near the shoreline. This is a good port of call if you want to explore some of the inland attractions of Long Island.

Old Bethpage Village Restoration (516) 572-8400 to the northeast of Freeport is a restoration of a pre-Civil War community. The only things you'll need are your sense of history and lots of time. Horticulturists shouldn't miss the hundred-plus acres of gardens, fields, and woods at Old Westbury Gardens (516) 333-0048. The grand house was built in 1906 and the original framed paintings and

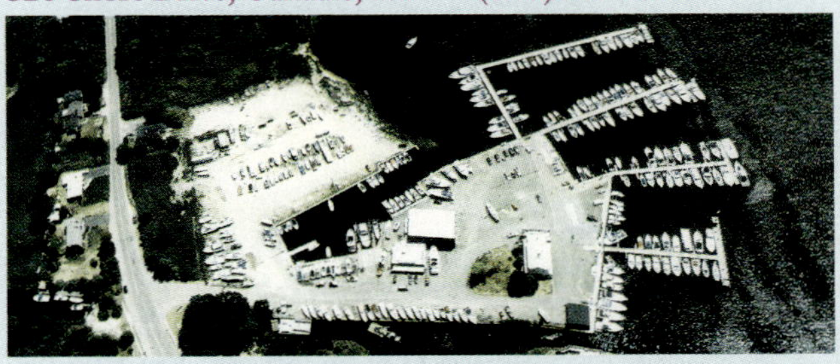

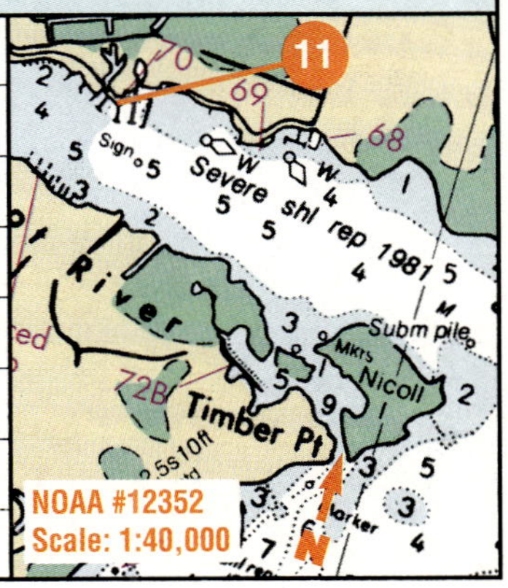

© Embassy

Looking at Shinnecock Canal and Bay from the north.

antiques remain. A full schedule of lectures and concerts is held year long, with lawn concerts in the summer.

Approaching Freeport from Jones Inlet, you will pass a bascule bridge as you head into Long Creek. The route is winding but well marked all the way to Freeport. The numerous marinas are located all along the waterfront. After your inland excursions in Freeport, rejoin Reynolds Channel by following the shoreline west around Middle Bay and follow the channel through Garrett Marsh and East Channel Islands.

Reynolds Channel follows the lower part of Hempstead Bay, running about 8nm between Jones Inlet and East Rockaway Inlet. **Long Beach** is the name of both the barrier beach itself and the resort town, which offers plenty of facilities for boaters that you'll see as you travel west along the channel. This is a good stop to unwind from your cruising; there are tennis courts, a five-mile boardwalk, a fishing pier, and a town made to serve out-of-towners.

East Rockaway Inlet is the westernmost inlet along the South Shore, and it generally has the safest conditions. While it is subject to changing conditions, depths are generally about 12 feet. The entrance is marked by red and white "ER" Morse (A) bell, and red gong "4" stands off from the eastern jetty marked by the 33-foot flashing red 4sec light. If you're in need of a temporary stopover, the Gateway National Recreation Area (718) 338-4200 operates a marina on **Barren Island** in well-protected Dead Horse Bay. The Plumb Beach Channel guides you northeast toward its entrance. Slips are available on a first-come first-served basis.

On your approach to New York Harbor, you'll see **Coney Island**, home to the world-famous amusement park, as well as Nathan's hot dogs. Its vast structures and rides are colorful by day and ablaze with sparkling light at night making for good landmarks. You can travel to Coney Island by car from Barren Island, or secure a berth in Sheepshead Bay, behind Coney Island, or Gravesend Bay, north of Coney Island.

Whether you arrived by way of the open Atlantic or left the inland waterway at Rockaway Inlet, the area just south of Coney Island marks a crossroads. You can turn north through The Narrows and up the Hudson River, head west into Raritan Bay, or swing south to Sandy Hook to pick up the inland route along the Jersey shore. Whatever route you choose, this confluence of rivers and bays shows the diversity of New York cruising.

Shoreside and Emergency Services

Ambulance, Fire, Police: 911

Coast Guard:

Montauk Station, Montauk Harbor, Long Island
(516) 668-2773 or VHF 16

Shinnecock Station, Ponquogue Point, Long Island
(516) 728-1171 or VHF 16

Moriches Station, Long Island
(516) 395-4421 or VHF 16

Fire Island Station, Fire Island, New York
(718) 661-9101 or VHF 16

Jones Beach Station, Jones Beach, Long Island
(718) 785-2995 or VHF 16

Rockaway Station, Rockaway Beach, Long Island
(718) 634-2848 or VHF 16

Sandy Hook Station, Sandy Hook, New Jersey
(908) 872-0300 or VHF 16

Shark River Station, Shark River, New Jersey
(908) 775-5029 or VHF 16

New York Station, Governors Island, New York
(212) 668-7935 or VHF 16

Fort Totten Station, Little Bay, Long Island
(718) 352-4422 or VHF 16

Eatons Neck Station, Huntington Bay, Long Island
(516) 261-6868 or VHF 16

New Haven Station, New Haven, Connecticut
(203) 773-2400 or VHF 16

New London Station, New London, Connecticut
(203) 442-4471 or VHF 16

Radio Telephone:

VHF Ch. 85; I.D. KZN 548, Bay Shore N.Y.
VHF Ch. 87; I.D. KMC 972, Central Long Island
VHF Ch. 86; I.D. WHU 638, New York, N.Y.
VHF Ch. 28; I.D. KLU 786, Riverhead, N.Y.

VHF Ch. 28; I.D. WHU 738, Staten Island, N.Y.
VHF Ch. 27; I.D. Bridgeport Marine Operator
VHF Ch. 26, 86; I.D. New London Marine Operator
VHF Ch. 87; I.D. KMC 972, Old Saybrook, Conn.
VHF Ch. 24; I.D. KQU 556, Navesink, N.J.

Tow Service:

Sea Tow Central Hudson, New Windsor
(914) 561-0933 or VHF 16
Troy Town Dock, Troy
(518) 272-5341 or VHF 16
Sea Tow Eastern Long Island, Southold
(516) 765-5300 or VHF 16
Sea Tow Moriches-Shinnecock, Westhampton Beach
(516) 653-6361 or VHF 16
Sea Tow Great South Bay - Fire Island, Oakdale
(516) 563-3391 or VHF 16
Sea Tow Freeport, Freeport
(516) 623-4183 or VHF 16
Sea Tow Lower New York, Brooklyn
(718) 338-3093 or VHF 16
Sea Tow Western L.I. Sound, City Island
(718) 885-0101 or VHF 16, 9
Sea Tow Harbor Marine, Mamaroneck
(914) 698-6523 or VHF 16, 9
Sea Tow Port Kar, Greenlawn
(516) 754-1545 or VHF 16
Sea Tow Port Jefferson, Port Jefferson
(516) 473-2869 or VHF 16
Sound Tow Co., City Island
(718) 885-3420 or VHF 16, 9
Douglass Marine Towing, Orient
(516) 323-2645 or VHF 16
Scopinich Towing & Salvage, Hampton Bays
(516) 728-2743 or VHF 16
Dickson Marine Inc., East Patchogue
(516) 475-1445 or VHF 16, 9
Wagner Marine Service, Patchogue
(516) 475-4343 or VHF 16
Reynolds Towing & Marine, Bayshore
(516) 667-9849 or VHF 16
Reliable Marine Towing, Babylon
(516) 587-8346 or VHF 16, 9
Marine Towing & Salvage, Pt Lookout
(516) 431-3179 or VHF 16

1 inch= 17 miles
Scale 1:1,077,000
0
17
34
51
68
NY
Hudson River
George Washington Bridge
1
Weehawken
Jersey City
Newark
Elizabeth
Long Island
Staten Island
NEW JERSEY
Joins page 158
Raritan Bay
Sandy Hook
U S C G (908) 872-0326
Delaware River
Long Branch
U S C G (908) 775-5029
Shark River Inlet
PENNSYLVANIA
Trenton
2
Pt. Pleasant
Manasquan Inlet
VHF-FM KIH-28 162.475 MHz
U S C G (908) 899-0130
Philadelphia
Camden
U S C G (609) 494-2661
Barnegat Bay
Delaware River
Manahawkin
Barnegat Inlet
C & D Canal
U S C G (609) 492-5751
3
Beach Haven
Little Egg Inlet
VHF-FM KHB-38 162.40 MHz
N
Bridgeton
Atlantic City
4
Absecon Inlet
U S C G (609) 344-6594
Great Egg Inlet
Delaware Bay
DE
5
ATLANTIC OCEAN
U S C G (609) 898-6981
Cape May Canal
Wildwood Crest
Cape May
Cape May Inlet
Joins page 190

The Garden State

While the face of New Jersey may gaze to the West, this state's back anchors it steadfastly to the East Coast. The spine of wetlands and barrier islands is bona fide Atlantic coastline. Its ocean waves attract an onslaught of tourists. Tourists require amusements and boardwalks and houses. What you've got is the Jersey Shore.

With its rugged Atlantic waters, New Jersey has a fascinating maritime background as well. There are some 200 sunken wrecks within a 50-mile radius of the Manasquan Inlet. And local museums are full of old photographs from the days when lifesaving stations dotted coastal beaches and surf boats were rowed out to rescue the tall-ship sailors who carried on the nation's commerce.

But there's a calmer face to this shoreline, too. There are the miles and miles of unspoiled wetlands lining the estuaries inside the barrier islands. Two decades of efforts by the state to clean up its waterways and stricter federal regulations on ocean dumping have greatly improved the quality of water in and around most of New Jersey. Local skin divers talk about the increasing variety of fish and the increasing numbers of striped bass and mussels around the jetties along the Jersey Shore.

Many boaters still take the offshore route following the sea buoys down the coast of New Jersey, a viable option particularly for large boats. There are a number of inlets to duck into along the coast, but some have been rendered impassable by the ever-changing topography. Others require careful navigation, especially due to large swells and strong currents. On the other hand, if you travel inside the barrier islands for much of the coast, keep in mind that some areas are shallow and boats with more than a 4-foot draft may have trouble at low tide.

Visitors enjoy the Victorian resort town of Cape May.

Courtesy New Jersey Division of Travel and Tourism

While the 110-nautical-mile ocean coast makes up the preponderance of the state's waters, there are two other important boating areas in New Jersey. In the south and west, Delaware Bay and the Delaware River line the

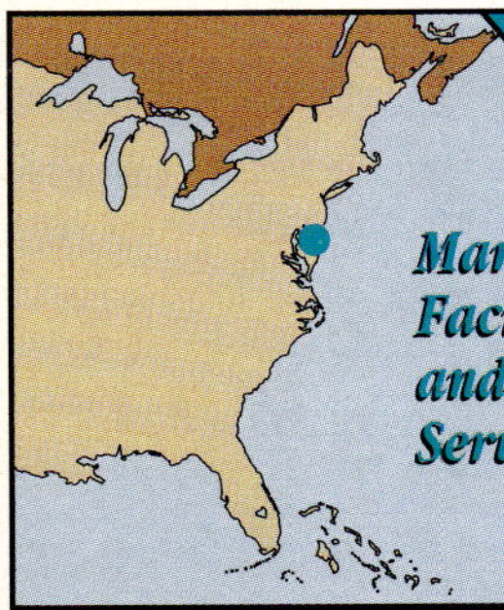

Marine Facilities and Services

See Adjacent Chart for Coast Guard Information

		DOCKAGE								SERVICES					SUPPLIES				AMENITIES			COAST GUARD
Facility	Phone	Monitors VHF Channel	Working Channel	Number of Transient Berths	Number of Transient Moorings	Seasonal / Year-round	Maximum Length Over-All	Approach Depth in Feet at MLW	Dockside Depth in Feet at MLW	Hookups: Fresh Water / Phone / Cable TV	110V★ 220V▲ 3 Phase■ Maximum Amps	Ramp / Dinghy Dock / Launch Service	Railway / Lift / Crane: Capacity (tons)	Repairs: Propeller / Sail / Electronics / Rigging / Gas / Diesel / Wood / Fiberglass	SCUBA: Air / Rentals / Equipment / Charters	Gas / Diesel Fuel	Fuel Brand	Marine / Groceries / Bait / Ice / LPG / CNG	Showers / Laundry / Pumpout / Restrooms	Hotel / Pool / Tennis / Golf	Public Phone / Restaurant / Snack Bar	MasterCard / Visa / Discover / AmEx
1 Port Imperial (p. 171, New York Chapter)	201-902-8787	88A	88A	50		Y	150	8	6	F	★▲■ 100		L35	PEGD		GD	Mob	BI	All	G	All	MVA
2 Crystal Point Yacht Club (p. 183)	908-892-2300	16	9	10		Y	100	9	7	All	★100	D		PEGDF	C	GD	Tex	BI	SPR	MP	PR	MV
3 Morrison's Marina (p. 184)	609-492-2150	68	68	12		Y	55	5	5	All	★▲50		L30	PEGDWF		GD	Gulf	MGBI	SLR		PR	MVA
4 Harrah's Casino Hotel & Marina (p. 186)	609-441-5315	9	16	107		S	130	35	9	All	★▲50							I	All	HPT	All	All
5 Two Mile Marina (p. 187)	609-522-1341	16	9/69	15		S	100	20	15	F	★▲■ 50							I	SPR		All	All

Information in these listings is provided by the facilities themselves.

coast. In the north, New Jersey's waters are linked to the New York metropolis. Cruise down the Hudson River, where there are scenic stops in Alpine and Englewood beneath the shadow of the Palisades Cliffs. Then, continue under the majestic George Washington Bridge, through New York Harbor, and into Raritan Bay. With the Statue of Liberty on one side and the New York skyline on the other, it's a journey you won't easily forget. (See preceding **New York** chapter.)

Northern Coast: Raritan Bay to the Shark River

Raritan Bay, the southern terminus of the Arthur Kill, spreads across the Bayshore of Monmouth and Middlesex counties. This is a popular spot with boaters, with plenty of room outside the shipping lane for racing, and there are a number of yacht clubs along its shores. Members of distant yacht clubs can get a mooring for the night at the Raritan Yacht Club near the mouth of Raritan River in **Perth Amboy**. The nearby Armory, a three-story former Naval building, has been converted and now houses restaurants and bars. Entering Raritan Bay from New York Harbor's Upper Bay, head for the Chapel Hill North Channel after passing through the Narrows. You can take the Chapel Hill South Channel all the way to Raritan Bay East Reach, or you may want to cut across the lower bay and pick up Raritan Bay West Reach. To do the latter, leave the Chapel Hill North Channel at flashing green 2.5sec "15" and head west leaving Old Orchard Shoal to starboard. Make sure to stay out of the red sector of Old Orchard Shoal Light. Pick up West Reach at flashing red 4sec "14." Follow the clearly marked channel all the way to Perth Amboy.

The Spy House Museum Complex (908) 787-1807 in **Port Monmouth** is a good place to explore the area's history. It features displays relating to sea and land farmers of the bayshore area. The museum is easily reached if you continue on to Atlantic Highlands and take the 10-minute cab ride to the museum. En route, at **Leonardo**, there is a state marina east of the Navy Pier, which you can't miss because it sticks out more than 1.5nm into the bay. The channel into Leonardo is clearly indicated by privately maintained markers.

At **Atlantic Highlands**, you can hop aboard the *Sandy Hook Lady* (908) 291-4354, an authentic 85-foot paddle wheeler that will take you along the Shrewsbury and Navesink Rivers farther south. Or have a good laugh by attending one of the comedy productions at the First Avenue Playhouse (908) 291-7552. You'll find a municipal marina with a restaurant and a yacht club located inside the breakwater here.

Be sure to round the lighted breakwater at the east end marked by a flashing white 4sec light. As you proceed west to the municipal marina, favor the breakwater side to avoid a spoil area to the south. A recently opened entrance at the west end has pilings close by and should be avoided, particularly at night.

Sandy Hook is part of the Gateway National Recreation Area, which sticks up like a thumb to border the east side of Sandy Hook Bay. Until 1920, America's Cup races were sailed off Sandy Hook in schooners not much different from those used by harbor pilots to board incoming ships. Today, you can spend the day swimming and fishing on the six miles of beaches. There are walking nature trails to the Sandy Hook Lighthouse, the oldest working lighthouse in the country, as well as the former army camp at Fort Hancock. A visitor center (908) 872-0115 at Spermaceti Cove provides literature and maps for touring Sandy Hook.

To enjoy Sandy Hook, anchor in Horseshoe Cove. From Atlantic Highlands, it's a straight shot across the bay. If you're coming from the ocean side, stay in the Sandy Hook Channel, which is well marked for shipping. Some boaters get into trouble by cutting "The Hook" too close. They end up in the shoals of "The False Hook." Pick up green flashing 2.5sec gong "17" as you round Sandy Hook. Head east again past the Sandy Hook Coast Guard Station for the anchorage at Horseshoe Cove. The shoals may extend farther from shore than charts indicate. Sticks in the water at Horseshoe Cove mark fish nets and the anchorage is outside of them. Many boats ride out storms in this safe anchorage.

At **Highlands**, as you exit the bay and enter the channel to the Shrewsbury and Navesink Rivers, the Twin Lights sit high atop a point of land. Aside from being a good navigational landmark when approaching from sea, they are a historic site housing a museum (908) 872-1814 that features equipment used in the coastal lifesaving stations that predate the Coast Guard. You'll find several marinas in Highlands. Slightly farther south you'll find marinas and good restaurants at **Sea Bright** on the eastern shore. Dock space is available for diners at some restaurants, but put out the fenders. The boat wakes and current can make for bumpy dockage.

To approach the channel to the Shrewsbury and Navesink Rivers, set a course for flashing red 4sec "2," which will be below the East Point Condominiums. Highlands will be to starboard. This is the start of a "No Wake" zone that extends to the bascule Highlands Bridge, where currents run up to 6 knots. Pass between green can "3" and red nun "4." The channel is straightforward from there.

Red Bank, on the Navesink, was once a steamship port from which New Jersey produce was shipped to New York City. It's now a hub of restaurants, bars, and shops. Attend a musical act or play at the Count Basie Theater (908) 842-9002. Inns, hotels, an antique center, and the Riverview Medical Center (908) 741-2700 are all located here.

Be particularly careful where the Navesink leaves the Shrewsbury River and you head west toward Red Bank. Although you can see some of the exposed bars, many boaters lose their stern-drives here at low tide. Heading west up the Navesink, pass the small island and turn back after flashing green 2.5sec marker "7" to head up the old channel for an anchorage above Upper Rocky Point. The channel in this lower end of the Navesink has been changed by winter storms and the markers have been moved, adding several doglegs that may not appear on your chart. At Red Bank, there are marinas near Marine Park, the town park with a pier. But close to shore at Marine Park itself, the water is shallow. Access to the Navesink beyond Red Bank is greatly limited by a fixed 8-foot-high bridge.

The Shrewsbury River runs west and south to **Long Branch**. At the river's southern leg, Long Branch was once the summer retreat of the rich, with casinos that attracted the likes of the famous gambler Diamond Jim Brady during its heyday in the late 1800s. The Seven Presidents Oceanfront Park (908) 229-7025 is a 33-acre oceanfront park, pavilion, fitness course, beach, and boardwalk. Because there are no facilities, it's best to take a cab from Highlands or Red Bank.

Monmouth Park (908) 842-4000, a thoroughbred race track that retains a grandeur of earlier days in the area, is in **Oceanport**, located at the more inland branch of the river. You can dine at the race track. Again, you should travel by car to the sites of this town.

Leaving Sandy Hook and Raritan Bay waters behind, you can exit the area and cruise just off the beach all the way to Cape May. The sea buoys are located one to two miles offshore and lead all the way down the coast. If you take this option, you'll want to give a wide berth to shoals that form around the seaward side of the inlets from Barnegat Inlet south.

Cruising close to shore, you'll see the changing face of New Jersey's coastal towns. Condominiums along the coast of Monmouth Beach and Long Branch often dwarf regal mansions that were once the summer homes of the rich. St. Michael's Roman Catholic Church, the large red structure at the south end of Long Branch, remains a landmark to mariners and navigators. Winter storms have left Long Branch without a beach and destroyed boardwalks in several New Jersey waterfront towns. Some of the boardwalks are not being rebuilt, but replaced, instead, with less expensive alternatives like red brick and asphalt. You'll also pass the oceanfront town of Asbury Park, once a grand amusement area featured on the front of singer Bruce Springsteen's "Welcome to Asbury Park" album. Although some rides still exist, the area has mostly lost its amusement appeal.

About 18nm south of the Sandy Hook Channel you'll find the **Shark River Inlet** and **Belmar,** a resort town that draws a large crowd of young people. If you anchor off or south of the municipal marina, you can row your dinghy to the gas dock and walk a block down 10th Avenue to the downtown area. You'll find a tavern that was a speakeasy during prohibition, now offering dinner cafe-style on the sidewalk. South on Main Street, it's two

Courtesy New Jersey Division of Travel and Tourism

Looking up at the Sandy Hook Lighthouse.

blocks to a Cajun restaurant with thoughtfully selected jazz and soft lighting. Enjoy Louisiana favorites like crawdads and shrimp and sausage etouffee. Or follow the road east along the inlet to Ollie Klein's Fish Market. Operated by the same family for 50 years, Klein's has a great local reputation for fresh fish, which you can take home or eat on a deck overlooking the water. Among the memorabilia hanging from the ceiling are the running lights from the *Morro Castle*, the cruise ship that burned off of Belmar on its maiden cruise in 1934.

The Shark River is the site of a technological development of significant importance to mariners, though few give it much thought anymore. Guglielmo Marconi sent the first wireless message from a small wooden tower on the river's south banks in 1896.

Marked by a red and white buoy "SI" with a whistle signal of Morse A, the Shark River Inlet entrance is straightforward. The mouth is bounded by jetties with flashing red 4sec light "2" atop a 33-foot tower on the north and a flashing green 4sec "1" on a 30-foot tower on the south. In heavy weather, a large swell can form at the mouth. A bascule bridge with 15-foot overhead clearance spans the inlet at the foot of the jetties and opens upon request year round. Inside the inlet, the channel is wide and heavily traveled by recreational boaters as well as a fleet of party fishing and charter boats. The Shark River Coast Guard Station is on the north side. There is a marina on the south side just before the Main Street Bridge.

Courtesy New Jersey Division of Travel and Tourism

Powerboats stream into Manasquan Inlet.

Current is strong in the inlet, particularly around the three bascule bridges before you enter the basin. Stay in the south channel through the bridges into the Belmar Basin. Head to the municipal marina or continue on to the park to anchor. The rest of Shark River may look inviting, but much of its bottom is exposed at low tide.

The Intracoastal Waterway: The Manasquan River to Cape May Inlet

New Jersey's real natives, the Lenni Lenape Indians discovered the resort quality of the **Manasquan River** long ago and feasted on summer clams and crab. In the mid-19th century, the towns around the Manasquan grew as a summer mecca. Along the river and oceanfront, they retain some of their resort character today. The Manasquan River is home to a fair-sized commercial fleet that docks just past the Coast Guard station along the south side of the river in Point Pleasant. There are a couple of fishermen's co-ops along the south side of the Wills Hole Thoroughfare where you can purchase freshly caught fish. There are also restaurants that offer dockage while you eat.

The river is home to a head boat and charter fishing fleet. Large tournaments for tuna and shark are annual events. Throw in a vast and varied population of recreational boaters, including a steady stream of jet skiers, large cruisers, and sailboats heading upriver to the Manasquan-Point Pleasant Canal, and a couple of imitation paddle wheelers and you have a summer weekend on the Manasquan.

The Manasquan River is the start of New Jersey's Intracoastal Waterway. It is well marked by a flashing red 4sec bell sea buoy about a half mile offshore. The water is deep all the way into the mouth of the inlet, protected by jetties on both sides and marked by a red flashing 4sec 30-foot tower on the north and a green flashing 6sec 35-foot tower with a horn on the south. Past the jetties, the channel narrows to about 150 feet, with an average depth presently at 6.5 feet. It's easy to run aground if you stray from the channel.

Although it is not difficult to enter in most conditions, swells at the mouth of this inlet can test your boat-handling skills. And as with most New Jersey inlets, severe storms can create new shoals outside the inlet faster than the chartmakers can make charts. But most complaints about the Manasquan Inlet involve heavy traffic on weekends, strong currents, and the railroad bridge about .75nm upstream. Constant attention is required when these three elements combine on a summer weekend afternoon as traffic backs up waiting for the slow-opening bascule bridge to free movement through the 48-foot-wide passage. Closed, the bridge allows only 3 feet of clearance.

Brielle and fishing go together like peanut butter and jelly. If you don't want to fish from your own boat, you'll easily find a charterboat to do the driving. Call Bogan's Brielle Basin (908) 528-5014, which has the largest party boat fleet on the coast. Or just take a cruise on the River Belle or River Queen (908) 528-6620 while enjoying a buffet lunch, brunch, or dinner. Located between the railroad bridge and the 30-foot high bascule bridge are several full-service marinas in Brielle.

We recommend a stop at **Point Pleasant**, which retains a strong feel of its early days when wooden boats were built at several yards along the water. Some of the early homes in Point Pleasant were built with lumber gathered by "wreckers" from stranded ships. There are plenty of restaurants, a supermarket, shops, and a ship chandlery for supplies. A boardwalk that offers a Tiki bar overlooking the ocean and several blocks of amusements and rides (908) 892-3274 draws a family crowd. Jenkinson's Aquarium (908) 899-1212 features a miniature railroad ride for the kids in addition to its exotic fish, mammal, and bird exhibits. And if you're in town on a summer Thursday, you'll enjoy the weekly fireworks.

Continue in the Manasquan River channel to flashing green "13." Just before the bascule bridge on the southern shore, you'll find **Crystal Point Yacht Club** which offers everything from a pool and 50-unit inn, to an on-site helipad where you can catch a quick ride to New York City or Atlantic City.

To continue on the ICW return to the **Point Pleasant Canal** at flashing red (2+1) 6sec. Point Pleasant Hospital (908) 892-1100 marks the south side of the canal. Current in the canal can run 4 to 6 knots. The waters are turbulent at the two lift bridges; boats that need 30 feet of clearance can get into trouble waiting for a bridge opening on a following tide. Large sailboats often wait to pass through the canal at slack tide or on a waning ebb.

The first of the series of estuaries that make up the intracoastal route in New Jersey is **Barnegat Bay**, a good place to ride out a storm. Even when the waves are being blown apart in the ocean, the chop in Barnegat Bay does not build much more than 2 feet. Barnegat Bay is another waterway where development and nature meet—and sometimes clash. But there remain large areas of protected marshland along the western shore. Like much of the Intracoastal in New Jersey, Barnegat Bay is best explored by small boat. Cruising down the bay in the summer, you may see some of the flat, wide sailing catboats and sneakboxes that follow the designs originated by duck hunters and fishermen to ply the bay's shallow waters and marshes a century ago.

The channel is narrow in many parts of the bay and traffic can make conditions choppy on summer weekends. Tides run less than a foot in the bay. A strong west wind for any length of time tends to push water out of the bay and can be a factor in getting into some marinas on the western shore. Boats drawing more than 4 feet of water should choose carefully which sections of this waterway to traverse. And keep in mind that winter storms often knock buoys off station or change the bay bottom around them.

From the Point Pleasant Canal, follow the channel west and then south to the long shoal that runs north from Herring Island. Red day marker "6" marks the narrow channel between Herring Island and the marshes along the western shore. The Metedeconk River runs west for about 3nm, offering minimum depths of 5 to 6 feet from side to side. Some boaters anchor for the night off the marsh near flashing green "1" on the south side inside its mouth. Should the wind die or come from the west, you'll need something to protect you from the greenhead flies and mosquitos.

Continuing south through the 14-foot bascule bridge, Barnegat Bay begins to open up. There's a marina immediately after the bridge on the west, and just past that is another old-fashioned sailboat marina where they still build wooden sailboats. It's also a sail loft, and a good haven for sailors needing repairs. On the east side, you'll find marine facilities in **Bay Head** as well as a gourmet restaurant and hotel. Visit the exquisite shops or art gallery: the Anchor & Pallet (908) 892-7776.

About a mile south of Bay Head on the opposite shore you'll see the first of a number of man-made osprey towers along the Intracoastal Waterway. Along the west side of the bay past Swan Point, keep an eye out for submerged clam stakes. Most of the clam stakes have been taken up and a pair of tugboats dragged the area with a chain, but local sailors still spot a few stakes broken off just below the water.

Farther south in **Lavallette** is Hankins Boat Works, where Charley Hankins, like his father before him, built the surf boats used by lifeguards up and down the coast. One of Hankins surf boats is on display in Smithsonian Institution's museum complex.

The channel in this part of Barnegat Bay is fairly wide and well marked. If you draw less than 4 feet you'll find plenty of sailing along the western side and avoid the traffic and resultant chop in the center of the channel. Steady onshore morning breezes generally shift to the south and come up the bay in the early afternoon. One caution: Don't cut the points of land along the western shore too closely. They tend to shoal out.

Sandy Hook Tides; High: +0:50, Low: +1:27, Mean 3.1 feet

$1.50 / ft per day
200 Berths, 100' Max LOA

Propeller, Electronics, Gas, Diesel, Fiberglass

Texaco
Gas, Diesel, Aviation Gas

Fresh Water, Phone, Cable TV
110V, 220V, 100 Max Amps

Nearby

Harpoon Willys 908-223-8880
Nautical, $8-15

MasterCard, Visa

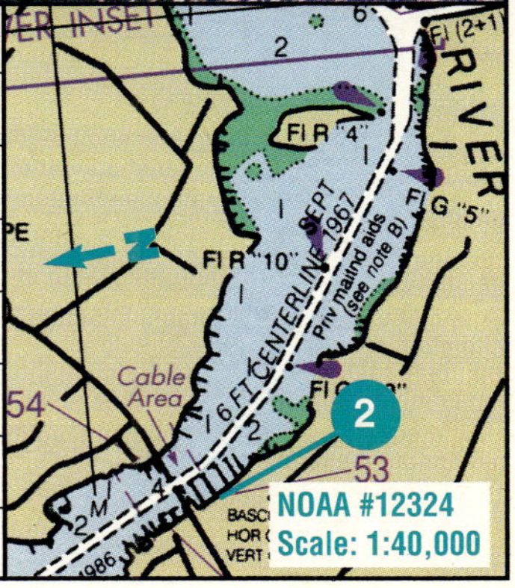

Cruisers frequently anchor in **Silver Bay.** A small, fairly well-protected bay surrounded by marshlands on the western shore with an entrance just north of quick-flashing red "22." Applegate's Cove, just south of Silver Bay and entered between flashing red 2.5sec "24" and red day marker "26," is also a good anchorage. There are shoals outside the channel off Silver Bay past Goodluck Point, south of the two Toms River bridges. The fixed bridge with a clearance of 60 feet is followed by a 30-foot bascule bridge.

Up the **Toms River** at **Pine Beach** you'll find Admiral Farragut Academy Maritime Museum (908) 349-4406 is open to the public, as is the Seaport Society Museum (908) 349-9209. Or you can carefully follow the channel upriver to the lobster restaurant, which has dock space available for diners with large boats.

Turn west at green day marker "31" to head up the Toms River. Stay south of flashing red "2" to avoid the shoal that runs well south off Long Point.

Across the bay from the Toms River is **Seaside Heights** and **Seaside Park**. At Seaside Park, the water stays 4 to 5 feet deep near the shore because there was once an inlet there. You can anchor, row to the beach or the public pier, and walk two blocks to some stores. By car, you can visit Seaside Heights with its large boardwalk packed with amusements, food stands, and games of chance.

Although too shallow for big boats, **Cedar Creek**, about 3nm from Goodluck Point, is a good place to go by dinghy for a picnic and explore the wetlands along the way. Berkeley Island County Park, running along the creek's north shore, has a fishing pier. The area just south of Tices Shoal is deep enough to have gained a reputation among coastal and offshore cruisers as a fine anchorage. It's also a popular spot for local weekend cruisers but usually not crowded on weekdays. Anchored here, you may get a look at the nesting pair of peregrine falcons in the marshes.

On the western shore is the town of **Forked River** on the river of the same name**.** Here, you'll find the JCP&L Energy Spectrum (609) 971-2100 with energy education center, interactive exhibits, and hands-on displays for the kids, as well as the Popcorn Park Zoo (609) 971-2100 operated by the Humane Society.

Flashing red "2" marks the beginning of the channel. Stay in the channel and head up the North Branch where there is a state marina for fuel and supplies.

Oyster Creek, to the south of Forked River, is navigable but offers no public facilities. The channel is marked by privately maintained seasonal aids.

The eastern shore of Barnegat Bay from Seaside Park to Barnegat Inlet is **Island Beach State Park**, eight miles of tree-lined, preserved open space and wide ocean beaches popular with swimmers and surf fishermen. Shoals run from most of the eastern shore well into the bay except for the noted anchorage at Tices Shoal.

If you want to duck in from the outside passage, you can use **Barnegat Inlet** and head north to Tices to ride out approaching storms or rest in the protection of the barrier island on the east and the Sedge Islands to the south. Long considered a dangerous inlet with shifting bars just outside its mouth, Barnegat Inlet has been tamed somewhat by the recent extension of its north jetty. However, an outgoing tide when swells are large still creates a good deal of turbulence, with waves breaking on both sides of the channel leading into the inlet. Barnegat remains a difficult inlet when the wind blows hard from the Northeast.

Some locals approach the inlet from the north, inshore through an unmarked channel in the shoal. For the unfamiliar, this is dangerous. Instead, enter from the east and follow the two pairs of buoys into the mouth of Barnegat Inlet between the flashing red 4sec "NB," 37-foot monument marking the end of the north jetty and the flashing green 4sec marker at the end of the south jetty.

Once inside the inlet, follow the Oyster Creek Channel through the marshes and Sedge Islands to the main Intracoastal channel in Barnegat Bay to get to Tices Shoal. Keep in mind that Oyster Creek Channel markers are moved around during the summer to keep up with the shifting sands inside the inlet.

Barnegat Lighthouse on the south side of Barnegat Inlet was renovated and reopened to the public recently. The Barnegat Inlet Coast Guard Station is south of the lighthouse. Climb the stairs of "Old Barney," with its 172-

Courtesy of New Jersey Division of Travel and Tourism

The casinos of Atlantic City glow at night.

foot tower (609) 494-2016 or have a picnic on the grounds. You'll also find Barnegat Light Museum (609) 494-9196 and shopping in the town of **Barnegat Light**.

If you follow the local channel just inside the inlet, turning south at Barnegat Lighthouse, you can buy fish on the dock from some of the commercial fishing boats along the cut known locally as Myers Hole. Boats sometimes anchor in this small but well-protected harbor. Marinas are located just south of lighthouse. Some transient slips are available in the spring and fall but are hard to come by during the summer.

The barrier island towns of **Long Beach Island** offer a number of attractions. **Beach Haven** offers the most variety along its main drag, Bay Avenue. Schooner's Wharf and Bay Village have several restaurants and gift shops, including a shop inside a replica of the schooner *Lucy Evelyn*. The Thundering Surf Water Slide (609) 492-0869 and Fantasy Island Amusement Park (609) 492-4000 both have miniature golf courses and are family oriented. Fantasy Island Amusement Park offers rides for kids and an arcade designed with a casino atmosphere for adults. Beach Haven is also home to the Long Beach Island Historical Society Museum (609) 492-0700. Exhibits include many photographs from the days when Long Beach Island was known to few people other than duck hunters, as well as photographs of some of the storms that raked the island over the years. Across the street from the museum is the Surflite Theater (609) 492-9477, where professionally staged plays are performed.

Morrison's Beach Haven Marina is located at red "106" in the heart of Beach Haven. The marina has deep water slips for boats up to 55 feet and is a full-service facility. Located just north of Little Egg Inlet, Morrison's also has a seafood restaurant, a ship's store, and laundry facilities. From here, you can explore Long Beach Island's barrier island towns.

Following the Intracoastal Waterway south, the channel narrows and runs along the eastern shore before turning back out into shallow Manahawkin Bay and under the 60-foot high fixed Manahawkin Bridge. The channel turns east again and comes back in along the shore at Brant Beach. Boats that require less than 15 feet of overhead clearance can take the local inshore channel that starts about 1.5nm north of the Manahawkin Bridge at flashing green "1." The channel rejoins the Intracoastal at Brant Beach at flashing red 2.5sec "78," and continues south along the shore.

At the end of Long Beach Island, you'll come to **Beach Haven Inlet** and **Little Egg Inlet**. Beach Haven Inlet, like **Brigantine Inlet** farther south, is not passable. Little Egg Inlet starts just south of Tuckers Island, which is just a sandbar. If you want to make time, Little Egg Inlet is a good place to head back outside. There is plenty of water in Little Egg Inlet, but you will see white water on both sides most of the way out.

The waterway turns west into **Great Bay** before turning south. It is very narrow as it winds through seven miles of wetlands before coming to Absecon Inlet at Atlantic City. If you stay on the inside route, you can run across Great Bay and up the **Mullica River**, which passes beneath the Garden State Parkway at a fixed bridge with 30 foot clearance that runs parallel to a 50-foot-high power cable. **The Bass River** runs northwest off the Mullica. There are several large marinas upstream as well as a major powerboat manufacturer.

Continuing down the Intracoastal from Great Bay, the channel passes through the Edwin B. Forsythe National Wildlife Refuge at **Brigantine**. The refuge includes more than 34,000 acres of preserved wetlands with some 200 species of birds. The U.S. Fish and Wildlife Service (609) 652-1655 conducts walking tours through parts of the preserve. The marshes along the west side around shallow Perch Cove are home to about 60,000 pairs of swans in the fall. The Marine Mammal Stranding Center, where you can see porpoise that have been saved from beach strandings, also is in Brigantine. If you want to visit these areas, you can take a cab from Atlantic City.

Roll the dice or pull the lever. Either way, you're in **Atlantic City**, one of the more unique East Coast cities with its 12 casino-hotel complexes and long oceanfront boardwalk. You can take a sightseeing tour in a rolling chair like those that have been pushed along the boardwalk since 1887. Shop at Ocean One, which offers a variety of stores and restaurants on the pier near Caesar's Atlantic City. You can rent a car at any of the hotels and drive to 14 golf courses within 30 minutes of Atlantic City or visit Atlantic City Race Track. Historic Smithville Village (609) 652-7777, established in 1787, is 20 minutes away by car. Atlantic City also is a good place to pick up and drop off crew. Atlantic City International Airport is a convenient cab ride from the marinas.

From the inside route, follow the cans up the inlet to green can "17." Just west of the Brigantine Bridge you'll find **Harrah's Casino Hotel and Marina**, a first-class facility with all the amenities. Harrah's has luxury accommodations and nightly entertainment. There are shower and laundry facilities, TV hook-up, and telephone at the dockside marina.

The entrance to **Absecon Inlet** and Atlantic City is easy to follow from flashing green 4sec "1" off the mouth northwest to red "2" and green "3," both 2.5sec flashing buoys marking the mouth of the inlet.

The towns south of Atlantic City are mostly residential with some seaside attractions. **Margate** has become known for Lucy the six-story elephant (609) 823-6473 built in 1881 and now a historic landmark. A marina is located there. A harbor at **Longport** also offers transient slips.

Ocean City, with shops and amusement piers, is a traditional summer haunt of people from Philadelphia. You

4 p. 178

Sandy Hook Tides; High: -0:27, Low: -0:35, Mean 4.1 feet
$1.15 / ft per day
107 Berths, 130' Max LOA
Fresh Water, Phone, Cable TV
110V, 50 Max Amps
Ice
(609) 441-5315
9 Restaurants 1-800-242-7724
Casual to Elegant, Prices vary
MasterCard, Visa, Discover, American Express

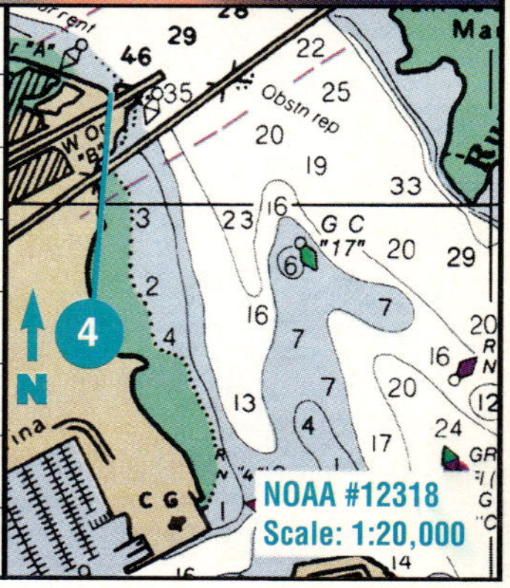

can see the sites by staying in nearby **Somers Point**. (The home of Harbormaster Richard Somers in Somers Point is the oldest in Atlantic County.)

To reach Somers Point, follow the ship channel. Enter the channel southwest of nun "250" giving it a wide berth to avoid a shoal area. This will lead you to Somers Point.

If you're coming in the Great Egg Inlet, use caution. It has been undergoing major dredging to align the channel as a straight shot from the sea buoy off its mouth through the shoals on both sides.

The narrow Intracoastal continues to wind through marshes down to **Corson Inlet**, which is not passable, and **Townsend Inlet**, which is passable by following the uncharted center channel buoys. The channel no longer runs beneath the part of the 23-foot-high bascule bridge that opens.

The Wildwoods are made up of **North Wildwood**, **Wildwood**, and **Wildwood Crest**. These barrier island towns offer an array of amusement rides for all tastes, as well as miles of beaches. The boardwalk is excellent for bike riding, and you'll begin to notice more open expanses this far south in New Jersey. Visit the George F. Boyer Historical Museum (609) 523-0277 to see its exhibits on the history of the Wildwoods. Or shop to your heart's content at The Boardwalk Mall (609) 522-2702.

Continuing south, a dredging project may cause some confusion at Great Channel. The narrow channel beneath the bascule railroad bridge just south of Grassy Sound is only about 4 feet deep at low tide. When cruising the Wildwoods-Cape May area, we recommend staying at **Two Mile Landing**, located on the Intracoastal Waterway adjacent to Cape May Harbor and Inlet, with dockage for vessels up to 100 feet. From here, you can see all of the Wildwoods and Cape May by car. If you follow the channel to just past green "479," you'll come to Lower Thoroughfare, which will lead you to the marina just inside the inlet.

Sunset Lake is a nearby anchorage with plenty of depth for anchoring. In the summer, however, it is a popular spot for jet skiers, so don't expect peace and quiet. To enter **Sunset Lake**, leave quick flashing green "465" to starboard and head south for flashing green "1." After flashing green "3," pick your anchoring spot in 15 feet of water.

Cape May has been a charming resort town for almost 200 years. Today, it is jam-packed with bed and breakfasts and its own year-round population of artisans and craftsmen. Not always so, however. A few of today's bed and breakfasts were brothels and gambling halls a century ago. If you seek a quiet place to enjoy some excellent food in an relaxed atmosphere, Cape May should be in your cruise plan. Take a tour of Victorian homes or an evening buggy ride. If you happen to be in Cape May for the last Sunday in August, belly up to the bar and try your luck in the local froth-blowing contest. People come from all over to watch the fall migrations of birds through Cape May. Attend one of the programs at the Cape May Bird Observatory (609) 884-2736, which include Cape May Spring Weekend and World Series of Birding, both in May, and Cape May Fall Weekend in September. Or call David Githens at Wildlife Boat Tours (609) 898-9631, and arrange for a trip up to Jarvis Sound or Hereford Inlet with a local guide.

The Emlen Physick Estate (609) 884-5404 is an 18-room Victorian-era mansion with tours available. And don't forget the Cape May Lighthouse (609) 465-3535. It's worth the 199 steps to the top. The attached museum includes photos showing the changing face of Cape May's shoreline over the years—indeed, near Cape May Point the sea has taken back 1,000 feet of beach since 1942.

Cape May Inlet is one of the safest along the Jersey Shore. And Cape May Harbor is clearly marked with day markers. Stay in the channel, though, as the harbor has shoaled to a couple of feet in places.

Delaware Bay and River: Cape May Canal to Trenton

Extreme care must be taken in rounding the Cape May Point. The point is rounded outside of Eph Shoal and Prissy Wicks Shoal—an area known as the "Rips" to local boaters. The shoals change frequently. You may see some local party boats coming from Cape May Inlet rounding the cape inshore of the Rips. Don't try it! These are fishermen who know the local waters well.

The best route to the Delaware Bay is through the **Cape May Canal**. The Cape May Canal was dug during World War II as a place to escape German U-boats and avoid rounding the exposed cape into Delaware Bay. Controlling depth is 6 feet for a good portion of the canal and two fixed bridges have vertical clearances of 55 feet.

At the head of Cape May Harbor, the channel forks. Take the fork to the north at flashing red 2.5sec "10" and head for flashing red 4sec "14" which marks the beginning of the canal. You'll encounter one of the bridges just as you enter the canal from Cape May. A railroad swing bridge .75nm farther along remains open forming a divider down the middle of the channel. The railroad crossing the canal is no longer in use. A third bridge, fixed with 55 foot clearance, is directly after the railroad bridge. Exiting the Cape May Canal into Delaware Bay you'll find a red and white striped whistle that emits Morse "A"—one short blast, one long—has replaced the red bell "8" shown on most charts. The ferry that goes to Lewes, Delaware, docks just inside the canal and uses much of the channel when entering.

The towns along the New Jersey side of the Delaware Bay grew up around oystering, which was done from sailing vessels until World War II. Then, power dredges quickly depleted the oysters and the mysterious MSX disease hit in 1957. By 1961, a $2.7 million industry had dwindled to an $18,000 harvest. Some oystering still goes on and the towns along the bay's navigable rivers have changed little.

Going up the bay, you can follow the shipping channel, which has the advantage of being well-marked and maintained but experiences a stronger current than the flats to starboard. Most of the creeks that run off the New Jersey side of the Delaware are navigable only by small boats. Watch for crab traps and gill nets. The currents in the Delaware Bay can run at 6 knots, especially during the ebb in the spring. To make the best time, start your trip up the bay an hour before the tide starts coming in.

Both commercial fishing and party boats work out of the **Maurice River** (pronounced, Morris). **Mauricetown** is a small hamlet of Colonial homes. Wheaton Village (609) 825-6800, a replica of 19th century glass making village with working glass blowers, is up the Maurice River just before Millville.

Sound your way across Egg Island flats to enter the Maurice River following the local markers, staying east of Fowler Island. The Maurice River Cove is shallow, but local boaters report 5 feet of water in the channel at the mouth of the river. The 43-foot East Point Lighthouse is a good landmark during the day. The channel is well marked up to Millville. There is a 25-foot-high fixed bridge at Mauricetown.

Legend has it that Captain Kidd buried some of his booty up the **Cohansey River**. The river is navigable up to the historic town of **Bridgeton**, which offers walking tours of some of the 2,000 Colonial homes there. **Greenwich**, about 3.5nm up the winding river, has homes dating back to 1740 and a memorial to the local tea party held before the Revolutionary War.

Entering the Cohansey, put the Ship John Shoal Lighthouse astern and steer for the inlet between the 24-foot quick flashing green "1" on the north and the flashing red 4sec 42-foot tower on the island that marks the south side of the river mouth. Red nun "10" marks the end of Dunks Bar. Give it a wide berth. Or enter the river from the Cohansey Cove through the deep channel behind the island, staying off the shallows of its southern tips. There is 36 feet of water inside the mouth of the river but, if you anchor there, you may spin around a bit when the current is strong. Some cruisers anchor up the river, around the bend past Laning Wharf.

Across from the Chesapeake and Delaware Canal, you'll find the **Salem River**. The town of **Salem** dates back to 1675. There is a village square and the Salem Oak, believed to be one of the oldest trees in New Jersey. The boat basin has transient slips and someone there will usually will give boaters a ride to the town.

Lighted buoys mark the Salem River up to the town of Salem. The channel is narrow with recent dredge spoils dumped just outside of it. Beyond the bascule bridge, power cables span the river at a charted height of 59 feet.

You can navigate the Delaware River all the way to Trenton by following the ship channel. The river becomes more industrial the farther up you go. North of the Salem River you will see Fort Mott State Park at Finns Point, constructed in 1896 to protect the port of Philadelphia. Passing beneath the Delaware Memorial Bridge above Pennsville you'll see the DuPont Plant. You'll pass **Camden**, once a thriving shipbuilding city where the U.S. Aircraft Carrier *Kitty Hawk* was built, and, of course, **Philadelphia** across the river. You can stop at **Burlington**, a major seaport in the 18th century, and visit the home of James Fenimore Cooper. At **Trenton**, the state capital, there are slips available for transients. The Trenton War Memorial, New Jersey State Library, and the Science Center are all located around the State House. Trenton is about as far north as you'll want to go as you will begin to encounter submerged rocks just upriver.

If you continue up the Delaware River to Trenton pay attention to the strong tidal currents and end your trip at the Trenton Marine Basin. Keep in mind that ship and barge traffic is heavy on the river.

You'll find that the Delaware Bay and River waters are as different from New Jersey's ocean coast as they are from the waters that border the northern side of the state. But, all in all, this varied state gives you a good taste of the Atlantic Coast.

Shoreside and Emergency Services

Fire, Ambulance, Police: 911

Coast Guard Stations:

Sandy Hook (908) 872-0326
Shark River Inlet (908) 775-5029
Manasquan Inlet (908) 899-0130
Barnegat Inlet (609) 494-2661
Beach Haven (609) 492-5751
Atlantic City (609) 344-6594
Cape May (609) 898-6981

Radio Telephone:

VHF 24; I.D. Sandy Hook Marine Operator
VHF 27; I.D. Toms River Marine Operator
VHF 25; I.D. Beach Haven Marine Operator
VHF 26; I.D. Atlantic City Marine Operator
VHF 28; I.D. Odessa Marine Operator, Delaware

Tow Service:

Sea Tow Northern, N.J., Highlands
(908) 872-1835 or VHF 16
Sea Tow of Central New Jersey, Point Pleasant
(908) 295-4466 or VHF 16
Island Boat Towing, Tom's River
(908) 929-1313 or VHF 16
Munro's Marina, Beach Haven
(609) 296-8202 or VHF 16
Sea Tow Atlantic City
(609) 266-1984 or VHF 16
Sea Tow Cape May
(609) 263-2222 or VHF 16
Sea Tow Philadelphia
(215) 824-3743 or VHF 16

Note: There is no commercial towing service located along the New Jersey side of the Delaware Bay. For assistance, call N.J. Marine Police, Bivalve Station, (609) 785-1330. ✦

Local Services

PA
MD
NEW JERSEY
DELAWARE
ATLANTIC OCEAN
Delaware Bay
1 inch=10 miles
Scale 1:570,240
0
10
20
30
40
Joins page 178
Joins page 196
Claymont
Wilmington
Penns Grove
New Castle
Delaware River
1
Chesapeake City
Delaware City
St. Georges
C&D Canal
Penns Grove
Chesapeake River
Reedy Island
Bridgeton
Vineland
Millville
Cohannesey River
Maurice River
Great Egg Harbor River
Port Mahon
Dover
Bowers Beach
Milford
USCG
(609) 898-6995
Hereford Inlet
Cape May
Cape May Canal
VHF-FM
WXJ-94
162.55
Cape Henlopen
Lewes
Lewes and Rehoboth Canal
Choptank River
USCG
(302) 227-2121
Seaford
Rehoboth Bay
Indian River Inlet
Indian River Bay
Monticoke River
N

The Diamond State

Tiny it may be, but Delaware boasts a shoreline bordered by four major bodies of water—Delaware Bay, the Delaware River, the Atlantic Ocean, and the Chesapeake and Delaware Canal. At 86.3 nautical miles long, the little state has waterfront along its entire length as well as its modest width.

With all that shoreline, it's no surprise that the residents of Delaware take to recreational boating. But that's only one part of the activity on the water. Every year, thousands of fishermen from other states haul their boats here, and thousands of out-of-state boaters use the Chesapeake and Delaware Canal to Chesapeake Bay.

The Chesapeake and Delaware Canal connects the two bays of its namesake forming a protected shortcut that is an important slice of the Intracoastal Waterway. It is 19 nautical miles stretching from Reedy Point in Delaware to Welch Point in Maryland. The C&D Canal, the Delaware River, and Delaware Bay all have strong currents, unlike the waters of Chesapeake Bay.

Whether you've just been up the Delaware River or you're cruising on Delaware Bay, there are several choices for your boating route in Delaware. One is to head down the Delaware River to the C&D Canal. The second is to pass the canal and travel down the entire coast of little Delaware to the Atlantic Ocean. And yet another is to use Delaware Bay as a route between New Jersey and the C&D Canal.

Whichever way you travel in Delaware, you should be aware that only select areas have facilities for recreational craft, although there are some additional choices for very small boats. The three principal areas are Delaware City and Summit both on the C&D Canal, and Lewes at Cape Henlopen on the Atlantic. So enjoy the Delaware shores as you cruise by, but plan your stays accordingly.

Also keep in mind that you are likely to meet ships while transiting the canal or the ship channel in Delaware River and Delaware Bay. In all cases, there's ample room to pass; just keep a good watch for large

© Mary Jane Hayes

Cruising on Delaware Bay.

commercial vessels and stay out of their way. Remember, you can maneuver much better than they can within the confines of the channel. Also, in areas other than the canal, you can sometimes go safely outside the channel while they don't have that option.

Delaware River and the C&D Canal

Coming from the Pennsylvania border on the Delaware River, the first town you'll see on the shores of little Delaware is **Claymont**, known for its steel mill. Stay in the channel heading south, keeping in mind that there is a separate channel on the New Jersey side and shallow water between the two. Next you'll see a DuPont plant at Edge Moor before you come to Wilmington at the mouth of the **Christina River**.

It's somewhat ironic that **Wilmington** is a common home port on transoms because there is little in the way of facilities for recreational boaters in the city (then again, no sales tax is a strong incentive for registering in the state). The city has been trying to make arrangements for a marina on government-owned land along the Christina River. Currently, the port hosts commercial ves-

Marine Facilities and Services

Marine Facilities and Services	Phone	DOCKAGE: Monitors VHF Channel	Working Channel	Number of Transient Berths	Number of Transient Moorings	Seasonal / Year-round	Maximum Length Over-All	Approach Depth in Feet at MLW	Dockside Depth in Feet at MLW	SERVICES: Hookups: Fresh Water / Phone / Cable TV	110V★ 220V▲ 3 Phase ■ Maximum Amps	Ramp / Dinghy Dock / Launch Service	Railway / Lift / Crane: Capacity (tons)	Rigging / Gas / Diesel / Wood / Fiberglass; Repairs: Propeller / Sail / Electronics	SUPPLIES: SCUBA: Air / Rentals / Equipment / Charters	Gas / Diesel Fuel	Fuel Brand	Marine / Groceries / Bait / Ice / LPG / CNG	AMENITIES: Showers / Laundry / Pumpout / Restrooms	Hotel / Pool / Tennis / Golf	Public Phone / Restaurant / Snack Bar	MasterCard / Visa / Discover / AmEx
1 Summit North Marina (p. 193)	302-836-1800	16	8	30		Y	125	8	12	All	★▲50	D	L50	All		GD	Mob	All	All	P	PS	MVD

COAST GUARD: See Adjacent Chart for Coast Guard Information

Information in these listings is provided by the facilities themselves.

sels bringing autos, bananas, and other cargo from foreign countries, so there are areas where you could tie up in an emergency.

Up the Christina past its confluence with the Brandywine, there is a floating dock owned by a nonprofit organization, the Kalmar Nyckel Foundation (302) 429-7447, dedicated to preserving the culture and maritime heritage of Delaware. The organization, named after a 17th-century Swedish tall ship, has a small museum on the premises. It is possible for visiting yachts to tie up temporarily, and volunteers there are likely to be helpful if needed.

Leaving the port of Wilmington and the Christina River, you'll get back in the channel just south of red and green bell "CR." After you pass beneath the twin Delaware Memorial Bridges (fixed with vertical clearance of 166 feet), the city of **New Castle** is just ahead on the Delaware side.

New Castle is a small city with a riverfront historic area that is being faithfully preserved. Residents will tell you the city turned down an offer of huge amounts of money to make it into a Williamsburg. They preferred not to create a replica, but to preserve the original. The result is colonial and early 19th-century homes and streets paved with cobblestones, yet no re-enactments of colonial life. Immanuel Church on the town green is a handsome colonial church with a 19th-century tower and steeple. Another prized historic building is the Old Court House on Delaware Street, Delaware's first state house built prior to Independence Hall in Philadelphia.

New Castle has many restaurants, antique shops, and stores. Although there is one very small marina with shallow water (as low as 3 feet), we recommend taking a taxi or the limousine service from the next stop at Delaware City.

Another site on the Delaware River is Fort Delaware, located on Pea Patch Island. You'll see the island to your starboard with the two-story fort. Built in 1859, the fort originally served to protect the ports of Wilmington and Philadelphia. During the Civil War, Confederate prisoners were held there. It's a brick structure with ramparts, soldiers' quarters, and gun emplacements. If you want to see the fort or the wealth of wading birds on the island, you'll have to tie up in Delaware City and catch the ferry to Pea Patch Island.

To get to **Delaware City** heading south from New Castle, stay in the main channel between flashing green 4sec "5D," marking the northern end of Bulkhead Shoal, and red flashing 4sec "4D." At flashing green 2.5sec "1D" you can use the short Bulkhead Bar Range to lead you to its intersection with the New Castle Range. This range will guide you to the east of Pea Patch Island. While it is possible to approach Delaware City passing to the west of Pea Patch, it is not recommended due to several unmarked shoal areas. South of Pea Patch Island, leave flashing green 4sec gong "3N" and red nun "4" to starboard. Turn to starboard red nun "2," and follow **Bulkhead Shoal Channel** northwest to the entrance of the branch channel marked by flashing green 4sec "1." There is a prominent white building near the north side of the branch channel entrance.

As you enter the branch channel, on the right is the dock used by a state-owned vessel that takes people back and forth to Fort Delaware (302) 834-7941. Beyond that first dock, used also by federal government vessels, is a wall on the right against which visitors may tie up to for three hours.

You'll find a marina a little farther in to starboard. It has transient facilities and monitors Channel 16. Fuel is available, and vessels up to 20 tons can be hauled. Some supplies are stocked at the marina; there is a pump-out facility and showers are available.

The state's launch ferries visitors across the river from Delaware City to Fort Delaware on weekends and holidays from the last weekend in April through September. For Civil War buffs, the fort is a must. For birders, on the other hand, the big attraction on Pea Patch Island is the marshes, where nine species of herons, egrets, and ibis nest during the summer. You can see them from an observation tower and a hiking trail. The heronry on the island is one of the largest nesting sites for wading birds anywhere in the east. Take along your camera with your longest lens. There are picnic tables and grills on the island; soft drinks, snacks, and ice water may be obtained at the fort.

Delaware City's downtown is nearby. Taxis and limousine service are available to travel downtown or to nearby historic New Castle. You can also travel about four miles north of Wilmington to see the Winterthur Museum and Gardens (302) 888-4600, a premier collection of antiques made or used in America from 1640 to 1840. Or you can take a scenic half-hour cab ride south to the Bombay Hook Wildlife Refuge (302) 653-9345.

Other sites you can get to by car from this area include Fort DuPont, another guardian of the river just south of Delaware City, and the Port Penn Interpretive Center (302) 834-0431, a charming small museum in the riverfront town of **Port Penn**, five miles south of Delaware City. The center has artifacts and exhibits demonstrating the culture of the small fishing village from 1870 to 1900, particularly the sturgeon fishing that used to supply large amounts of caviar. (As the Delaware is cleaned up, the sturgeon are returning.) The center similarly shows how turtles were captured for making snapper soup, and how muskrats were trapped in their wetland environment.

As you depart Delaware City, you will see the eastern end of the well-known **Chesapeake and Delaware Canal** to the south. Your approach to the canal may be directly from the **Cape May Canal** if you're coming from southern New Jersey rather than the Delaware River. The distance between the two canals is 48nm. In that case, in your trip up Delaware Bay, you have the choice of crossing to Brandywine Shoal Lighthouse to pick up the ship channel or taking the shortcut to Miah Maull Lighthouse.

Whether traveling down from the Delaware River or straight from the Cape May Canal, there are anchoring options near the entrance to the C&D Canal. Keep in mind that currents are strong, and you'll be within earshot of the jetty horn at either location.

One of these places is just north of the Reedy Point jetty on the north side of the canal opening, and west of the channel leading to Delaware City between green flashing 2.5sec "3" and nun "4A." The area is marked on the chart as "General Anchorage No. 4" with a depth of 20 feet. Closer to shore the depth is 15 feet. This area is open to the northeast. It's well within earshot of the horn at the end of the jetty; if you're looking for a quiet anchorage, move on.

Another option sometimes used by local boaters is to anchor between **Reedy Island** and Augustine Beach, just south of Port Penn. There is room to anchor in 19 feet of water here. This is just south of the canal entrance and not far from the channel. If you're traveling from Cape May, beware. Boaters run aground and suffer damage every year south of Reedy Island. There's a concrete dike here that protects the channel at a curve in the river. It is clearly marked on the chart and has warning markers on it at intervals—white with orange bands and the word "DIKE"—*but it is submerged at high tide!* None of the warnings prevent chartless or inattentive boaters from smashing into it.

If your trip is taking you through the C&D Canal, you should be aware of certain guidelines. You must go through the canal under power; sailing is prohibited. You must yield to all commercial vessels, and you cannot anchor in the canal. You don't have to report to the canal dispatcher, but you will be visible in the canal office on a television monitor.

The Army Corps of Engineers, with a dispatch office at Chesapeake City on the Maryland side of the canal, monitors Channel 13 to communicate with ships. They do not welcome radio calls from pleasure boats because it's important to keep the radio channel open, but you may contact them by telephone at (410) 885-5621.

The 19.2-mile-long, 450-foot-wide canal, dredged to 35 feet, began as a 10-foot barge canal in 1829. The original canal was privately financed and had four locks. The federal government bought the waterway in 1919 and converted it to a sea-level canal in 1927 by removing the locks. Today, more than 10,000 deep-draft commercial vessels pass through it annually.

The bridges over the canal in Delaware traveling east to west include the Reedy Point Bridge, two bridges at St. Georges, the Conrail Bridge, and then the Summit Bridge. Once in Maryland, there's also the Chesapeake City Bridge. There's ample clearance for recreational boats under the road bridges—about 135 feet. The Conrail railroad bridge allows 45 feet when down and 133 feet when raised. This bridge is usually open and the dispatcher monitors Channel 13.

As you enter the canal, there are jetties on both sides at Reedy Point. The one on the north side has a quick flashing red "2" horn; the one on the south is marked by quick flashing "1." There is a traffic light at Reedy Point. If it is red, do not enter the canal; call canal control. Canal control's call sign is WB33. (The counterpart to this light at the western end of the canal is at Old Town Point Wharf.)

Just under 2nm from the jetties, you will come to the branch channel. It was part of the original canal before it was straightened and deepened and is currently known as the **Delaware City Branch Channel**. A marina is at the other end of this short waterway. Only runabouts and other low craft can use this shortcut to the marina because of a fixed bridge with only about 6 feet clearance at high tide.

Proceeding west in the canal, about 2.5nm from the branch waterway and before the railroad bridge, you'll pass the town of **St. Georges**. There are no facilities for transient yachts here. But just after the railroad bridge, there is **Summit North Marina**. It was built in a part of the canal that was abandoned during work in 1966 to further straighten and widen the canal so it is totally protected. The marina has fuel service day and night, a pumpout facility, haul-out equipment, and some supplies.

Recreational facilities at the marina include a swimming pool, paddle boats, volleyball and basketball courts, horseshoe pits, and picnic areas. There are plans for a laundry. Yachts up to 125 feet can be accommodated; the marina monitors Channel 16.

Taxi and limousine services are available in this area, but it's best to call in advance if you need them. Famous Longwood Gardens in Kennett Square, Pa. (215) 388-6741, with 350 acres of outdoor gardens and woodlands, is about a 40-minute cab ride. Or you can go back to the Delaware coasts to the sites of New Castle and Wilmington from here.

Proceeding through the canal, you'll soon come to **Chesapeake City,** with a high bridge connecting the divided town, both parts of which cater to visiting yachts. On the north side is a well-known restaurant and marina on the banks of the canal. Fuel and water are available at the dock. You'll see pump-out equipment, but the town of Chesapeake City has shut down the equipment refusing to accept outside contributions to its sewage disposal system.

The launches you will see at the east end of the dock are used by pilots, climbing on and off ships. An eastbound ship drops the Maryland pilot and takes on a Delaware pilot. Westbound, of course, it's reversed. On shore, just east of the restaurant, the pilots have a motel-like building in which they wait for ships.

The town on the southern side of the canal has tasteful gift shops and historic buildings, one of which is the 1844 house in which the author Jack Hunter wrote "The Blue Max." There is a basin on this southern side that belongs to the U.S. Army Corps of Engineers. You may anchor in the basin for up to 48 hours for free, but the entrance is shallow, sometimes as low as 4 feet in the center. Inside, the depth is 11 feet.

One of the most alluring features of Chesapeake City is the Canal Museum on the Engineers' property (410) 885-5622. You can find a place to beach a dinghy nearby. The museum shows barges being pulled through the original canal by mules. There's also a lift wheel that was used beginning in 1850 to add water that was lost in opening and closing the locks. The wheel, 37 feet in diameter, could raise 130 tons of water 14 feet each minute. Two big steam engines drove the huge wheel, made of cypress. And there are fossils found during excavation of the canal—even a dinosaur bone. The museum is open 8 a.m. to 4 p.m., Monday through Saturday. Admission is free.

Just past Chesapeake City, the canal runs into Back Creek, and channel bouys, mostly nonexistent in the canal, reappear. With no locks, all is at sea level. However, the sea level at any given moment is not the same at the Delaware end as it is at the Elk River, or Chesapeake end, because of differences in the tide. The canal is flooding when current is moving east toward the Delaware River, where the normal tide range is 5.4 feet as compared with 2.6 in the canal at Chesapeake City. The normal high and low tides in the canal are greatly affected by strong winds. But don't worry—the normal depth at low water is 35 feet.

Delaware Bay: C&D Canal South to Lewes

You may chose not to travel to Chesapeake Bay and instead head down Delaware's coast to the Atlantic. In this case, when you're at the canal's end at Reedy Point, stay clear of the south jetty, and head east to the ship channel, marked by quick flashing green gong "1N" and red nun "2N." This is the intersection of the New Castle Range of the channel and the Reedy Island Range. It is not necessary or desirable to stay in the channel. Instead run just close enough on either side to read the numbers on the buoys. West of the channel and northwest of flashing green 2.5sec "5R" is an old pier extending out from Reedy Island. Also, watch for crab pot buoys in this area.

The channel turns more to the south at the next pair of buoys, green can "1R" and quick flashing red bell "2R." The next turn is opposite the Salem Nuclear Power Plant, which dominates the landscape in this region on the New Jersey side. The turn is marked by quick flashing green "1B" and flashing red 2.5sec "2B," where the channel moves from the Baker Range to the Liston Range, and heads generally southeast.

Along the Liston Range the bay widens and the channel buoys are not in adjacent pairs. After passing red flashing 6sec "4L," it is wise to get into the channel or just west of it because of an obstruction beyond the east side of the channel before the next marker, "42," a lighted beacon (flashing red 2.5sec).

On the Delaware side of the bay is Bombay Hook Point, marked by a tower. This is a 15,000-acre paradise for birds, such as heron, ibis, yellow legs, eagles, and snowy egrets. Federal and state wildlife areas line the Delaware coast for miles in this region.

Proceeding down the bay, you'll come to a series of lighthouses, all unmanned and with horns. The first is Ship John Shoal, a brown octagonal dwelling with a steel walkway. The flashing 6sec light is 50 feet high. After three more buoys, you come to a structure on three legs marked "32." This is the end of the long Liston Range, and the channel turns into the short Cross Ledge Range, where you will find the second of the lighthouses, with its name "Elbow of Cross Ledge" printed on it. The structure is 61 feet high and has a rusty orange base. Local fishermen congregate near the lighthouse.

East of flashing green 2.5sec "27" and the channel are the ruins of an abandoned lighthouse. Opposite flashing green 4sec "25" and a little east of the channel, only 4.3nm from Cross Ledge, is the third lighthouse, a round red house with gray bottom, named Miah Maull Shoal, 59 feet high. If you're going to Cape May Canal, turn southeast and leave the ship channel. The distance is 16nm. It's at least 9 feet at normal low tide on this route, but there are no aids to navigation.

On the Delaware side of the bay 8.7nm from the lighthouse is **Bowers Beach**, a popular Delaware fishing port at the mouth of the Murderkill River. Headboats occupy most of the docks, and there is a strong current, but one could tie up there in an emergency.

Those who wish to stay in or near the channel heading for Cape May, Lewes, or the Atlantic Ocean will find two more lighthouses. First comes Fourteen Foot Bank, a white house on a black base, west of the channel, 59 feet high. Quick flashing green "19" precedes the lighthouse. Near flashing red 4sec bell "12" and well out of the channel is Brandywine Shoal Lighthouse, 60 feet high. From here, Cape May Canal is almost directly east, 7nm. The Cape May Canal is readily visible as you approach on a clear day because there are two black stone jetties that contrast with the white sand beaches on both sides. The canal entrance is 2nm north of **Cape May Point** and is marked by flashing green 2.5sec light "11" with a horn on the left, and by flashing red 4sec "10" on the right.

South of Brandywine Shoal Light is flashing red 4sec nun "10" and west of that is the entrance to the Mispillion River. This is a narrow, winding river with some facilities that can be valuable in an emergency, but it is difficult to enter and should be avoided at low tide. The entrance (with a controlling depth of 2 to 4 feet) is between two jetties. The one on the left has deteriorated and the outer part is not visible, but there are day markers on it. Boats entering stay in the center heading for Mispillion Lighthouse, 67 feet high, with a flashing 4sec light. There's a restaurant on pilings near the lighthouse, and marinas up the narrow, winding river, as well as on Cedar Creek, near the river mouth.

Cape Henlopen, on the Delaware side of the mouth of Delaware Bay, is almost directly south of Brandywine Shoal Lighthouse. Brown Shoal Light, 23-foot white flashing 2.5sec "WR9" with a diamond-shaped daymark, marks the end of the channel for those heading toward the sea. Cape Henlopen is 7 miles south. Roosevelt Inlet, the entrance to the **Lewes and Rehoboth Canal**, is about 3 miles west of the cape. If you're going to Lewes, pronounced "Lewis," you may head directly for the inlet, for Lewes, an important little port for pleasure craft and fishermen, is on the canal.

Roosevelt Inlet is protected by jetties, with lights at the outer ends. The tide range here is 4.4 feet and the current is about 0.9 knots. On your right, entering the inlet, is a 20-foot tower with a flashing red 4sec light, marked "2." On the left side, the tower is 21 feet high, marked "1R," and has a flashing green 4sec light. The dredged entrance channel is about 6 feet deep at low water. To the right, inside the inlet, is the mouth of the shallow Broadkill River. The canal is on the left side. Facilities of the University of Delaware's College of Marine Studies and of the Lewes Yacht Club are just inside the inlet, along with a U.S. Coast Guard substation.

The canal, which extends 8nm to Rehoboth Bay, leads to restaurants and docks in **Lewes**, where full services are available—fuel, repairs, hoist and railway, and supplies. Lewes is also a pleasant little city to visit, with shops and restaurants. The city was founded in 1631 as a Dutch settlement called Zwaanendael. A museum of that name deals with the early history.

Across the canal along the bay shore is **Lewes Beach**, one of a string of resorts in lower Delaware. East of the beach area is the Delaware terminal of the Cape May-Lewes Ferry. Recreational boats have no reason to go near the ferry terminal. The terminal is behind a breakwater, one of three in the area. The other two are an outer and an inner breakwater.

Breakwater Harbor is between the inner breakwater and the shore. It offers anchorage with protection from all directions, except the northwest. The depth is 7 feet. There are piers along the south shore. To the west is Cape Henlopen, with shallow water near the cape.

The **Harbor of Refuge** is between the inner and outer breakwaters. There is a gap of deep water between the tip of Cape Henlopen and the south end of the outer breakwater. That end is marked by the Harbor of Refuge Light, 72 feet high, with a horn. A light 40 feet high is at the northern end.

Returning to the Lewes and Rehoboth Canal, beyond the facilities at Lewes, the canal is shallow at several points so that it is useful mainly to runabouts. The same is true of the Assawoman Canal, farther south, connecting Indian River Bay with Little Assawoman Bay, the most southerly of Delaware's series of three inland bays in the coast resort area.

There is access to **Indian River Bay** from the Atlantic Ocean by **Indian River Inlet**. The inlet is marked by flashing green 2.5sec gong "1," and there are marinas just inside the inlet.

As you leave Delaware, you may find deeper ports and more facilities elsewhere, but don't forget, Delaware probably helped you get there.

Shoreside and Emergency Services

Ambulance, Fire, Police: 911
Coast Guard:
Indian River Inlet, Del. (302) 227-2121
Cape May, N.J. (609) 898-6995
Radio Telephone:
VHF Ch. 28; I.D. Wilmington Marine Operator
VHF 27; I.D. Lewes Marine Operator
Tow Service:
Sea Tow Philadelphia
(215) 824-3743 or VHF 16
Sea Tow Delmarva, Millville
(302) 537-2305 or VHF 16
Delaware City Launch & Towing
(302) 834-4172 or VHF 16

✦

Joins page 206

The Old Line State

"There is no life west of the Chesapeake Bay." So says a bumper sticker common on Maryland's Eastern Shore, and while to some it may seem a jab at the west, it also succinctly illustrates how, for so many people, the Chesapeake Bay defines Maryland. The state's history is indebted to the bay, its culture inexorably wedded to it.

Seventeen of Maryland's 23 counties touch the bay or its tributaries, and lawmakers have designated the skipjack, striped bass, and blue crab as the state boat, fish, and crustacean, respectively. This is not to diminish Maryland's Atlantic Coast and inland waters. But cruising the Chesapeake reveals, in infinite ways, much of Maryland.

Formed more than 10,000 years ago as the glaciers melted and the Atlantic rose and flooded the Susquehanna River valley, the bay spans 180 miles from the tidal Susquehanna to Cape Charles, Va., and ranges between 3 and 30 miles wide. The waters that enticed early settlers to establish port cities on the Chesapeake today carry more than 10,000 ocean-going commercial vessels each year. Fish, crabs, clams, and oysters gathered by bay watermen have a commercial value of more than $1 billion annually. And its myriad tributaries—home to osprey, bald eagles, great blue heron, dozens of fish species, oysters, and blue crabs—provide thousands of miles of cruising grounds, a gunkholer's paradise.

Its estuarine nature defines much of the bay's character. The tides and currents are fairly lenient; the gentle daily cleansing of bay waters usually produces a tidal difference of about 2 feet. In some areas, the current runs more strongly (river mouths, the Chesapeake & Deleware Canal, the deepwater main channel that is the Susquehanna's ancient riverbed), but mostly currents aren't critical. Depth, however, is. With an average depth of 28 feet, the bottom ranges from sand to mud, and though there are few rocks to worry about, shoaling can be a major problem. Your depth sounder will be your friend as you negotiate the creeks and inlets, especially along the Eastern Shore. The bay's shallow waters make for a quick, steep chop in foul weather. And don't let the more forgiving bottom and countless safe harbors lull you; it's just as easy to get in trouble in the middle of a ferocious bay thunderstorm two miles off Annapolis as it is anywhere else.

Chesapeake Bay skipjacks were designed for oystering on the bay.

The best time to visit Maryland depends in part on what you want. Powerboaters may prefer the heat of summer, when less wind means less chop. Sailors may seek the strong breezes of spring and fall. The boating season lasts from April through October. Still, many keep their boats in year-round; some people can't get enough of Maryland and the Chesapeake. You'll see why.

Eastern Side of the Northern Bay: C&D Canal to Worton Creek

You can approach the northern Chesapeake Bay via the **Chesapeake and Delaware Canal**, which connects the

Marine Facilities and Services	Phone	DOCKAGE: Monitors VHF Channel	Working Channel	Number of Transient Berths	Number of Transient Moorings	Seasonal / Year-round	Maximum Length Over-All	Approach Depth in Feet at MLW	Dockside Depth in Feet at MLW	SERVICES: Hookups: Fresh Water / Phone / Cable TV	110V★ 220V▲ 3 Phase ■ Maximum Amps	Ramp / Dinghy Dock / Launch Service	Railway / Lift / Crane: Capacity (tons)	Rigging / Gas / Diesel / Wood / Fiberglass	Repairs: Propeller / Sail / Electronics	SCUBA: Air / Rentals / Equipment / Charters	SUPPLIES: Gas / Diesel Fuel	Fuel Brand	Marine / Groceries / Bait / Ice / LPG / CNG	AMENITIES: Showers / Laundry / Pumpout / Restrooms	Hotel / Pool / Tennis / Golf	Public Phone / Restaurant / Snack Bar	MasterCard / Visa / Discover / AmEx	COAST GUARD: See Adjacent Chart for Coast Guard Information
1 Sassafras Harbor Marina (p. 199)	410-275-1144	16	71	20		Y	110	15	10	All	★▲■ 100	R	L50	In '94		C	'94		MGBI	All	P'94	PR	MVD	
2 Annapolis Landing Marina (p. 201)	410-263-0900	9	72	25		Y	100	12	10	FPC	★▲■ 50	D		SGD			GD	Amo	MGI	All	PG	PS	MVD	
3 Lippincott Marine (p. 203)	410-827-9300	68	68	15		Y	70	8	7	F	★▲50	DL	L25	All					MIL	All	P	P	MV	

Information in these listings is provided by the facilities themselves.

© Maryland Office of Tourism

An aerial view of Oriole Park and Baltimore.

two bays. The 19.2-mile-long, 450-foot-wide canal is controlled by the Army Corps of Engineers, which operates a dispatch office in Chesapeake City. Canal control's call sign is WB33, and it monitors VHF 16 and 13.

Keep in mind that it is a busy canal, which means two things: one, you'll have a lot of fun seeing big ships up close and personal, and two, keep a weather eye for them and give them plenty of room. Canal control recommends staying to the starboard side of the channel. Also, be aware that the current runs through the canal more strongly than most places on the Chesapeake, about 2 to 4 knots. To save some time, try to gauge your travel time so that it runs with the current.

Plan to spend a few hours in **Chesapeake City**, a small town full of charm, Victorian architecture, and small shops and antique stores. As you might expect, fine seafood restaurants are superb and abundant in Chesapeake City. But as you might not expect, marine facilities are relatively scarce. Instead, there's a basin maintained by the Corps of Engineers where boats may anchor for free. The entrance is shallow in the middle (4 to 5 feet) but if you stick to either side, it's about 7 feet. (For more information on the C&D Canal, see the preceding chapter on **Delaware**.)

Once you leave the C&D Canal and turn south, you're actually entering the **Elk River**, which feeds into the Chesapeake. You'll find that the shipping channel markers are a dependable guide. And as you head down the river, you'll begin to see why people love boating in this area. The scenery is splendid. On the eastern shore, you'll come to the **Bohemia River**, known for pleasant anchorages, beautiful bay atmosphere, and nettle-free swimming. The channel marker off the river's mouth is flashing red 4sec "16." Head for the middle of the river, keeping Town Point well to port since there's some shoaling on that side. **Veazey Cove** is a good anchorage on the southern shore as soon as you enter the river.

After the Bohemia, try the **Sassafras River,** a longer and more spidery vein that you'll encounter as you head south. One of the bay's prettiest rivers, it has several secluded anchorages great for dropping a lunch hook or spending a quiet night. The yachting area of **Georgetown**, located 7nm upriver, with its neighbor, **Galena,** offers some good antique shops, and the natural beauty of the Sassafras makes it well worth the trip.

Use channel marker flashing red 6sec "58" as an entrance point, and keep to mid-river. Travel east about 4nm to reach flashing red 4sec "2." As you approach Ordinary Point, don't stray too far inside green can "3" as it becomes shallow quickly. From Ordinary Point, follow about 4nm of the winding, well-marked channel that will bring you into the Georgetown-Fredericktown area. About 7 miles from the mouth of the river, you'll find **Sassafras Harbor Marina,** a newly renovated marina with many amenities and fine restaurants within walking distance.

Once back out the Sassafras, head down the bay's eastern side. If you don't plan to travel up the Chester River to historic Chestertown (which we'll describe in detail a little farther south), you can travel there by car from **Worton Creek**. The buoy off Worton Creek's entrance is yellow nun "42B." Head straight in to quick flashing red "2," followed by green cans "3" and "5." As you enter the narrow channel, give the sandy point to starboard a healthy berth.

As you leave the creek, this is a good time to decide whether to travel down the bay's western or eastern shore. You can break away west toward Baltimore after passing Poole's Island (3nm southwest of Worton Creek), or continue south along Tolchester Channel toward Rock Hall. Admire Maryland's Eastern Shore for its wildlife, small towns and beautiful rivers; enjoy the bay's western shore for its energetic cities packed with things to do and see.

Western Shore: Baltimore to the Potomac River

The tendency when approaching the west shore from the C&D Canal is to head straight for Baltimore and all its activity. But before you do, if you take the time to turn north and head to **Havre de Grace**, you won't be disappointed. Steeped in history and colonial charm, this waterfront town will be a peaceful memory to look back on when you are in busier, more urban ports. Visit the Concord Point Lighthouse, a granite structure built in 1827 and decommissioned in 1975. Everything in the historic district (which is just about the whole town) is within walking distance of the water, including The Decoy Museum (410) 939-3739, with over 2,000 carved decoys. The Millard F. Tydings Park is a popular spot, featuring tennis courts, picnic tables, barbecue grills, and a playground. The park is host to numerous festivals including the Classic Boat Show in June, the Seafood Festival in August, and the Havre de Grace Arts & Crafts Show during the third weekend of August. The Decoy Festival is held during the first weekend of May at local schools.

Several facilities can be found along the shore. At the red and green "A" buoy off Turkey Point, the channel winds around Spesutie Island and turns north toward the **Susquehanna River**. After flashing green 4sec "17," a channel heads due west between Park Island and the mainland.

As you head south toward Baltimore, you'll pass **Bush River, Gunpowder River, Middle River,** and **Back River.** Note restrictions on the Bush and Gunpowder due to the army's Aberdeen Proving Grounds. Middle River is filled with facilities and boats to go along with them. And Back River can provide a peaceful trip before heading into the hub of the city.

Baltimore will keep you busy for days, whether your interests are eating, sightseeing, history, or music and entertainment. It's a hard-working, industrial town that's also home to some spectacular galleries and libraries, as well as Johns Hopkins University. The Inner Harbor is ringed by Harborplace (a food and shopping paradise), the Maryland Science Center (410) 685-5225 (which includes an Imax theater and a planetarium), the World Trade Center, Pier Six Concert Pavilion, and the National Aquarium in Baltimore (410) 576-3800, one of the finest of its kind. Delving into the city's history is as easy as a trip to Fort McHenry (410) 962-4290, which the British attacked on Sept. 12, 1814, a battle that inspired Francis Scott Key's words for the national anthem. Don't miss the B&O Railroad Museum (410) 752-2490, the Museum of Industry (410) 727-4808, the Peabody Conservatory of Music (410) 659-8110, and the Peale Museum (410) 396-1149.

Just blocks outside Inner Harbor is Little Italy, and past it, Fell's Point. This small neighborhood is enchanting with its small shops, and it also offers excellent music, entertainment, and wonderful food. The Federal Hill section, a few blocks from Inner Harbor, also has good nightspots and restaurants. And for you sports fans, Oriole Park (410) 685-9800 at Camden Yards is just a few blocks from Inner Harbor. Oh, and don't forget the Babe Ruth Museum (410) 727-1539.

To visit Baltimore, look for quick flashing red "2B" as your entrance to the Patapsco River. This is the junction of the Brewerton and Craighill channels, so keep your eyes peeled for commercial traffic. Enjoy the well-marked shipping channel that will lead you past Sparrows Point off to starboard, past an abandoned fort, under the Key

Sassafras Harbor Marina

P.O. Box 68 Georgetown, MD 21930
410-275-1144 FAX 410-275-9966

1
p. 196

Completely renovated for the 1994 season, Sassafras Harbor Marina is located on the west side of Route 213 in Fredricktown, Maryland, seven miles from the mouth of the Sassafras River. Offering 30' to 45' floating docks and 45' to 60' fixed docks, covered and uncovered, Sassafras Harbor Marina can accommodate vessels up to 110' at competitive rates.

Amenities such as cable TV, telephone service, metered electric, ice, air conditioned bath houses, and washers & dryers will be available at the marina.

Fine dining is available at the Kitty Knight House and the Granary Restaurant, both within walking distance from the marina. A 50-ton travel lift with a full-service department will be available for the 1994 season. Swimming pool facilities, an activities center and a marine store are planned for 1994.

Baltimore Tides; High:+2:05
Low: +2:05, Mean 2.0 feet

$ 1.00 / ft per day + elec.
20+ Berths, 110' LOA

Fresh Water, Phone, Cable TV
125V, 250V, 110 Max Amps.

Kitty Knight House, (cas.) 410-275-2000
Granary Restaurant, (fine)410-648-5566

MasterCard, VISA, Discover

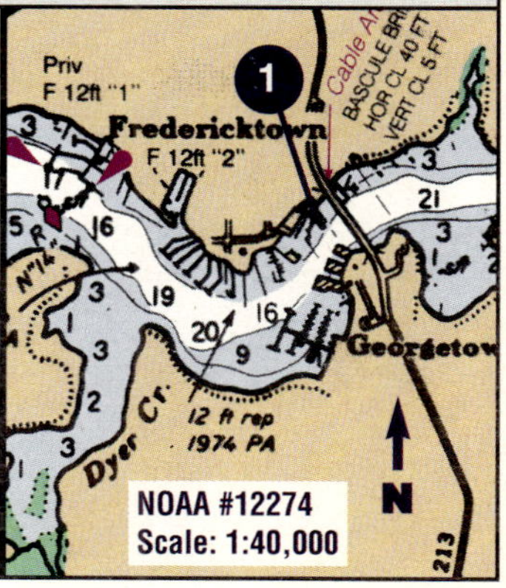

© Maryland Office of Tourism

Houses line the waterfront in historic Chestertown.

Bridge, and straight on to Fort McHenry. To reach the Inner Harbor, turn to starboard at Fort McHenry up the Northwest Branch, and keep following the channel. If you're arriving by night, take it slowly; the city's bright lights and abundance of flashing channel markers make it tricky. Don't worry about depth though, since ships and tugs move all around this working harbor. Once you reach Inner Harbor, you may contact the dockmaster on VHF 16 or at (410) 396-3174 for information about anchoring or the local marinas.

The next major stop south of Baltimore is **Annapolis**, Maryland's capital. It's also the self-proclaimed sailing capital of the nation, and for good reason. Spend a Wednesday night during summer and watch scores of boats compete in Annapolis Yacht Club's Wednesday night race series, which often ends with a spectacular spinnaker finish at Eastport Bridge. This is home to some of the country's best sailors and top boats. Still, there's plenty to do on land, much of it closely linked to Annapolis' history as a colonial port. With its cobbled main streets and tree-lined, brick sidewalks, Annapolis breathes history and charm. Take a walking tour led by guides in colonial costume, or buy a tour on tape, which will guide you to historic buildings and homes. Be sure to tour the State House (410) 974-3901, the oldest in continuous legislative use in the country, and visit the U.S. Naval Academy (410) 267-6100, where you can watch midshipmen go through their paces. When you tire of walking, stop at the restaurants and bars lining City Dock and throughout town. No matter what your appetite is like, you're likely to find a restaurant to suit it. Once you've revived, take your pocketbook to shops and stores that offer everything from fine art to blue jeans to Maryland crafts. Annapolis is a wonderful town that offers a little something for everyone, whether you want to kick up your heels or relax with a quiet evening on the water.

After passing beneath the William Preston Lane Memorial Bridge (commonly called Bay Bridge), green can "1" is a target where you can make a wide turn to starboard and head into the Severn River. Red nun "6" and green can "7" mark the inner channel entrance to the Severn; from here, it's 1.5nm to Annapolis Harbor. Leave the 40-foot 4sec flasher marking Greenbury Point to starboard. On the western shore at Back Creek, you'll find **Annapolis Landing Marina**, a full-service marina with showers, a laundry, pool, picnic area, badminton, and volleyball.

If you're going to Spa Creek, you can continue to the Naval Academy seawall, keeping flashing green 4sec "11" and the Horn Point Shoal pole, a 6sec flasher, well to port. Turn to port as you near the seawall (the marks are green can "1" and flashing red 2.5sec "2"), and you're in Spa Creek. Here you'll find the Eastport Bridge straight ahead. A sign informs boaters of the bridge's schedule; the operator monitors VHF 13. The Eastport section of town is to port. Annapolis Harbor does have moorings and docking; call the harbormaster on VHF 16 or (410) 263-7973.

Once you leave Annapolis, on your way down the western shore, linger over Thomas Point Light (leaving it to starboard), one of the remaining working screwpile lighthouses on the Chesapeake. Then head west-southwest toward the marks leading you to **Galesville** on **West River,** and neighboring **Rhode River**. West River is further evidence that good things come in small packages. It won't take you long to stop there, but Galesville is a charming small town with several good restaurants, and the river is quiet and lovely.

From Thomas Point, head 3nm southwest to flashing green 2.5sec "1A," and 15-foot flashing red 2.5sec "2." To West River, travel 2nm southwest to 14-foot flashing red 4sec "4." Turn south into Galesville, leaving green "5" to port and looking for 15-foot flashing red 4sec "6." On the right you'll find several marine facilities. To find Rhode River, head northeast to 16-foot flashing red 4sec "2" marking Dutchman Point, then turn north into the river, leaving green "3" to port. Up here you'll find a marina.

Your next destination is **Solomons Island**, a longtime fishing community that's working hard to take on Annapolis as a sailing mecca. With gorgeous water views nearly everywhere you turn, just about everything is within walking distance, including Calvert Marine Museum (410) 326-2042. Founded in 1970, this museum focuses on marine paleontology, estuarine biology, and maritime history, linking them to the **Patuxent River** and its past. The museum offers tours in historic vessels, and you might get to watch some model skipjack racing. Solomons has several fine restaurants and bars—indoor and outdoor—with plenty of nightlife. It's an excellent fishing area as well, and every August it's home to *Yachting* Race Week, part of a national sailboat racing series that attracts top boats.

Head south down the western side of the bay, using Drum Point light as your turning mark into the Patuxent River. As you look toward Solomons on the right, avoid The Flats ahead, an extremely shallow shoal that has a channel on either side. Both channels have good water, but the inside channel along the shoreline is quicker. From Drum Point, travel northwest .75nm to red "2" and green "1," which marks the innermost edge of The Flats.

From there, the 12- to 23-foot-deep channel is well marked into Solomons proper. As you enter, you'll see a small, manmade island called Molly's Leg; turn to port at the junction buoy here and you'll find a place to dock directly behind Molly's Leg on Janes Point. You can leave Molly's Leg to either side as you head into Back Creek. In Back Creek, more marina facilities can be found.

Solomons is also an ideal jump-off point for cruising the Patuxent River. Named after the Patuxent tribe that the English settlers found on the river's banks, this is, on average, the deepest of the bay's rivers. Yellow and white perch spawn here, while striped bass, alewife and other species make it their seasonal home. Travel upriver and enjoy some great gunkholing once you round the point just past the Route 4 bridge. Follow the channel to St. Leonard Creek. Broomes Island, just past St. Leonard Creek, is also a pleasant anchorage.

After the Patuxent River, travel south to the last major river on the bay's western shore and one of its biggest tributaries: the **Potomac River**. The enormous Potomac travels 400 miles from its source in West Virginia's Allegheny Mountains. While exploring it in 1608, Capt. John Smith wrote that the fish were "lying so thicke with their heads above water, for want of nets, we attempted to catch them with a frying pan." The Potomac offers beautiful scenery and, of course, it's the only way by water to Washington, D.C. It's also the way to historic St. Mary's City, where two ships landed in 1634 and established the colony of Maryland. St. Mary's City was the state's first capital. An 800-acre living history museum, St. Mary's City is also home to the *Maryland Dove*, a full-scale replica of a typical 17th-century square-rigged merchant ship. Seeing the *Dove* tied along the river's pristine banks makes it easy to imagine being those early river travelers. This is also home to St. Mary's College of Maryland, founded in 1840 and now producing some of the country's top collegiate sailors.

To enter the Potomac, round Point Lookout, then stop in **Smith Creek.** There's a fishing center, marinas, and restaurants. **St. Mary's City** is a brief car ride. You can also stay in St. Mary's River. From Point Lookout, travel

P.O. Box 3350
980 Awald Drive
Annapolis, MD
(410) 263-0090

2

p. 196

ANNAPOLIS LANDING MARINA

All are welcome at Annapolis' finest and friendliest marina, easily accessible by both land and water. Slips are equipped with water, 30 or 50 amp service, telephone and cable TV hook-up, and a de-icing system. Our 165' fuel dock offers competitively priced Amoco diesel and gas, pump-out, a dock shop, and we have a courtesy vehicle!

Amenities include clean showers, laundry, a pool (coming in '94), picnic area, grills, badminton, volleyball, bike rentals, and more. Our deck features a cafe serving during the season, and a year-round meeting room for all events.

A canvas shop, marine, engine repair shop, charter boat service and "Boat and Breakfast" are on location.

Baltimore Tides; High: -1:38, Low: -1:49, Mean .9 feet

$.75 - 1.50 ft/day
109 Berths, 100' LOA

Amoco
Gas/Diesel: Volume discount

Fresh Water, Phone, Cable TV
110 & 220 V, 50 Amps

Several nearby - courtesy vehicle available

Several nearby - courtesy vehicle available - various prices

MasterCard, Visa, Discover

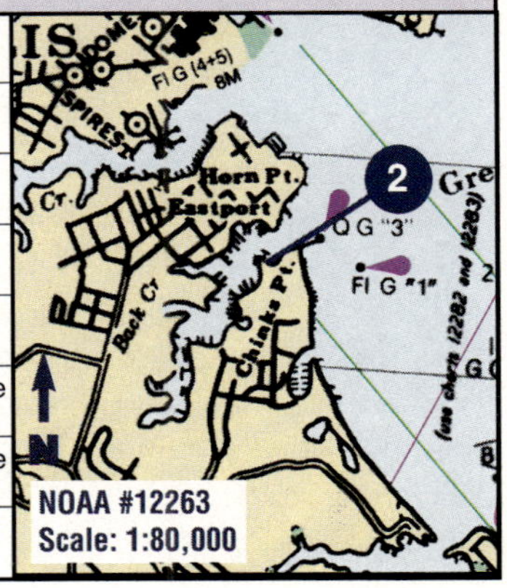

northwest 6nm to a flashing red (2+1) 6sec junction buoy, then head north into the river. About 5nm up is **Carthagena Creek**, marked by green "3" and flashing red "4." Follow the channel and you'll find a marina, or travel 3nm upriver and drop the hook in the beautifully quiet anchorage off historic St. Mary's City.

Because it's a long trip up the Potomac, you may want to stop in one of the marine facilities along the way: in **Saint Patricks Creek,** at the mouth of the **Wicomico River**, or in **Piscataway Creek,** just a few miles south of the city. Plan to spend some time in Washington D.C. The Metro is clean, inexpensive, and easy to ride, and it will take you all over the city. The best in music, entertainment, food, museums, and exhibits are easily accessible for visiting boaters. The Smithsonian "village" of museums is a must (202) 357-2429.

As you approach **Alexandria, Va.**, and **Washington, D.C.**, stay well within the channel, as depths come up precipitously in some places. The channel narrows as you approach Woodrow Wilson Memorial Bridge. This bridge (vertical clearance 50 feet) has a restricted schedule (which Congress mandated), so contact the operator at (202) 727-5522 for information. Once through, Alexandria will be to port. Follow the channel, leaving the junction buoy marking the entrance to Georgetown Channel to starboard. About .5nm further is a second junction buoy marking entrances to the Anacostia River and Washington Channel. There are marinas and access to the capital by Metro.

Eastern Shore: Tolchester to Crisfield

After visiting Maryland's western shore, you'll find a comfortable blend of activity and quietude traveling the Eastern Shore. Begin at **Rock Hall**, across the bay from Baltimore. This small watermen's community, built on fishing, crabbing, and oystering as a way of life, has recently undergone some major changes. But it still retains its local flavor, and the recently opened Waterman's Museum (410) 778-6697 will help you appreciate its history. Rock Hall has excellent seafood at several restaurants, as well as bars, antique stores, and shops that make it worth an overnight. It's recommended, however, that you have a detailed chart for the approaches to Rock Hall Harbor and Swan Creek. The area is tricky with shoals, and a wrong move can run you aground quickly. Likewise, although charts show a northern shortcut across the shoals at Swan Point, it's only advisable for those with local knowledge.

The best and safest way into Rock Hall is to locate flashing red 2.5sec "8" marking the end of Tolchester Channel. Then look for flashing green 6sec "7," the entrance to Swan Point Channel, and from there travel 2.5nm southeast to green can "1." Turn north, leaving green can "3" to port, and look for flashing red 4sec "4," Huntingfield Point light. From here you have two choices: head right into Rock Hall Harbor or straight ahead into Swan Creek. For Rock Hall Harbor, after passing flashing red 4sec "4," look for green can "1," then flashing red 2.5sec "4" marking the jetty entrance. Once in the harbor, mind the channel markers closely.

If you continue north toward Swan Creek from Huntingfield Point Light, look for green can "5" and then 15-foot flashing red 4sec "6" along Gratitude Point. Follow the channel. After red "10," you'll find numerous marinas.

Leaving Rock Hall, head south into the **Chester River**. This is one of the loveliest rivers on the Eastern Shore, with wonderful tributaries and anchorages. Ducks, geese, bald eagles, and other birds make their homes here, and the waters are spawning areas for shad, alewife, and striped bass. The Chester River has two other major advantages: Chestertown and the entrance to Kent Narrows.

Chestertown dates from 1668, and with its brick, tree-lined sidewalks and quiet atmosphere, the town invites you to relax, take in some terrific architecture, and contemplate the past. Designated a "port of entry," it was a thriving trade center for the Eastern Shore's agriculture industry. The Customs House, dating from the 1740s, and many of the stunning brick homes and buildings built in that era are open for tours. Several beautiful inns and taverns serve up history as well as delectable dining. Chestertown is also home to Washington College (410) 778-2800, founded in 1782 and now the oldest liberal arts college in the country. If you'd like to stretch your legs, the Kent County Chamber of Commerce (410) 778-0416 in town publishes maps and descriptions of several bicycle trips, all in luscious Eastern Shore farmland. If a drive is more your speed, you can rent cars downtown.

To reach Chestertown, follow the channel down past Eastern Neck Island National Wildlife Refuge and around the big elbow, turning northeast as the river starts to narrow. On your way up you can pull into **Langford Creek**. Look for green and red junction buoy "LC" marking the creek entrance, then follow the markers. You'll find a place to dock on the creek.

Approaching Chestertown, you'll find facilities that have a courtesy car so you can visit Chestertown 3 miles away. Or you can anchor in the river and dinghy to a small park at the foot of High Street, where sail and steamships once tied up to unload their cargo.

When you head back down the river for **Kent Narrows**, get ready to move at a slightly faster pace. Kent Narrows is an up-and-coming destination, especially with the powerboat crowd. The restaurants dish out a wide range of seafood, and several dock bars offer nightlife, live music, and entertainment. On the western side, shop till you drop at the factory stores within walking distance, or take a cab to top-notch golfing 5 miles away. If you visit at the end of July and early August, you'll see the Kent Narrows Powerboat Challenge, an annual weeklong event that features offshore hydroplane races. Still, Kent Narrows remains an important port for watermen. Check out the watermen's dock to see some beautiful working boats.

While Kent Narrows is a fun stop and the shortest way to Eastern Bay and Saint Michaels, we advise keeping a chart handy and arriving during the day. The channel heading in on the north side is narrow, winding, and leaves little room for error. The red and green 6sec flasher (2+1) "JC," at the bottom of the Chester River's big elbow, marks the entrance. From here, follow the channel religiously. The Army Corps of Engineers recently finished dredging the channel, but some local boaters have reported skipping hard across the bottom with a 7-foot draft. (If you've any doubts, you can go around Kent Island, which is a much longer trip to the Narrows, Eastern Bay, and Miles River, but with no questions about depth.) The Narrows also has a drawbridge that opens every half-hour between 6 a.m. and 9 p.m., May 1 to Oct. 31. Off-season, it's on demand, and the bridge monitors VHF 13.

Once in the Narrows and through the bridge, we recommend docking at **Lippincott Marine,** which is on the port side of Marshy Creek. Look for red "2" and green "1" to guide you to Marshy Creek's entrance, then follow the channel markers (it's about 8 feet deep) till you see Lippincott to port. The marina has a pool and cabana, laundry, and restaurants within a mile.

Once you've taken in the hustle and bustle of Kent Narrows, get ready to slow the pace a little. **Saint Michaels** is a popular place, and it's especially wonderful by boat. After docking, you can spend hours walking up and down small, tree-lined streets, sampling restaurants, antique shops, clothing stores, and more mundane but necessary places like the grocery and drug store. You can stop at the seafood restaurant located on the water next to the Chesapeake Bay Maritime Museum (410) 745-2916 (its red building and white sign are impossible to miss). Boaters can dock for free for two hours to enjoy a variety of seafood inside or on the deck. Don't neglect the maritime museum, either. This is one of the best on the bay, featuring dozens of historic boats, a building entirely devoted to the evolution of the steam engine and the history of steamships on the Chesapeake, a display on waterfowling on the bay, and a gift shop with a wonderful maritime book selection. Saint Michaels is lovely year-round, whether during the hectic July 4th holiday, or on a cool, quiet autumn weekend.

To get there, head south out of the Narrows into Eastern Bay, and look for flashing red 4sec "4" off Tilghman Point as your starting point to navigate the Miles River. Find flashing red 4sec "2" and green can "3" marking the entrance to Saint Michaels. To starboard is Hooper Straits Lighthouse, a screwpile lighthouse now a land-bound exhibit at the maritime museum.

From Saint Michaels you'll want to head for another charming Eastern Shore town called **Oxford.** Settled along the banks of the **Tred Avon River**, delightful Oxford has managed to stave off some of the tourism that

Lippincott Marine is a full-service marina located in the prime yachting area of Chesapeake Bay on the Eastern Shore of Maryland. This fine marina is open all year round and offers many extras such as a new pool & cabana, laundry, tile showers, vending machines, ice, dinghy racks, store with basic maintenance supplies, ample parking, dock carts, fresh water & electric at every slip, pumpout station, excellent restaurants, and a deli & liquor store within one mile.

Their full service yard provides a 25-ton Travelift, highly trained mechanics for diesel & gas engine repairs, outdrive repairs, elec. & electronic repairs, wood & fiberglass repairs to hull, deck & interiors, specialists in the latest "State of the Art" blister repairs and blister preventive coatings, awlgrip & imron painting, and repairs and alterations to rigging.

The Lippincott Marine is an authorized sales & service dealer for Northern Lights generators & Cruisair A/C. The same attention is paid to little jobs as well as large ones. Lippincott Marine is proud of the work they do and they welcome you to stop by and enjoy all of their many services and fine facilities.

Baltimore Tides; High: -1:40, Low: -1:28, Mean 1.2'

$1.25 / ft per day
200 Berths, 70' LOA

Prop., Sail, Elec., Rigging, Gas, Diesel, Wood, Fiberglass

Fresh Water
Elec: 110V, 220V, 50 Max Amps

1.5 miles/ 0.5 to Quick

Several Seafood Restaurants in the Area - Pick-up Serv. Aval.

MasterCard, Visa,

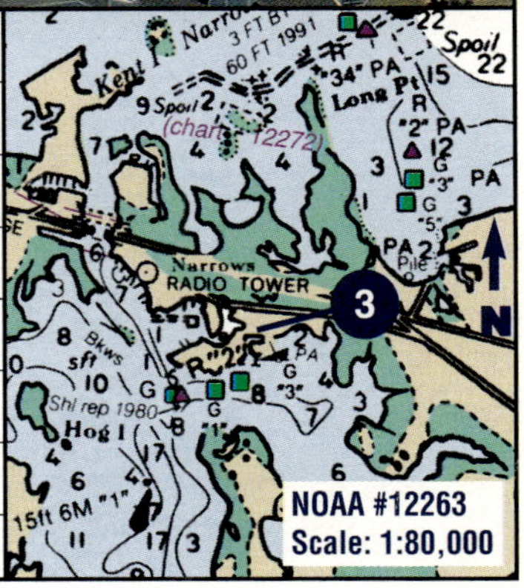

The Chesapeake Bay Maritime Museum in Saint Michaels.

sometimes taints other harbor towns. Here, you can walk along The Strand and picnic on the grass under tall trees overlooking the river and perhaps see some log canoes out sailing. Wander among some of the boatyards, for they are full of history; during World War II, Oxford Boatyard (410) 226-5101 employed 300 people and built patrol boats. Ride the Oxford-Bellevue Ferry, which started in 1683 rowing people across, then switched to sail, steam, gas, and now diesel. It's the country's oldest privately owned, continuously operated ferry, and it's also a shortcut by land to St. Michaels, just 7 miles from Bellevue. Oxford is also home to some excellent restaurants and old taverns.

From Eastern Bay, head south to the Chesapeake "proper," where you'll leave Poplar Island well to port. There is a shortcut to Oxford just south of Poplar Island via Knapps Narrows, the channel that separates Tilghman Island from the mainland. Shoaling and narrow, however, the shortcut should be reserved for those with local knowledge.

The safest way is to head for Sharps Island Light off Black Walnut Point (note the lighthouse is listing rather severely due to the force of ice pushing against it during the winter). From Sharps Island, travel 2nm east into the Choptank River to green can "9" and over to flashing red 2.5sec "10." It's 5.8nm to the Choptank River Light, a 35-foot 4sec flasher that's surrounded by riprap marking the entrance to Tred Avon River. Turn north, leaving 15-foot flashing green 2.5sec "1" at Benoni Point well to port. Travel 2.5nm to flashing red 4sec "2," then continue past The Strand to the entrance of Town Creek to starboard.

Once you leave Oxford, it's a short hop back into the Choptank River and up the river to **Cambridge.** The 68-mile Choptank, the eastern shore's longest river, is named for an Indian word describing a place where the current runs strongly. Harriet Tubman was born here on a river plantation, and she later used the river's marshes and banks as part of her Underground Railroad. Cambridge is the largest community on the river, and visitors can find many restaurants, tours, and activities. Take a walking tour of the town's historic buildings and federal-style architecture. Visit the James B. Richardson Foundation Maritime Museum (410) 221-1871, dedicated to the memory of the shore's craftsmen and boatbuilders. For a unique look at the bay's history, call the Brannock Maritime Museum (410) 228-1245, which opens by appointment only. Visit the construction site of a new skipjack, the *Nathan of Dorchester.* And don't forget a trip to the Wild Goose Brewery (410) 221-1121 for a taste of their classic ales.

From Choptank River Light, follow the well-marked channel toward the southeast. The last creek to starboard before the bridge is Cambridge Creek, where you'll find marine facilitites within walking distance of town.

After leaving Cambridge, our travels continue toward Crisfield, but make a stop in **Slaughter Creek** off the Little Choptank River. There you'll find a marina as well as a restaurant with a floating bar that's made from an old Coast Guard station.

Though far down the bay (just north of the Virginia border), **Crisfield** is a must-see all year round, but even more so if you can make it on Labor Day weekend. That's the date of the annual National Hard Crab Derby (now in its 46th year), which is three days of festivities that energize this tiny community. Surrounded by more seafood than you can eat, you'll watch crab-picking contests, the crowning of "Miss Crustacean," live entertainment, and be witness to the crab race. A small town with a true waterman's atmopshere, Crisfield is an appropriate way to end a trip down Maryland's Eastern Shore.

From the Little Choptank, head down the bay, passing the Honga and Nanticoke rivers and Hooper Straits. Kedges Straits brings you into Tangier Sound. Mind the channel markers through here, there are some shoals, wrecks, and obstructions outside. Look for Janes Island Light, a 37-foot 4sec flasher, and green can "1." Turn northeast into Little Annemessex River. Head for the quick flashing red "2" (there's a six-foot bump behind it), then 15-foot flashing red 4sec "4." After red "8" and flashing green 4sec "7," turn northeast and head straight for Crisfield. Ahead you'll see a city dock and marina within walking distance of grocery stores, restaurants, bars, and shopping.

Maryland's Other Shore: Ocean City

The Chesapeake Bay is such a large part of Maryland it's sometimes easy to overlook another fine boating and fishing area: **Ocean City**. Located on the Atlantic Coast barrier island, Ocean City is a popular jump-off point for charter fishing. The bays behind it—Isle of Wight, Sinepuxent and Little Assawoman—are prime boating grounds. Ocean City is bordered to the north by Fenwick Island and to the south by Assateague Island and is connected to the mainland by the Route 50 bridge. Like many beach communities, its attractions are plentiful: a boardwalk, countless restaurants, bars, and stores. But it also has some unique features, such as the Ocean City Life Saving Station Museum (410) 289-4991, offshore powerboat races, and several top-level sportfishing tournaments (more than 200 boats were expected at the most recent 19th Annual White Marlin Open in August).

Approaching the Ocean City Inlet, look for flashing red 4sec "4" marking the northern jetty and flashing green 4sec "5" marking the south. It's best to approach the inlet from the north where there's deep water; to the south are shoals and rocks, and part of the jetty continues underwater. This inlet is not as ferocious as some, but a northeast or southwest wind can kick up a good chop. Once inside, follow the well-marked channel to the junction buoy (flashing red (2+1) 6sec and green "OC"). From here, if you turn to starboard toward the drawbridge, you'll find marinas and a fishing center. The drawbridge has about 18 feet of clearance at low tide and monitors VHF 13.

Whether you visit Maryland via the Atlantic Ocean or Chesapeake Bay, your trip south will take you into Virginia waters. And as you look back, you're sure to agree that no matter how you approach it, Maryland is a boater's paradise.

Shoreside and Emergency Services

Ambulance; Fire; Police: 911

Coast Guard:

Ocean City (410) 289-7559
Still Pond (410) 778-2201
Curtis Bay (410) 789-8050
Annapolis (410) 267-8107
St. Inignoes (301) 872-4344

Radio Telephone:

VHF Ch. 25 & 26; I.D.: Baltimore Marine Operator
VHF Ch. 26; I.D.: Ocean City Marine Operator
VHF Ch. 28; I.D.: Cambridge Marine Operator

Tow Service:

Sea Tow Maryland Central, Annapolis
(410) 267-7650 or VHF 16
Sea Tow Solomons, Solomons
(410) 326-6801 or VHF 16
Sea Tow Lower Potomac, Ridge
(301) 872-5144 or VHF 16
Georgetown Yacht Basin, Georgetown
(410) 648-5112 or VHF 16, 9
Deckelman's Marine Towing
(410) 391-6482 or VHF 16
Boat Tow, Essex
(410) 780-4121 or VHF 16, 9
Abraham & Rebecca Towing and Salvage, Tolchester
(410) 778-3404 or VHF 16, 9
Baltimore Marine Recovery, Pasadena
(410) 440-3319 or VHF 16
Baker Charters & Marine, Kent Narrows
(410) 827-8204
Ann Bay Towing, Inc., Annapolis
(410) 263-1260 or VHF 16, 9
Casa Rio Marina, Mayo
(301) 261-7111 or VHF 16
Conlyn Marine Services, Inc., Tracy's Landing
(410) 261-5151 or VHF 16
Chesapeake Charter & Service, St Michaels
(410) 745-3233 or VHF 16, 9
Tow Jamm Marine, Neavitt
(410) 745-3000 or VHF 16

Local Services

Joins page 196
MARYLAND
Chestertown
DEL
(See page 190)
Annapolis
Washington D.C.
Arlington
Alexandria
Manassas
Easton
MARYLAND
Choptank R.
Cambridge
Port Tobacco
VHF-FM
KHB-36
162.55 MHz
Patuxent River
Salisbury
Fredericksburg
Colonial Beach
Potomac River
Port Royal
Point Lookout
Pocomoke City
Crisfield
VHF-FM
WXM-57
162.40 MHz
Smith Point
U S C G
(804) 336-6511
Chincoteague
VIRGINIA
Irvington
Chesapeake Bay
Rappahannock River
Onancock
Richmond
Deltaville
U S C G
(804) 787-9527
U S C G
(804) 725-2125
VHF-FM
WXF-65
162.475 MHz
Mobjack Bay
Hopewell
York R.
Williamsburg
Gloucester Pt.
Atlantic Ocean
Yorktown
U S C G
(804) 331-2000
Newport News
James River
Hampton
U S C G
(804) 464-9371
VHF-FM
KHB-37
162.55 MHz
Norfolk
Portsmouth
Virginia Beach
U S C G
(804) 421-6200
I.C.W.
Dismal Swamp Canal
1 inch= 19.5 miles
Scale 1:1,235,000
0
19.5
39
58.5
78
Joins page 218
N
1
2
3
4
5
6
7

Old Dominion

Many Atlantic Coast states hold their share of historic sites, but few lay claim to the past as does Old Dominion. The Indian heritage lingers in the rivers and streams that cut through the land with names like Piankatank, Wicomico, and Nansemond. In 1607, Captain John Smith sailed into the Chesapeake Bay. Admiring its beauty and bounty, he decided that this would be a good place to establish the first colonial settlement. He graciously named it and the adjacent river in honor of his patron, King James of England. Six generations later, the settlers (now calling themselves Americans) fought off the British at the Battle of Yorktown, and won independence for their new nation. After some debate, they laid out the plan for the nation's capital at the head of the largest river on the Bay and named the new city for the Democracy's first President: Washington. Four score and more years later, the world's first ironclad warships, the Yankee *Merrimac* and the Confederate *Monitor* dueled (to a draw) in Hampton Roads for control of the still-young nation. And ever since, the bounty of the land, the seafaring tradition, and the power of the growing nation have shaped the development of Virginia. Indeed, to travel Virginia is to take a trip into our nation's history.

Nowhere is the power of the nation and Virginia's seafaring tradition more evident than at the Newport News shipyards and the Norfolk Naval Station, home to the world's largest collection of warships. It's an awesome sight when a 1,100-foot carrier enters the harbor, its flight deck lined with thousands of sailors eager to be home after months at sea. But equally impressive is the brilliant array of stars to be seen on a clear night in a secluded anchorage off Mobjack Bay or Onancock Creek or the Piankatank River. And there are thousands of other enchanting sights on Virginia waters—a mecca for working watermen, sailors, fishermen, and cruisers.

Boats line the Hampton waterfront.

Marine Facilities and Services

Facility	Phone	Monitors VHF Channel	Working Channel	Number of Transient Berths	Number of Transient Moorings	Seasonal / Year-round	Maximum Length Over-All	Approach Depth in Feet at MLW	Dockside Depth in Feet at MLW	Hookups: Fresh Water / Phone / Cable TV	110V★ 220V▲ 3 Phase■ Maximum Amps	Ramp / Dinghy Dock / Launch Service	Railway / Lift / Crane: Capacity (tons)	Repairs: Propeller / Sail / Electronics / Rigging / Gas / Diesel / Wood / Fiberglass	SCUBA: Air / Rentals / Equipment / Charters	Gas / Diesel Fuel	Fuel Brand	Marine / Groceries / Bait / Ice / LPG / CNG	Showers / Laundry / Pumpout / Restrooms	Hotel / Pool / Tennis / Golf	Public Phone / Restaurant / Snack Bar	MasterCard / Visa / Discover / AmEx
1 Windmill Point Resort (p. 209)	804-435-1166	16	68	20		Y				FC	50	All			C	GD	Ex	All	All	All	All	MVA
2 Dozier Marine Center (p. 210)	804-776-6711	16	72	6		Y	60	6	6	All	★▲ 100		L50	All		G	Tex	All	All	P	P	MV
3 The Tides Inn (p. 211)	804-438-5000	16	68	19		S	100	10	6	All	★▲■	D				GD	Ex	MGI	All	All	All	All
4 The Tides Lodge & Marina (p. 211)	804-438-6000	16	68	42		S	100	10	6	All	★▲50	D				GD	Ex	MGI	All	All	All	All
5 York River Yacht Haven (p. 213)	804-642-2156	16	69	30		Y	185	8	16	F	★▲■ 100	DL	L35	All		GD	Tex	MGILC	All	P	All	All
6 Waterside Marina (p. 214)	804-625-2000		68			Y	300	45	25		50								SL	H	R	MV
7 Tidewater Yacht Agency (p. 215)	804-393-2525	16	68	100		Y	120	10	8	All	★▲■ 100	D	L60	All		GD	Tex	MGI	All	HP	All	All

Column groups: DOCKAGE, SERVICES, SUPPLIES, AMENITIES, COAST GUARD. See Adjacent Chart for Coast Guard Information.

Information in these listings is provided by the facilities themselves.

Trophy bluefish, trout, and flounder as well as stripers, croaker, and spot make the Lower Chesapeake Bay attractive to anglers. The famed Chesapeake blue crabs are everywhere, which you'll quickly realize when forced to dodge the numerous floats tied to crab pots. Because of the floats and the occasional wooden-staked fish traps, night traveling on the Lower Bay and its rivers can be a challenge. Navigating during the day is fairly straightforward, but has gotten more difficult because the Coast Guard seems to rearrange the buoys rather frequently, and some of the unlighted buoys that used to be channel markers have been removed. These buoys are shown as black and white or yellow nuns on older charts.

There are several things to keep in mind when cruising the Lower Chesapeake Bay and its tributaries in Virginia. In particular, the Bay's notorious summer thunderstorms can be hazardous to cruising vessels. The Lower Bay is always busy with freighters, colliers, container ships, tankers, tugboats, and Navy vessels, so keep a sharp lookout at all times. And finally, there's a good chance you'll run aground if you stray from the channels. The waters shoal quickly, but the bottom is soft. As one veteran sailor put it: "If you haven't got mud on your bottom, you haven't been anywhere on the Bay." If you do run aground, tides in the Lower Bay will offer little help, rarely varying more than 3 feet.

In addition to Chesapeake Bay, Virginia has an ocean coastline east of the Delmarva Peninsula and east of Virginia Beach. This area between Chincoteague Bay and Cape Charles is mostly marshland behind a string of uninhabited islands. Twisting waterways between the islands are shallow and are generally plied only by small fishing boats. Chincoteague Inlet gives access to Chincoteague Bay of Maryland, but most cruisers bypass this area for the greater options of the Chesapeake Bay—a body of water as richly laden with history as the land that surrounds it.

Western Shore: The Potomac River to Virginia Beach

The northernmost body of water to border Virginia (though actually belonging to Maryland) is also one of the Chesapeake Bay's biggest tributaries, the **Potomac River**. The enormous Potomac travels 400 miles from its source high in West Virginia's Allegheny Mountains, splits Maryland from Virginia, and spills over the treacherous Potomac Falls and into the bay.

From the Bay, the entrance to the Potomac is guarded on the north by Point Lookout Light, a 39-foot flashing 5sec horn, standing almost a mile south into the Bay, and by Smith Point Light, a 52-foot flashing 10sec horn, two and a half miles east of the point in the Bay. It is 96nm from Smith Point Light to the Jefferson Memorial in Washington, so either plan on traveling fast, or go slow and enjoy the scenery and a night's layover along the way. There are literally dozens of choice anchorages, marinas, and attractive towns along the river. Minimum depth to Washington is 18 feet, so don't fret a grounding unless you aren't paying attention.

Your first anchorage is Virginia's **Coan River**, about 10.5nm upriver from Smith Point Light. The first major anchorage on the Maryland side of the river is in **Smith Creek** at nautical mile 12 where you will find a fishing center, marinas, and restaurants. More waterfront restaurants and marinas can be found on the **Yeocomico River** at mile 14 on the Virginia side.

The entrance to the **St. Mary's River** and **St. George Creek** is at nautical mile 15. It is here that the square-rigged merchant ships the *Ark* and the *Dove* landed in 1634 to establish the colony of Maryland and build its first capital. Beautiful St. Mary's City is now an 800-acre living history museum and home to a full-scale replica of the *Dove* (301) 862-0990.

Nomini Bay, (Va.) at 26nm is used primarily by commercial fishermen and has little to offer recreational boaters. On the other hand, **Breton Bay** (Md.) at 27nm is a favorite anchorage for yachts. Depths are generally about 6 feet, but deeper draft vessels visit the area frequently. Marinas and supplies can be found up the bay on Combs Creek.

The **Wicomico River** (Md.) at 32nm offers many fine anchorages and marinas, as do many of the smaller tributaries along the Potomac as you head north. Dockage, supplies, and repairs can be found in Mattox Creek (Va.) at 38nm. The U.S. Marine Corps Base at Quantico, Va., at 70nm is a major landmark as you head upriver.

Soon you will reach Mount Vernon (703) 780-2000, George Washington's Plantation on the banks of the Potomac. You will have a much better appreciation of this gracious landmark if you arrive by boat instead of being stuck with the crowds of cars in the parking lot. Limited docking facilities are available for visiting yachtsmen.

Keep a close eye on your charts and depth sounder as you get near the Woodrow Wilson Bridge at 91nm as the river shoals quickly outside of the channel. The bridge has a vertical clearance of only 50 feet and a restricted opening schedule, so contact the bridge tender at (202) 727-5522 for more information. The old port town of **Alexandria,** Va., with lots of good restaurants and boutiques, is immediately to port after the bridge.

Hains and Greenleaf points are about 95nm upriver. Here the river splits, with the mainstream of the Potomac heading to port (northwest); the big marinas in the Washington Channel basin are straight ahead, and the Annacosta River bears off to starboard (northeast). Much of the waterfront area, including all of East Potomac Park, is low and therefore subject to spring flooding. Although a world-class city, all of the buildings in **Washington D.C.** are also low. The skyscrapers common to other large cities are absent. As a result, you can see the Capitol Building and Washington Monument standing high over the rest of the city.

The George Mason Bridge (at 96nm) from Washington, D.C., to Arlington, Va., and the Pentagon marks the end of the buoyed part of the river and the end of the tidal flood currents. The height of the water is still affected by the tides (2.9 feet) for several miles farther up river to the fall-line. Just past the bridges on the Virginia side is a large marina. Thanks to excellent taxi, bus, and metro train service, all the wonders of Washington are easily accessed from any of the marinas. Admission is free at all branches of the Smithsonian Institution (202) 357-2429, including the Zoo, the Air & Space Museum, the National Gallery, and a dozen others. As you would expect, crowds are heaviest on the weekends and in the summer, but all the public buildings are worth the wait in line. There are dozens of special events that take place year-round, so call the Chamber of Commerce & Visitor's Bureau (202) 638-3222 for a list of happenings while you are in town. And, thanks to the number of foreign dignitaries in town, Washington has the best selection of ethnic restaurants in the world (outside of the Big Apple).

NOAA chart coverage extends to the Chain Bridge, 101nm upriver from Smith Point Light. This is a very beautiful section of the river, with impressive public buildings including the Kennedy Center and Georgetown University, along with many interesting commercial and residential buildings lining the banks and clinging to the 300-foot bluffs. The current in the river can be very swift, and there are plenty of rocks about, so watch your position carefully. After heavy rains it is unwise to navigate this section of the river in a cruising boat due to the very fast currents and the amount of floating debris. Small boats are no safer under such circumstances, as evidenced by the number of fatalities here and at the falls.

Back out on Chesapeake Bay, and 52 feet above the waves, Smith Point Light marks a crossroads in your cruise—here you have to decide whether you want to run down the popular western shore of the Bay or challenge the less-traveled route through Tangier Island and along the eastern shore. The western shore is by far the more interesting option with numerous ports and facilities. It's particularly charming if you like to spend the night swinging on an anchor in a lonely cove or creek. And although it's only about 60nm from Smith Point to the big-city offerings at Norfolk, you'll be glad you took time to explore the beauties of the Rappahannock, Piankatank, and York rivers, as well as Mobjack Bay.

It's a short 6nm hop from Smith Point to the 42-foot 6sec flashing Great Wicomico River Light. Stay south of the light before heading west into Ingram Bay. Leave quick flashing red "6" to starboard, then make a hard turn north into Cockrell Creek. A day trip is the best way to go in this area. Leaving the Great Wicomico, you'll be tempted to swing south when you return to the Chesapeake, but don't—there's only 3 feet of water in spots south of the Great Wicomico River Light, so travel a mile east and avoid an unnecessary grounding. The town of **Reedville** on **Cockrell Creek** is home to Virginia's menhaden industry. Many of the early menhaden moguls' mansions have been restored to their former grandeur and make for a pleasant stroll through town. Menhaden is a kind of minnow, trapped in large quantities and sold as fertilizer. The oil is used in cosmetics and paint. Visit the Reedville Fishermen's Museum (804) 453-5371 for a detailed look at the rise and fall of this once-prosperous town.

Between the Great Wicomico and the Rappahannock, there are several great streams that provide good protection and anchorages. **Dividing Creek** lies 5.5nm to the south, the entrance marked by a flashing green 4sec 15-foot "3," and although there are no transient facilities, there's plenty of water for overnight anchorages and a great place to seek refuge if foul weather is forecasted. Another 2.5nm to the south, 19-foot flashing 2.5sec "B" leads the way into **Indian Creek**, with a deepwater channel to a marina which has transient slips.

Beginning in the mountains far to the west, the **Rappahannock River** is one of Virginia's biggest and most beautiful waterways. Depending on the winds and the tide, the entrance to the Rappahannock can be rough as its waters pour into the Bay. We suggest making your run during a moderate tide. From Indian Creek, about 5nm to the southeast is the crab-like Windmill Point Light. Wise boaters will stay within half a mile of the 31-foot flashing 6sec structure before swinging into the

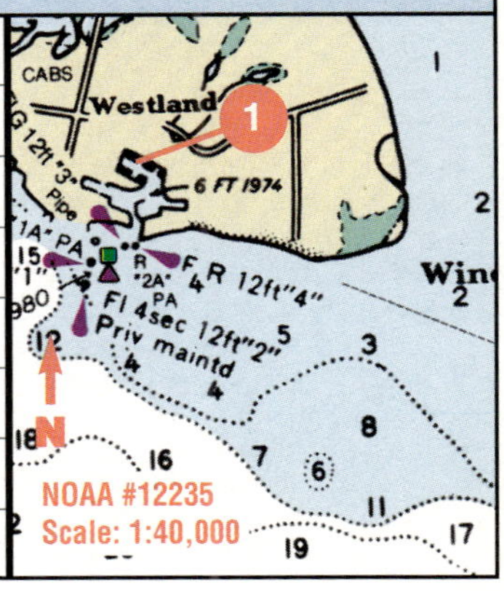

Rappahannock. Just around the bend is the **Windmill Point Resort** in a protected harbor on Fleets Island. The new owners are busily upgrading the facilities, which already include two pools, a golf course, tennis courts, a fitness center, fishing boat rentals, a waterfront restaurant, covered and open slips, and a ship's store. It is a good place to relax and meet fellow floaters.

Directly across the Rappahanock from Windmill Point, and sitting at the tip of a river of land, is **Deltaville**. This charming little town of 2,000 residents boasts a fleet of 3,000 boats and 17 marinas. Although a long way from anywhere by road, Deltaville is a very convenient stopover for boats transiting the Lower Bay. This strategic position has not been lost on the local residents who have made their living building traditional Chesapeake deadrise boats, and now welcome visiting yachtsmen to the many fine marinas and service yards. Southern hospitality isn't a myth in Deltaville.

Broad Creek, the main harbor for Deltaville, is 2nm west of Stingray Point Light, a 34-foot flashing 4sec structure that stands near the spot were Captain John Smith was stung by a stingray while exploring the Bay in 1608. The entrance to Broad Creek, marked by 15-foot flashing green 4sec "1," is dredged to 7 feet. It's narrow but easy to follow for the mile it takes to reach Deltaville and the **Dozier Marine Center**, a full-service, family-owned and operated marina with a clubhouse, pool, and complimentary bicycles among its many amenities. Excellent service and a well-stocked marine store are also available at the Center. Other facilities and anchorages can be found in Fishing Bay on the Piankatank River side of Deltaville.

Another 6nm up the Rappahannock, carrying Highway 3 traffic, is the Rappahannock River Bridge with 110 feet of vertical clearance. Another 2nm takes you to Carter Creek and the tiny village of **Irvington** on the north shore which is dotted with varied boutiques, antique shops, and art galleries. Just to the north of Irvington is the beautiful Christ Church (804) 438-6855. Built in the Georgian style among the towering pines by a wealthy plantation owner in 1732, it was an architectural wonder of the new world.

Leave 15-foot flashing green "1" well to port as you enter **Carter Creek**. A short run will bring you to numerous transient facilities in Irvington, including **The Tides Inn** and **The Tides Lodge**. Owned by two brothers, The Tides Inn is very formal (jacket and tie required after six o'clock), while The Tides Lodge is informal. A Scottish motif is carried through the lodge, which is located across from the Inn on the heavily wooded Carter Creek. Tied up at the Inn is the 127-foot *Miss Ann*, a yacht built in 1926 and now available for charter and cruises after an 18-month long restoration. Slips are available at the Tides Inn and even more at the Tides Lodge, as well as golf, tennis, sauna, bicycles, evening entertainment, and excel-

3
p. 206

The Tides Inn . . .

804-438-5000 Dockmaster 438-4453

P.O. Box 480

4
p. 206

and Lodge!

P.O. Box 309 • Carter's Creek
Irvington, Virginia 22480
804-438-6000 Dockmaster 438-6400

★★★★★

Monitor VHF Channel 16/68

Hampton Roads Tides; High: +2:29, Low: +2:46, Mean: 1.4 feet	Freshwater, Phone, Cable TV Elec: 110V, 220V, 50 Max Amps
$1.35 / foot per day 42 Berths, 100' LOA	On-site and local delivery
All Nearby	Casual and more dressy at both resorts
Exxon Gas, Diesel	MasterCard, Visa, American Express

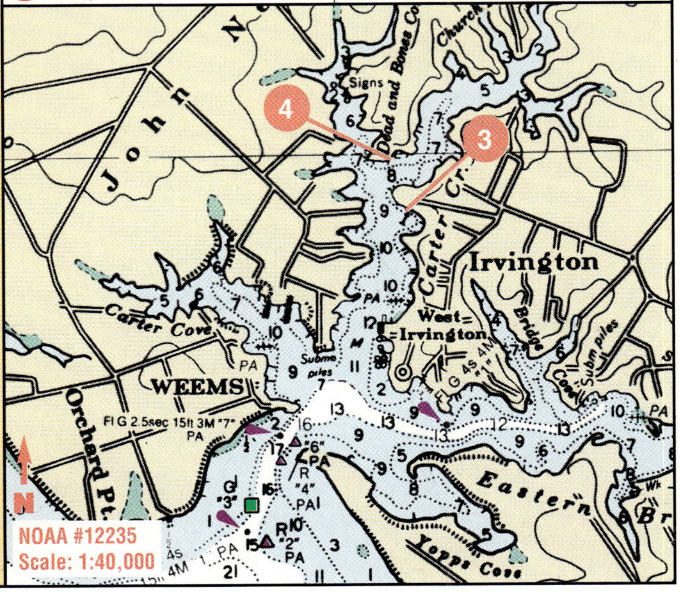

THE TIDES LODGE
Irvington, Virginia 22480

When you cruise the Chesapeake Bay, treat yourself to a few days at The Tides Lodge. It's tucked away on its own private peninsula in the heart of the historic Tidewater region.

Here you set your own pace. Relax and enjoy. How about 45 holes of challenging, uncrowded golf, or complimentary tennis? Rental clubs and racquets are available. There's no charge for many amenities: exercise room, sauna, bicycles, playroom, playground, shuffleboard, putting, heated fresh and dockside salt water pools, oyster roasts, dancing and evening entertainment.

Alternatives to the galley include both a snackbar and sandwich bar poolside, lunch and dinner in the casual Binnacle, cocktails at The Pub, or a five-course meal in the main dining room.

Toss us a line and we'll welcome you ashore to the only resort on the Bay to win the Atlantic Cruising Club's highest award, "Five Bells!"

THE TIDES INN
IRVINGTON, VIRGINIA

For more than 46 years, The Tides Inn has welcomed yachtsmen to its elegant facility - poised on a bluff overlooking the mouth of Carter's Creek in the Rappahannock. The marina at The Tides Inn has become a tradition for generations of yachtsmen transiting the eastern seaboard.

Whatever your needs, the Inn can accommodate you: the finest golf on the Chesapeake Bay, Best Fancy Dining Room (Chesapeake Bay Magazine, July 1993), an acclaimed children's program, swimming pool with beach, small boats, bicycles, tennis, Wednesday night owner's reception - whatever you wish! As a yachtsman, you will particularly enjoy cruises aboard the 1926 classic yacht, the 127' Miss Ann.

People travel from all over the world to visit The Tides Inn, year after year. Don't miss the opportunity to enjoy "One of the World's Finest Small Resorts." Start your own tradition!

lent dining. Since many guests return year after year to these four-diamond resorts, reservations are suggested.

If you treasure more open waters, keep cruising west 5nm from the high bridge to the broad **Corrotoman River**, whose shores are rapidly becoming residential. Many out-of-towners have found Virginia's Northern Neck to be the perfect place to seek refuge from the cold, cruel world. The Corrotoman offers many anchorages: It's as easy as picking a spot and dropping a hook. Entering the Corrotoman, leave flashing red "2" just to starboard. Head northeast a mile to red "4" where you can decide to explore the Eastern or Western branches.

Another option from the high-rise bridge is a visit to the town of **Urbanna**. This town plays host to the annual Oyster Festival (800) 453-6167 in late October or early November. Protected by a man-made breakwater, the well-marked entrance to Urbanna is about 8nm west of the bridge on the south shore of the Rappahannock. Although there's deep water and gorgeous scenery far up the Rappahannock, there are few amenities for transients above Urbanna.

Heading out of the Rappahannock, don't try to cut inside Stingray Point Light because it's surrounded by shoals. And keep your charts handy as you head to the **Piankatank River** —it's a lovely river but a navigational challenge. Leave the fish traps to starboard as you enter, and run 2.5nm to flashing red 6sec "6." Head south toward flashing green 2.5sec "7," leaving flashing red 2.5sec "8" off Stove Point to starboard.

On the south shore of the mouth of the river lies your next destination: **Gwynn Island**. Follow the markers to a swing bridge, leaving the restaurant and marina to port. The current can run 3 or 4 knots here, and it's a main route for working watermen, so stay alert. A few shops and restaurants are within walking distance, but the island is not a tourist attraction. So if an isolated island atmosphere is what you're after, pay a visit to Gwynn Island. The Gwynn Island bridge opens on demand, and the bridge tender monitors VHF channel 13. The waters separating the island from the mainland are known as **Milford Haven** and are home to a Coast Guard station. Half a mile east of the bridge to port is a cove with a pleasant anchorage for shoal-draft boats.

To go farther up the Piankatank, head for flashing red 4sec "10." Then head north into **Fishing Bay** on the south side of Deltaville for anchorages and marinas. The scenery is marvelous, and the anchorages are fine another 5nm upstream, but there are no facilities for transients beyond Fishing Bay.

Leaving the Piankatank, it's a 10nm run southwest to Wolf Trap Light (52-foot flashing 15sec) and another 7nm to New Point Comfort Light, a towering white silo at the entrance to **Mobjack Bay**. The light, now abandoned, is surrounded by shoals so stay a mile to the east and south as you head into Mobjack's pristine waters. It's 5nm from the New Point Comfort Light northwest to the entrance of the **East River**, marked by flashing green 4sec "1." The hamlet of Mobjack lies to port, and an abandoned wharf is to starboard as you enter. The river is wide but shallow, so follow the markers. Off to the east of green "7" is an abandoned tidal water mill once owned by the late Beatle, John Lennon, and his wife, Yoko Ono. To the west, the best anchorages in the **North River** are about 4nm up past red "6," but it's not necessary to travel that far to spend a peaceful night. You can drop a hook just about anywhere in these waters. On the **Ware River**, go past another Windmill Point, leaving 16-foot flashing green 2.5sec "3" and green "5" to port, and anchor south or west of quick-flashing red "6." Less open but with plenty of anchorages is the **Severn River**. Enter the Severn at 16-foot flashing green "1" and go carefully upriver to anchorages off Stump Point, School Neck Point, Bar Neck, and Cod Point.

Mobjack Bay is a popular weekend cruise with few facilities, only safe anchorages and rustic farmhouses dotting the undeveloped shores. It's the beauty of these rivers, untouched by urban sprawl that makes Mobjack Bay a jewel. Blue crabs can be caught at almost every anchorage in the Lower Bay by tying a scrap of meat on a string and dropping it near the bottom. When a crab bites, pull it slowly to the surface, and scoop it up with a net. The clear waters of most anchorages tempt many boaters to take a swim, but beware of jellyfish when the waters are warm.

From Mobjack Bay, it's an easy run to the historic **York River**, whose bottom still holds sunken warships from the battles of the Revolution. Bars and shoals make taking shortcuts dangerous, and tankers ply the waters to the Yorktown refinery. The York River entrance is marked by quick-flashing red gong "18," 5nm southwest of New Point Comfort Light. The river's channel is wide, deep, and well-marked, but make sure you stay well to the north of the flashing white 6sec Tue Marshes Light in the flats off the Goodwin Islands.

The lower areas of the York River often experience heavy boat traffic, usually transients making a stopover. What many boaters don't realize is that the rest of the York River has plenty of water, enough for even deep-draft vessels, and is a great cruising ground. Facilities can be found in a number of creeks at the mouth of the river, and there are several creeks farther north that offer beautiful anchorages.

North of Tue Marshes Light is the **Perrin River**, an easily-entered stream with a marina but very limited anchorages. You'll pass the refinery and wharves on the south shore of the river, and fish traps along the north bank as you head upriver to **Sarah Creek**, a popular destination for weekend travelers and your best bet for dockage while visiting Yorktown or Williamsburg. To enter the harbor, swing north, leaving 15-foot flashing red 4sec "2" to starboard. The entrance is busy and narrow. Able to accommodate up to 300 vessels, **York River Yacht Haven** offers full service and has on-site car rentals. The management knows that the attractions of the area are on land *and* water, and make it easy for you to get to all of them.

The river narrows and is crossed by a swing bridge with a vertical clearance of 60 feet at **Yorktown**, a must-see for history buffs. The trenches in the bluffs have been restored so you can see where American and British soldiers fought in the last great battle of the Revolutionary War, which led to Cornwallis' surrender to Washington. The Yorktown Colonial National Historical Park (804) 898-3400 and Yorktown Victory Center (804) 253-4838 offer films and exhibits illustrating this historic area. There's also the Watermen's Museum (804) 887-2641 with exhibits on the role of watermen throughout history. Unfortunately, there are no marine facilities at Yorktown.

After exploring the York River, head back to the river's mouth at red gong "18." To the south lies the **Poquoson River**, a waterman's haven with numerous anchoring spots in **Chisman Creek**. It's tempting to take shortcuts, but once again, don't—the open waters of the Poquoson Flats have lured many a stranger to a hard grounding. Travel 3nm east to York Spit Light, and head south. Continue 7.5nm to flashing green bell "1" marking the entrance to the very limited waters of the Back River.

Two miles south of bell "1" are the wetlands of **Salt Pond**, which is now lined with houses and waterfront development. Markers quide you easily a mile west from red "2" to the rock-lined cut that takes you 100 yards through the dunes. Anchoring has become difficult because of the heavy traffic. But if darkness or a storm is approaching, consider spending the night at the marinas in Salt Pond.

The city of **Hampton** is the next major port of call and is one of the most frequented cities on Virginia's coast. Numerous markings will help to guide you into the waters of Hampton Roads at the junction of the James and Elizabeth rivers. On the west side of the entrance to Hampton Roads is the square, brown brick Chamberlin Hotel (804) 723-6511, an eight-story structure on the grounds of Fort Monroe (804) 727-2111 where Confederate President Jefferson Davis was held prisoner after the Civil War. Also of interest is the Hampton Carousel (804) 727-6347. Built in 1920 and recently restored, it's open for rides from April to December. But the big attraction in Hampton is the Virginia Air and Space Center (804) 727-0800, featuring space craft, 12 full-sized aircraft, and a 280-seat IMAX theater.

As you enter **Hampton Roads**, keep a close watch because these waters are very busy. The shipping channels are well-marked and well-used. Prudent travelers usually sail just outside them. The surrounding waters are quite deep, and there are no hazards. The entry to the harbor is nearly a mile wide, marked on both sides by concrete islands. The island on the east abuts Fort Wool where in

Adjacent to the historic triangle of Yorktown, Williamsburg, and Jamestown is the most comprehensive marina in Virginia, the York River Yacht Haven. With accommodations for over 300 vessels this marina offers facilities and amenities unsurpassed anywhere. It's complimented by a management and staff dedicated to making a yachtman's stay enjoyable.

Dockside amenities include: spotless air conditioned showers and restrooms, on site car rentals, complete ship's store, full-service boatyard, pool, airport transportation, dry storage, picnic tables, barbeque grills, and coin operated laundry.

The Rivers Inn restaurant's main dining room offers a full service menu at lunch and dinner, while the outdoor deck provides a more casual seafood, raw bar fare.

Commonwealth Yachts and Coastal Yacht Sales offer brokerage and new boat sales for both power and sailing yachts.

- Hampton Roads Tides; High: + 0:26,Low: + 1:01, Mean 5.5 ft
- $.95 / ft per day, 300 Berths, 185' LOA
- Prop., Sail, Elec., Rigging, Gas, Diesel, Wood, Fiberglass
- Texaco Starport, Gas, Diesel
- Fresh Water, Phone, Cable TV, Elec: 110V, 220V, 100 Max Amp
- Ship's Store, Courtesy Car to Local Stores 804-642-2156
- River's Inn 804-642-2156, Leisure Dinning, $7 to $19
- MasterCard, Visa,Discover, American Express, Texaco

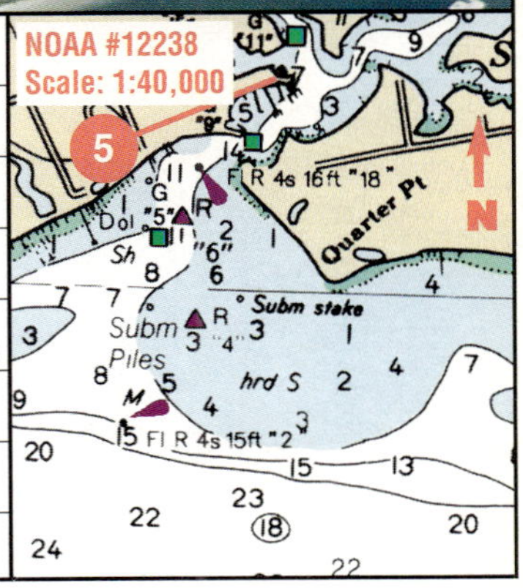

1862 President Abraham Lincoln watched Union artillery men lob cannon shells at Confederate troops in Norfolk. The fort, an unremarkable patchwork of concrete added bit by bit over a century, was abandoned for decades but has been partially restored. If you've an unquenchable hankering for history, boat tours to the fort are available at the visitors center in Hampton (804) 727-1102.

From the mouth of the **James River**, follow the fairway between flashing green 2.5sec "9" and flashing red 4sec "10." This leads to the **Norfolk Harbor Reach**, which turns into the **Elizabeth River**. As you go up the Elizabeth River, be wary of two submerged rocks a few yards west of flashing red 4sec "20," just outside the channel. The rocks are marked with white warning boards. A mile farther up at Lamberts Point are coal piers where train cars dump coal into the holds of colliers.

Stick near the channel as you head for the downtown skylines of **Norfolk** and **Portsmouth**. After you pass the coal piers, swing to the southeast and leave the Western Branch of the Elizabeth River to starboard. You'll cruise between shipyards and loading terminals as you go about 1.5nm from Lamberts Point past Scott's Creek to Hospital Point and quick-flashing red "36." Just southeast of Hospital Point is a well-used anchorage. Across the river on the Norfolk bank is the ship-like bluish-gray building, Nauticus—The National Maritime Center, which will open as a marine science attraction in 1994.

Both **Portsmouth** and **Norfolk** are historical cities that played key roles in the Revolutionary and Civil wars, and are replete with museums and treasured buildings. Ferries ply the river between the cities, and cabs are available at any hour. As in most metropolitan areas, strolling at night in quiet neighborhoods is unwise. This area also marks the official start of the ICW. The piers of the Norfolk Navy Base (804) 444-0000, the world's largest naval base, start 2nm southwest of Fort Wool, identified by the towering silhouettes of aircraft carriers. Stationed at the base are 90 Navy ships, including 6 aircraft carriers and 26 submarines, although half of them are often away patrolling the oceans. Tours of the station are available seasonally.

Among **Norfolk's** downtown attractions is the recently enlarged Chrysler Museum (804) 622-1211, home to impressive paintings and sculpture from around the world and a renowned collection of glass work. Also in downtown is the Douglas MacArthur Memorial (804) 441-2965, where the famed general who called Norfolk home is buried amidst artifacts from his illustrious military career. A few blocks away is St. Paul's Episcopal Church (804) 627-4353, a 1739 structure that survived the battles of the Revolution but still has a cannonball embedded in one of its walls. Other Norfolk attractions include a zoo and a botanical garden.

In Norfolk, Waterside is the focus of the city's waterfront development efforts, with a collection of indoor and

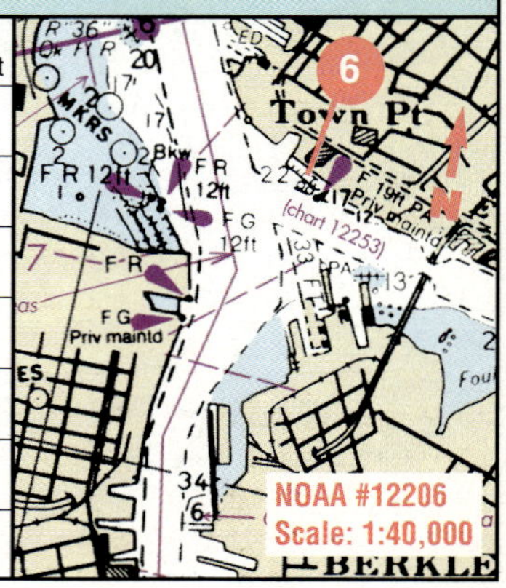

outdoor restaurants, an open-air marketplace and **Waterside Marina** with tour boats. The guides aboard the tour boats are incredibly knowledgeable about the Navy and commercial traffic in the harbor and well worth it. As they say: "Enjoy the view and leave the driving to someone else." Town Point Park, with 150 events per year, is next door to the marina. The dockmaster at Waterside is eager to make a reputation for his facility and can get anything you may need including a courtesy car to the grocery store or theater, marine supplies delivered dockside, or bargain prices of bulk fuel purchases. Across the river in Portsmouth are many museums including the Lightship Museum (804) 393-8741 and the Naval Shipyard Museum (804) 393-8591. Services are still held at nearby Trinity Episcopal Church (804) 393-0431, erected in 1762.

In Portsmouth, **Tidewater Yacht Agency** has over 350 slips in a protected basin and is located in the heart of the city and at the unofficial start of the Intracoastal Waterway at Mile Marker 0.0. Owing to their position, Tidewater pays particular attention to the needs of transient yachtsmen and offers all repairs and supplies onsite. The deck at Amory's Wharf restaurant at the marina is the best place in town to see the skyline and watch the world go by. Other attractions in town include the Pokey Smokey Steam Railroad (804) 538-4102, a 36-gauge miniature coal-fired steam train and amusement park, and the Children's Museum and Art Center (804) 393-8983.

Make the trek up the **James River**, as Captain John Smith and his settlers did in 1607. Leaving Hampton River, travel southwest for 4nm to a mile west of Middle Ground Light, a 52-foot red plug not far from where the world's first ironclad warships fought in 1862. Then head west up the Newport News Channel through the rock islands marking the entrance into the James River. The current here can be feisty. The bridge pilings are home to the stripers and draw fleets of fishermen during the fishing season.

Coal piers line the north shore of the river, supplanted as you go upstream to **Newport News** by the sprawling facilities of Newport News Shipbuilding Company, the nation's only builder of aircraft carriers. Usually there's a carrier or two under construction in the yards, and just the sight of these enormous vessels is breathtaking. Just west of the shipyard towers the James River Bridge, which has a vertical clearance of 60 feet and can be raised to 145 feet. Anyone cruising Virginia waters will want to spend an afternoon at the recently expanded Mariners Museum (804) 595-0368. The museum features vessels and artifacts from the Chesapeake Bay and the battles of the Revolutionary and Civil wars. Other museums, including the Peninsula Fine Arts Center (804) 596-8175 and the War Memorial Museum of Virginia (804) 247-8523, are scattered throughout the Newport News.

Continuing northwest up the James River, your next stop is **Jamestown Settlement**, the site of America's first

Hampton Roads Tides; High: +0:18, Low: +0:15, Mean 1.5 feet
$1.10 / ft per day
100 Berths, 120' LOA
Prop., Sail, Elec., Rigging, Gas Diesel, Wood, Fiberglass
Texaco
Gas, Diesel
Fresh Water, Phone, Cable TV
110/220V, 30/50/100 Max Amps
Ship's store with groceries
804-393-2525
Amory's Wharf 804-399-0991
Casual $8 - 15
MasterCard, Visa, Discover, American Express, Texaco

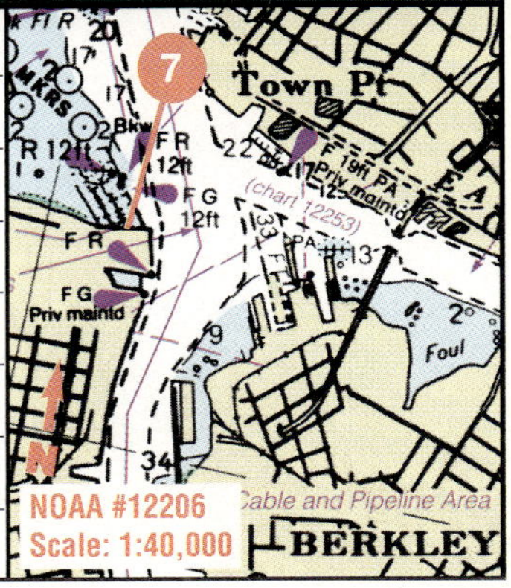

NOAA #12206
Scale: 1:40,000

permanent English colony and home to replicas of the *Susan Constant*, the *Godspeed*, and the *Discovery* which brought Captain Smith and his settlers to the New World. Here you can tour the modest remains of the first permanent English settlement, and colonial Williamsburg is only a short cab ride away. This restored colonial city—one of Virginia's most popular tourist attractions—will keep you busy for days with unique 18th century houses, a 173-acre outdoor living history museum, art centers, exhibitions, demonstrations, and a host of other sights. For more information, call the Colonial Williamsburg Foundation (804) 220-7645. Jamestown Island is located at the flashing green 4sec gong "55." Unfortunately, marine facilities on the island are limited by a 12-foot bridge. We suggest anchoring across the channel from Glass House Point and taking a dinghy across.

About 20 miles east of Norfolk is the Atlantic Ocean, where **Virginia Beach** boasts a 28-mile resort beach as the big attraction. Home to Seashore State Park, a campground and park dripping with Spanish moss, this area teems with tourists in warm weather. On Fort Story, at the entrance to the Chesapeake Bay, are two lighthouses. The first is the Old Cape Henry Lighthouse, America's oldest government-built lighthouse dating to 1791 and open to the public (804) 460-1688. The second is the new Cape Henry Lighthouse built in 1881 and currently in operation as a Coast Guard station. The Old Cape Henry Light is open to visitors from March through October. The Virginia Beach Marine Science Museum (804) 437-4949 features hands-on exhibits. Sea-side honky-tonk attractions like the Royal London Wax Museum (804) 491-6876 can be found on "the strip," just off the beach. For more information on things to do in this resort city call the Virginia Beach Visitor Information Center (800) 822-3224.

To reach Virginia Beach and Lake Rudee, pick up the Thimble Shoal Channel at Old Point Comfort. Head southeast past **Cape Henry**, then look for green can "1." From here, head south for 5nm to red and white "RI" just off the Rudee Inlet. Stay between flashing red 2.5sec "2" and flashing green 4sec "1." Once past the inlet, keep an eye on your depth sounder as the channel is unmarked. At the fork, head north into Lake Rudee, and look for facilities off to starboard, just beyond the fixed bridge (vertical clearance 28 feet).

Eastern Shore: Tangier Island to Great Bridge

We've explored Virginia's western shore of the Bay, so let's go back to Smith Point Light and run through Tangier Island and down the isolated waters of the Eastern Shore.

From the Smith Point, it's 9nm to the 15-foot flashing green 2.5sec "1" and the entrance to **Tangier,** a flat island that is home to 650 people whose livelihood is the blue crab. As you stroll around town you'll hear an Old English accent, a charming reminder that Tangier is still isolated from the rest of the world. Although tour boats bring visitors from Crisfield, Onancock, and Reedville every day, the pace is slow. No alcohol is sold on the island, and there are only a few cars.

It's easy to find the entrance to the harbor if you steer for the silver water tower visible from the west. Stay clear of the prohibited area to the southwest where Navy jet pilots sharpen their bombing skills. The shattering roar of jets can be startling as you cruise through what seem like undisturbed waters.

The channel from green "1" past the island to Tangier Sound has been dredged to 8 feet, but it silts continuously. Deep-draft vessels should check with the Coast Guard before attempting the passage. If you skip the island, run south 3.5nm to the 45-foot Tangier Sound Light, and head for Onancock Creek. If you go into the island, follow the channel, and swing to starboard between crab-boat pilings to a very small marina.

As you carefully follow the straight channel northeast to Tangier Sound—leaving red markers to port—you'll see scores of shacks on stilts where natives pen up and harvest soft-shell crabs. When you reach 15-foot flashing green "1" about a mile from the marina, swing to the southeast for a 9nm run across Tangier Sound to the entrance of Onancock Creek, staying well south of the marshy flats of Watts Island.

Another 3.5nm from flashing green 4sec "1" brings you to the town of **Onancock**, home to a city marina and a dandy anchorage a few hundred yards to the south. It's an easy stroll to the small-town atmosphere of Onancock, which has a bakery and an Italian restaurant. Visit Kerr Place (804) 787-8012, a 1799 period house open for tours.

The next refuge for transients is 35nm to the south at the railroad ferry town of **Cape Charles**, which has a marina and fishing facilities but little else to attract tourists. Six miles south of the town of Cape Charles, two huge concrete ships provide protection for an anchorage off Kiptopeke State Park (804) 331-2267 and its beaches. The ships were sunk there to act as breakwaters.

Another 4.5nm takes you to the north end of the Chesapeake Bay Bridge-Tunnel, a 13nm-span that carries Highway 13 traffic from the Eastern Shore to Virginia Beach. A 75-foot high-rise bridge allows boat traffic to head for the ocean about a mile southwest of Fisherman's Island, 4.5nm from Kiptopeke. Open waters over tunnels at the Chesapeake Channel and the Thimble Shoals Channel provide unfettered passage between the Bay and the Atlantic.

Two miles east of the south end of the bridge is Lynnhaven Inlet, crossed by a fixed bridge with a 35-foot vertical clearance. The entrance is 4nm west of Cape Henry Light, the historic 164-foot beacon marking the south side of the ocean entrance between the Virginia capes. The Cape Charles Light stands on the Eastern Shore, about 14nm to the northeast.

Enter the Lynnhaven Inlet at flashing green 2sec "1," run a mile under the bridge, and swing left, passing restaurants and marinas to port. Then run a mile through a narrow channel under another 35-foot bridge. The channel takes you on to Broad Bay and Linkhorn Bay, but be wary of water skiers. The Lynnhaven Channel was dredged to 7 feet in 1988, but silts, and the current can be wicked in narrow channels.

Two miles west of the south end of the bridge tunnel lies Little Creek, home to numerous marinas and restaurants and about 20 Navy ships. Enter at flashing green 4sec "1," follow the straight channel a mile to 15-foot flashing red 6sec "8," and swing to starboard into Fishermans Cove. A westward run of 7nm from Little Creek takes you to the entrance to the Hampton Roads harbor at Fort Wool.

Back in Norfolk and Portsmouth, quick-flashing red "36" marks the beginning of the Intracoastal Waterway which runs through the suburban city of **Chesapeake**. The Chesapeake Planetarium (804) 547-0153 is a favorite place to visit, especially among children and astronomy buffs. On the right just past the Jordan Bridge are dozens of old Navy ships, part of what is known as the "mothball fleet"—a reserve fleet tied up and waiting to be used or sold. Another 2nm to the south is the Gilmerton Bridge, which opens on demand. Water traffic can be slowed here also by two railroad bridges, but they'll be open unless a train is passing.

At statute mile 7.1 is the Interstate 64 high-rise bridge. Here you have to choose your route down to North Carolina. To starboard is Deep Creek which becomes a canal at the Deep Creek Lock and runs about 30nm through the **Great Dismal Swamp** to Elizabeth City, N.C. The lock opens four times a day, (at 8:30 a.m., 11:30 a.m., 1:30 p.m., and 3:30 p.m.). At the North Carolina end of the canal is the South Mills Lock.

South of the high-rise bridge runs the popular route for Waterway travelers (via **Albemarle and Chesapeake Canal**) who stream north by the thousands in spring and south by the thousands in fall. A run of 2.5nm brings you to the Great Bridge Lock. The openings are timed with those of the bridge at **Great Bridge** just to the east. Great Bridge is a regular stopover for cruisers who stay at the marina just east of the bridge or tie up to pilings on the south side of the canal just west of the bridge. Either way, Great Bridge is a convenient stopover where you can stock up on provisions or eat at a restaurant.

The scenery and wildlife can be breathtaking as the waterway swings through swampy areas and into the North Landing River, which flows into Currituck Sound in North Carolina.

Whether you take the popular western route down the bay or travel the lonely waters of the Eastern Shore, you'll leave Virginia with a better understanding of the challenges that faced our founding fathers when they came to the New World. And you'll know why coastal cruisers consider the Chesapeake Bay among the world's most enjoyable and diverse cruising grounds.

Shoreside & Emergency Services

Police, Fire, and Ambulance: 911

Coast Guard:

Chincoteague Station (804) 336-6511 or VHF 16
Parramore Beach Station (804) 787-9527 or VHF 16
Cape Charles Station (804) 331-2000 or VHF 16
Milford Haven Station (804) 725-2125 or VHF 16
Little Creek Station (804) 464-9371 or VHF 16
Portsmouth Station (804) 421-6200 or VHF 16

Radio Telephone:

VHF 25, 26, 27, and 84; I.D. Point Lookout (Md.) Marine Operator
VHF 25, 26, 27, and 84; I.D. Norfolk Marine Operator

Tow Services:

Marine Services, Norfolk, VA
(804) 483-2673
Rebel Marine, Norfolk, VA
(804) 588-6022
Sea Tow Lower Chesapeake Bay, Gloucester, VA
(804) 642-6800 or VHF 16
Sea Tow Hampton Roads, Virginia Beach, VA
(804) 496-1999 or VHF 16 ✦

Local Services

Joins page 206
VIRGINIA
NORTH CAROLINA
U S C G
(919) 453-3411
Elizabeth City
Albemarle Sound
Belhaven
Greenville
Washington
U S C G
(919) 987-2311
VHF-FM
KIG-77
162.475 MHz
U S C G
(919) 745-3131
Pamlico Sound
Hatteras Inlet
Cape Hatteras
U S C G
(919) 928-3711
U S C G
(919) 986-2175
New Bern
Oriental
Ocracoke Inlet
VHF-FM
KEC-84
162.40 MHz
Raleigh Bay
U S C G
(919) 354-2462
Havelock
Beaufort
Morehead City
Jacksonville
Bogue Inlet
Beaufort Inlet
Cape Lookout
New River Inlet
N
U S C G
(919) 343-4881
Onslow Bay
Wilmington
VHF-FM
KHB-31
162.55 MHz
U S C G
(919) 278-5945
ATLANTIC OCEAN
Cape Fear
Long Bay
1 inch= 23 miles
Scale 1:1,457,000
0
23
46
69
92
Joins page 232

The Tar Heel State

They're often described as a necklace—the barrier islands of North Carolina. But perhaps to the boater they are better seen as a shield, their inner rim providing a sheath of protection, their outer rim projecting barbs. For while they form the protective rim of the Albemarle and Pamlico Sounds, the islands also veer into the Atlantic in a series of sweeping arches to form three distinct capes—Cape Fear, Cape Lookout, and Cape Hatteras. These submerged deltas extend out to sea for several miles and create a navigational nightmare for any skipper skirting the coastline. They make up what is commonly called the "Graveyard of the Atlantic."

Fortunately, inside the Albemarle and Pamlico Sounds, in the shadow of towering sand dunes and ancient marine forest where pirates and pilgrims gained fame and misfortune, lie some of the most exciting cruising waters on the East Coast. The vast waters of the Albemarle and Pamlico will lead you to such scenic and attractive ports as Belhaven, Oriental, Edenton, and Manteo. The seemingly untouched National Seashores of the Outer Banks offer comfort from the open seas and refuge from the world, while the more active ports of Washington, Beaufort, and Wrightsville Beach offer a less-reclusive backdrop.

This diversity can also be seen in North Carolina's waters, where you'll find deep cuts as well as some dangerously shallow areas where you'd expect generous waters. A propensity for sandbars, shoals, and unexpected obstructions can wreak havoc on the careless skipper, so keep an eye on your chart and your course. Those conditions aside, the beauty of North Carolina is that boating is a year-round activity, with autumn being the ideal season.

© North Carolina Division of Travel & Tourism

***Henrietta II* glides past the Wilmington waterfront.**

Although most cruisers choose to stick to the Intracoastal Waterway, there are a few navigable inlets from the outside route. Shoaling has been a problem with the inlets along this stretch of the Atlantic. When an inlet was passable, communities arose to take advantage of the opportunities for commerce. When the inlets shoaled, the communities disappeared. Currently, the inlets for Beaufort, Wrightsville Beach, and the Cape Fear River are the best choices, although the Bogue and New River inlets can be used with local knowledge. If you stick to the inland route from Virginia, you'll get a much more varied look at North Carolina cruising.

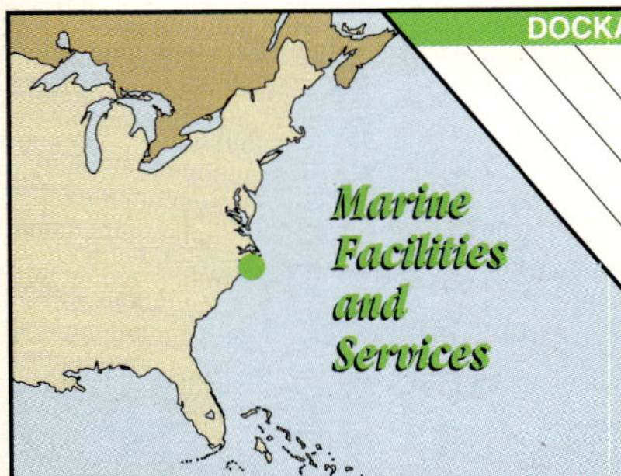

DOCKAGE: Monitors VHF Channel through Dockside Depth · SERVICES: Hookups through Repairs · SUPPLIES: SCUBA through Fuel Brand · AMENITIES: Marine/Groceries through Public Phone · COAST GUARD: See Adjacent Chart for Coast Guard Information

Marina	Phone	Monitors VHF Channel	Working Channel	Number of Transient Berths	Number of Transient Moorings	Seasonal / Year-round	Maximum Length Over-All	Approach Depth in Feet at MLW	Dockside Depth in Feet at MLW	Hookups: Fresh Water / Phone / Cable TV	110V★ 220V▲ 3 Phase ■ Maximum Amps	Ramp / Dinghy Dock / Launch Service	Railway / Lift / Crane: Capacity (tons)	Repairs: Propeller / Sail / Electronics / Rigging / Gas / Diesel / Wood / Fiberglass	SCUBA: Air / Rentals / Equipment / Charters	Gas / Diesel Fuel	Fuel Brand	Marine / Groceries / Bait / Ice / LPG / CNG	Showers / Laundry / Pumpout / Restrooms	Hotel / Pool / Tennis / Golf	Public Phone / Restaurant / Snack Bar	MasterCard / Visa / Discover / AmEx
1 Dowry Creek Marina (p. 222)	919-943-2728	9	16	36		Y	100	8	8	FP	★▲50	D				GD	Phil	All	All		PS	MVA
2 River Forest Marina (p.223)	919-943-2151	16	68	35		Y	175	9	9	All	★▲50	All	L40	All	All	GD	Tex	MGBIL	All	HPT	All	MVA
3 Marlen C. Robb & Son Boatyard (p.225)	919-943-2110	16	69	25	12	Y	140	13	10	FC	★▲■ 50	All	L75	PRGDWF		GD	Phil	MI	All		All	MV
4 Deaton Yacht Service (p.226)	919-249-1180	16	68	6		Y	65	8	7	F	★▲50	D	L25	All				All	SPR		P	MVA
5 Beaufort Docks (p. 227)	919-728-2503	16	10	81		Y	250	15	20	F	★▲■	D	C25	All	C	GD	BP	All	All	H	All	MV
6 Bock Marine (p.228)	919-728-6855	16	9	4		Y	100	14	8	F	★▲50 50		L60	All			Mob	MGBIL	SLR		P	MV
7 Morehead City Yacht Basin (p.229)	919-726-6862	16	71	15		Y	150	12	8	All	▲50			PGD		GD	Exx	MI	All		PR	MVA

Information in these listings is provided by the facilities themselves.

Virginia Border to the Albemarle Sound

From Great Bridge, Va., there are two access routes to North Carolina. If you choose the trek through the **Great Dismal Swamp Canal**, Elizabeth City at the head of the Pasquotank River will be your first port of interest. The Dismal Swamp Route has plenty of water and should pose no problems. There are two locks to contend with in South Mills, so call ahead for the latest opening schedule (919) 771-5906.

If you choose to stick to the Intracoastal Waterway and the Virginia-North Carolina cut (a.k.a. the **Albemarle and Chesapeake Canal**), you'll pass through the upper reaches of the Currituck Sound and down along the Intracoastal Waterway to Coinjock. The channel through this section of the Currituck Sound has experienced shoaling in the past, so be aware of your surrounding depths. The rest of the Sound is rather shallow and overgrown with vegetation—perfect for fishing and bird hunting, but treacherous for cruising boats. **Coinjock** is a refreshing stop along this route on the waterway. And a stop is all it is. Most would agree that Coinjock is not a destination, but rather a convenient place to fill your tanks, rest up a bit, and move on. There are several marine facilities that will cater to the needs of the temporary transient.

From Coinjock, the temptation is to hurry down the Alligator River on a course for Belhaven, but this would deny you one of the best-kept cruising secrets on the East Coast: the **Albemarle Sound**. The Albemarle is one of those pristine cruising grounds, renowned for its long stretches of open fresh water and undeveloped shoreline. The historical significance of the Sound invites you to spend at least part of your cruise examining Manteo, Columbia, Edenton, and Elizabeth City, for it is here that our country was first settled and thereby established its roots in the New World.

Transient boaters are viewed as an extension of the Albemarle's own sea-faring heritage, and few communities express this adoration more eloquently than **Elizabeth City**. Nestled in a bend on the **Pasquotank River**, Elizabeth City was first charted by explorers from the English colony on Roanoke Island in 1585. The town prospered as an agricultural, lumber, and fishing community until George Washington foresaw the value of an inland waterway link between the Chesapeake Bay and Albemarle Sound and signed on with the Canal Corporation. Not only did he invest financially in the project, but he helped survey the area. The strategic importance of the Dismal Swamp Canal during the War of 1812 brought new growth to Elizabeth City, and though the town's influence has diminished through the years, it is home to the first marine facilities on the Dismal Swamp route and has a reputation for its hospitality.

Elizabeth City's walking tour is peppered with homes from the neoclassical era. Within the historic districts of Elizabeth City you will find over thirty buildings listed on the National Register, many of them clustered along the city's waterfront.

If you approach from the north through the Dismal Swamp Canal, your first mark will be to starboard at flashing green 4sec "9," just east of Knobbs Creek. Continue through the bascule bridge at Elizabeth City, keeping red "8" to port, then look on either side of the river for appropriate dockage.

Approaching from the Albemarle Sound, you'll sail up the Pasquotank River. Keep flashing red 4sec "4" off Miller Point to starboard and quick flashing "5" and flashing green 4sec "5A" off Brickhouse Point to port. Another two miles will bring you abeam of Cobb Point on your port side and quick flashing green "7." From here, the town begins to open up before you, and dockage can be found on either side of the river.

The town has several fine marine facilities that cater to overnight transients, and each of the facilities is within easy walking distance of the neo-classical homes, brick streets, drug stores, groceries, and fine restaurants.

As you head south from Elizabeth City along the Pasquotank River, you'll want to turn west and run along the north shore of the Albemarle Sound. Both **Little River** and **Perquimans River** offer tranquil, pastoral settings, but little else. Neither can compare, however, with the vast **Chowan River**, the largest of the streams feeding the Albemarle. Limited facilities and even fewer navigational hazards make this area perfect for a day trip.

One of the highlights of any cruise on the Albemarle Sound is a visit to **Edenton** on the Chowan River. Here, the celebrated Village Green, with its famed teapot, manicured lawn, and ancient elm trees, represents a show of patriotism and intolerance to excessive British rule. In a show of rebellion against British rule, the women of Edenton vowed to abstain from drinking British tea. This act of defiance marked the beginning of our young nation's struggle against British authority.

Prior to the Revolution, Edenton was a flourishing seaport, with over eight hundred ships sailing to England loaded with tar, turpentine, and timber. Today, it's a quiet town: every house has a garden, every lawn clipped and pruned with precision. The grand old water oaks shade the streets from the summer heat and the ancient cypress trees funnel the summer breeze through town.

The city dock is the most convenient berth, situated as it is in the heart of the historic district, but it can only be recommended in light or northerly winds due to its exposure to the prevailing southerly breeze. Flashing red 2.5sec "2" marks the outer edge of the channel into town, and the dock is just north of red "8." The **Pembroke Creek** anchorage lies off the marked channel 1.5nm west of the town dock, and while it presents a long dinghy ride, it's a better choice.

Continuing along the south shore of the Albemarle, **Roanoke River** and **Scuppernong River** are certainly accessible by boat, but many skippers choose to bypass these streams for the busier ports farther along. The Scuppernong is the access route for the town of **Columbia**, where fishing is the mainstay and the river is its provider.

The prized jewel of the Albemarle Sound is **Roanoke Island** and the town of **Manteo**. Here, nestled between the mainland and the barrier islands, English colonists led by Sir Walter Raleigh planned to establish the first permanent settlement in the New World. A rather curious footnote in history, the colony vanished completely. Today the story of those early pilgrims is told through "The Lost Colony," a symphonic outdoor drama presented nightly during the summer season at the Waterside Theatre at Fort Raleigh (800) 488-5012. Reservations for this highly acclaimed drama are recommended. During your stay in Manteo, you should make plans to visit the Elizabethan Gardens (919) 473-3234, Fort Raleigh, and the *Elizabeth II* (919) 473-1144. Other attractions in the area are the Wright Memorial at Kill Devil Hills (919) 441-7430, Nags Head Woods, Jockey's Ridge State Park (919) 441-7132, and the Cape Hatteras Lighthouse. Automobile rentals are available in town, and some of the marinas have courtesy cars for patrons.

Manteo has three well-equipped marinas, two of which are located near the downtown historic district. Good protection is the key asset of these facilities—a breeze from any direction would be hard pressed to disturb this bight.

The route to Manteo from the north is shallow yet it is also simple. As you leave flashing red 2.5sec "36" to port, turn south and begin your approach down Roanoke Sound, leave red "34" to port and green "33" to starboard. When you leave red "32" and "30" to port, turn west into Shallowbag Bay. Just a word of caution: the flashing green 6sec "MC" marker you see does not indicate mid-channel, rather it introduces you to "Manteo Channel" and should be left to starboard. You are now returning to port, so leave red "2" to starboard and green "1" to port. Once you have cleared flashing red 4sec "4" and red "6" to starboard, Manteo Harbor is directly before you.

Belhaven and the Pamlico Sound

From Roanoke Island, you may choose to retrace your course through the Albemarle Sound, then into the Alligator River on your way down the waterway to Belhaven. This is the preferred route, due to its security and convenient facilities. If you choose this course, however, you miss the opportunity to skirt the Outer Banks and visit the charming village of Ocracoke. This course takes you through either the **Croatan Sound** or the **Roanoke Sound**, both generally hospitable for coastal cruisers, offering good facilities and anchorages and well-marked passages. The Croatan cuts between the mainland and the west coast of Roanoke Island—a scenic trek bordered mostly by undeveloped shores and forest. The Roanoke Sound is a bit more shallow, but has a generous channel and allows access to the Outer Banks and the east shore of Roanoke Island.

Ocracoke, located on the barrier island of the same name, is about as close as you'll get to island cruising in North Carolina. The anchorage is well protected, bordered by beautiful beaches, a marine forest, and an ancient lighthouse. The village offers an assortment of dining opportunities from pizza to gourmet selections, but marine facilities are limited. Groceries and supplies can also be purchased on the harbor. The village exudes a certain Cape Cod ambiance, peppered with pirate legends and a rich nautical heritage, for it was here that Blackbeard the Pirate met his death off Teach's Hole, the channel that now bears his true name (Edward Teach).

Cycling is the preferred means of touring the village, and you can rent bicycles from several shops. The British Cemetery and the lighthouse are two popular attractions. If you have the time, visit Portsmouth Island across the Ocracoke Inlet. This village is part of the Cape Lookout National Seashore and was once the largest settlement on the Outer Banks. Now, it serves as a tribute to the men and women who have struggled against the shifting sands and tides of the Outer Banks.

Your approach from Pamlico Sound is straight and deep, provided you stay in the channel. The most common error is to confuse the daymarkers identifying Nine Foot Shoal Channel to starboard with your next mark. If you maintain a southerly heading, leaving the flashing red 4sec "14BF" to port and flashing green 2.5sec "13" to starboard at the channel's entrance, you'll have a straight shot down the channel. As you reach the quick flashing mid-channel mark, which will come into view just after red "4," turn to port and begin your approach to Silver Lake.

If you choose to bypass Ocracoke and follow the waterway on your passage across the Albemarle Sound, you should set your course for the **Alligator River**. Although the river entrance is broad, we suggest you enter the channel at flashing 2.5sec "1AR" and make for flashing green 4sec "3" off the shoal at Middle Ground, keeping the beacon to port. Vessels which draw 5 feet or less may carry this shoal, but we suggest sticking to the channel—a stiff blow from the north can lower the water in the river by as much as 2 feet. From Middle Ground, pass between flashing red 4sec "6" and green "5." Once through this narrow gate, you can stray from the channel as the depths in the river increase. Clear the Alligator Bridge, which opens on demand, and your next port will then be Belhaven.

The name means "beautiful harbor," and the delightful southern town of **Belhaven**, with its old forests and clear streams, remains one of the most popular stops in North Carolina. It's a village framed by large oaks and homes constructed in the simplistic architecture of eastern North Carolina. This is the site of the Belhaven Memorial, a museum displaying the Eva Blount Way collection of antiques and artifacts.

1 p. 218

Come discover the Dowry Creek Marina just 3 miles north of Belhaven. After exiting the 20-mile Alligator River-Pungo River Canal, our marina is your first stop for pleasure. Our facility is all new so everything is sparkling clean and fresh.

Bear 350° at Channel Marker 15 and cruise through our well marked channel. We're located on the ICW at Mile Marker 132. You won't find a more convenient access.

The Dowry Creek Marina offers 72 protected deep water slips on beautiful Upper Dowry Creek. You can enjoy the fabulous views from the porch of our new ship's store. Or you can stroll across our foot bridge over the marshland as you unwind from your travels.

We'll lend you a car to go into town or you can enjoy lunch in our landscaped picnic area. If you need to conduct business, we offer mail service and a fax machine on the premises.

We've planned carefully so your visit with us will be just right. You might plan only to tie up for the evening but then decide to drop anchor in our beautiful waterfront planned community. Our pool and tennis courts should be opening in '94 for your enjoyment.

We hope you'll let our friendly staff at Dowry Creek Marina make your waterway trip a special memory that can happen again and again.

P.O. Box 341 Belhaven, NC 27810

Tel: (919) 943-2728

Pamlico Sound; High: +1.45 min, Low: +2.24 min, Mean: 0.5 feet	Freshwater, Phone Elec: 110V, 220V, 50 Max Amps
$0.75 / ft per day 36 Berths, 100 LOA	Grocery/general supplies (919) 943-2728
	Belhaven Restaurants Approx. 3 miles
Phillips 66 Gas, Diesel	MasterCard, VISA, American Express

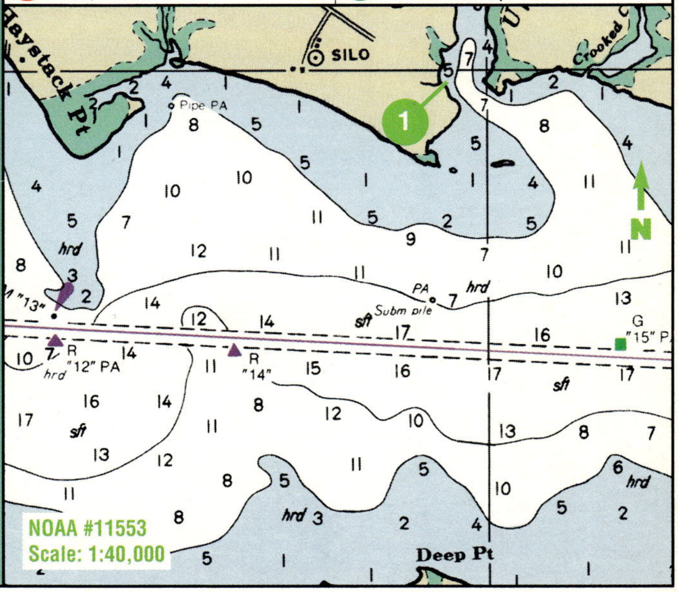

RIVER FOREST MARINA

2 p. 218

(919) 943-2151
Reservations/Information
1-800-346-2151
Monitor Channel 16

Welcome... to the only marina in Belhaven with a restaurant and bar as well as a full-service shipyard and travelift! River Forest Marina is the first marina inside the Pantego Creek breakwater, located in a protected yacht harbor.

Enjoy our lovely antique waterfront inn, featured on the cover of Country Inns of America. Extensively decorated with stained glass, crystal chandeliers, oak wainscoting and several period antiques, you'll soon discover why it's our favorite place to relax!

Enjoy dining at River Forest Manor, where the tradition of serving Southern-style food continues into its fourth decade. Try our World Famous Southern Smorgasbord nightly, or stay for Sunday brunch. House specialties include a delicious crabmeat casserole, oyster fritters, homemade sausage, and a large assortment of mouthwatering desserts. Our restaurant has received the Gourmet Diners Club of America Silver Spoon Award for the past three years, and we're sure you'll agree that we deserve that award this year as well!

With a new swimming pool, tennis courts and a hot tub jacuzzi, you're sure to find something to please every member of your crew at River Forest Marina!

Hampton Roads Tides; High: -1:38, Low: -1:41, Mean: 1.9 feet	Fresh Water, Phone, Cable TV 110/220V, 50 Max Amps
$0.75/ foot per day 35 Berths, 175' Max LOA	Foodland Grocery Store 919-943-2737
Prop, Sail, Elec., Rigging, Gas, Diesel, Wood, Fiberglass	River Forest Manor 919-943-2737 *Casual* $6.95 - 12.95
Texaco Starport Gas, Diesel	MasterCard, Visa, American Express

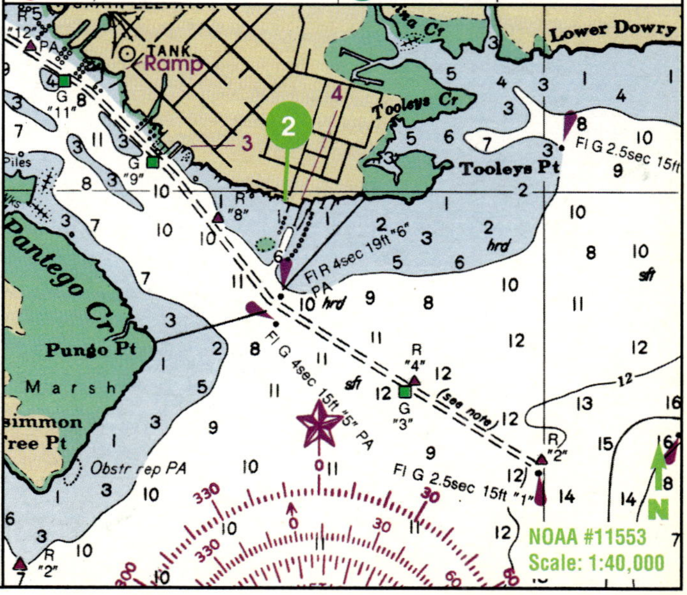

Much of Belhaven's fame can be ascribed to the numerous facilities to be found there. Follow the Alligator River—Pungo River Canal to the Pungo River. The channel turns sharply to the west at green "19." At green "15," bear 350° for **Dowry Creek Marina**. The new marina has views of the marshlands, a fully-stocked ship's store and shower facilities as well as a landscaped picnic area, and a courtesy car for traveling to town.

Inside the breakwater at Pantego Creek, you'll find another excellent facility, **River Forest Marina**. The marina features an antique waterfront inn, award-winning restaurant, swimming pool, jacuzzi, and tennis courts. The family-owned boatyard is a full-service facility offering travelift and towing services.

Not quite a half mile farther into the creek, at green "9," is the next marina, **Marlen C. Robb & Son Boatyard**. Formerly Belhaven Marine, the marina recently expanded to accommodate a fully-stocked chandlery, a visitor's lounge with shower facilities, a 75-ton travelift, and engine, propeller, and shaft repairs on premises. They also operate a 24-hour towing service.

As you depart Belhaven and continue along the Intracoastal Waterway, you'll head south through the **Pungo River** which offers safe anchorages in the bight of several creeks, but there's little in the way of facilities south of Belhaven. Farther south, the Intracoastal cuts across the **Pamlico River** to the **Hobucken Cut** and heads to the Neuse River. But don't let this route distract you from heading west through the Pamlico River. You won't be disappointed.

The towns of Bath and Washington both hold significant historical and cultural importance. While it seems every town in coastal North Carolina can lay claim to at least one story of Blackbeard the Pirate, it was the small village of **Bath** that enticed Edward Teach to give up his life of plundering and make this village his final home. Prior to the summer of 1718, his bold acts of piracy and pillaging had reached heroic proportions, fueling contempt and outrage from the citizens of Virginia and South Carolina. The governors of these two neighboring states expanded their efforts to capture Blackbeard, and for the first time in his career, he began to fear for his life. This new and real threat of hanging led Blackbeard to accept Royal Governor Charles Eden's condition of pardon. And he gave up pirating.

Teach married a sixteen-year-old girl and settled into the life of a gentlemen merchant engaged in legitimate trade. He purchased a home on Plum Point, across the mouth of Bath Creek, and from this bluff overlooking the Pamlico River, he watched the many ships sailing into Bath. These very ships, loaded with goods, would become his downfall. In September 1718, he claimed to have discovered a richly laden French vessel drifting crewless at sea. Fearing his return to piracy, his neighbors grew restless and began to fear for their safety. Although he still called Bath home, his heart yearned for the sea, and by November that year, his passion for adventure cost him dearly. Requested by merchants fearing Blackbeard's wrath, naval Lieutenant Robert Maynard sailed to Ocracoke Inlet and slew the pirate in a bloody battle.

While in Bath, dockage can be obtained at the town docks on the eastern shore, but there is no water, electricity, or restroom facilities. A little south of the town docks is the Bath Guest House, a bed and breakfast which offers dockage for its patrons only. The best anchorage is in **Bath Creek**. The creek is protected by a high bluff that overlooks the half-mile wide harbor.

The entrance to **Bath Creek** is wide, but the channel is narrow and tucked back off the Pamlico River near the shoreline. Pass flashing green 2.5sec "1" to port off Archbell Point, then clear red "2" to starboard off Plum Point. The creek opens up as you approach the town. Be sure to pass green "3" to port and flashing red 4sec "4" to starboard if you are continuing up Bath Creek.

Farther west up the Pamlico River is **Washington**, a perfect place for a day trip. Located at the head of the river, Washington is the final destination on the Pamlico. Many cruisers choose to dock elsewhere, spend a day cruising out to Washington, and then heading back.

Heading east through the Pamlico River, leaving Washington behind and making your way back to the Intracoastal Waterway, the combined effect of the river current at your back and the lack of rough water will make this as pleasant a cruise as you'll find in North Carolina. Navigating the cut to the **Neuse River** is also an easy sail and offers several fine anchorages.

Oriental is a village of nostalgic enchantment, untarnished by the modern trappings of urban sprawl, noisy interstates, and cheap tourism. This humble fishing community embraces the cruising sailor with celebration and reverence. Framed by the rivers and streams of the lower **Pamlico Sound**, Oriental looks towards the Outer Banks facing the sea with a soul linked to the traditions of coastal exploration. This is a bike and walk-about town where people wave and smile and offer their company for a spell.

The settlement lacks many of the conveniences of a modern city, and that's much of the charm. There's Red's, a gas station turned lunch grill, which still sells burgers and hot dogs for twenty-five cents. The offerings are small, but the price is right, and the meal is as much a novelty as it is basic nourishment. There are no traffic lights in Oriental. You move on at your own pace—or not at all.

The approach to Oriental is straightforward. The Intracoastal emerges from the Hobucken Cut and cuts across the mouth of the **Bay River**. From here, you will enter the Neuse River as it empties into the lower basin of the Pamlico Sound. You'll find plenty of water in the river on this southwest heading, provided you keep the respective markers of Maw Point, Piney Point, and Gum Thicket Shoal to starboard.

We may be the second marina in from the breakwater, but we're SECOND TO NONE when it comes to meeting YOUR needs!

Low, low, LOWEST! Diesel price on ICW for purchases over 150 gallons. Clean, fresh, water-free PHILLIPS 66 fuels!

TOWING! On call 24 hours a day. Channel 16 or **1-800-234-2869**.

Transient Marina - All tie-ups alongside in protected basins, plus moorings.

Magnificent new cedar planked showers, heads & laundry facilities.

Golf carts highway licensed for shopping convenience.

Dine at our new barbeque grilles & picnic tables or walk two minutes to ***The Helmsman*** - Belhaven's finest restaurant.

ON PREMISES:

75-ton Travelift - Extra high & extra wide on deep water. North Carolina's LARGEST!

Propeller Shop - Repair, reconditioning and balancing.

Welding Shop - Welding of all metals.

Machine Shop - Shaft repairs & fabrication, strut straightening & alignment, cutless bearing replacement & much more.

Engine & Gear Shop - Maintenance, repair and rebuilds by our factory trained technicians using only genuine factory parts.

Paint, Fiberglass & Carpentry Shops - For your needs & desires.

Chandlery & Parts Department - Genuine factory parts and other supplies.

Factory Authorized Dealers for:

DETROIT DIESEL, CAT, Perkins, YANMAR, WESTERBEKE, Universal Marine Power, KOHLER, LUGGER, ZF, TWIN DISC, RACOR, NAIAD

Hampton Roads Tides; High: -1:38, Low: -1:41, Mean: 1.9 feet	Freshwater, Cable TV Elec: 120/240V, 30/50 Max Amps
$0.75 / foot per day 25 Berths, 140' LOA	Highway licensed golf carts available at marina
Propeller, Rigging, Gas, Diesel, Wood, Fiberglass	The Helmsman 919-943-3810 Casual $5.50
Phillips Gas, Diesel	MasterCard, Visa, Phillips 66

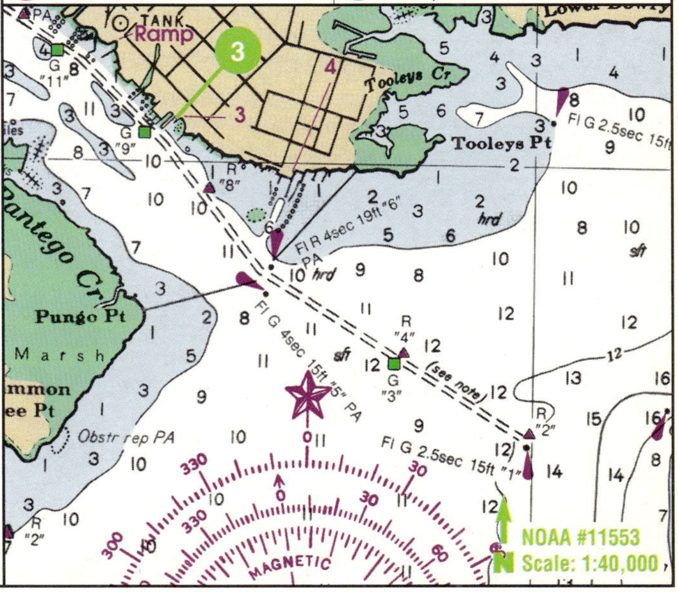

From flashing green 4sec "7," off Garbacon Shoal, look west towards the semi-high rise bridge, and set a course for flashing green 4sec "1." As you enter the channel, you will leave to port green markers "3" and "5." Flashing red 2.5sec "6" should now be to starboard. Set a course for flashing red 4sec "8," and once you clear the stone breakwater, you're in the harbor.

The main anchorage is behind the breakwater. The bottom is soft, black mud in 6 to 8 feet of water. Try to avoid anchoring close to the bridge, since this area was once an old logging site and some of the timbers may still remain submerged. Public dockage may be found on the waterfront along the bulkhead by continuing farther into the harbor past the seafood plant and shrimp boats. **Deaton Yacht Service** is a full-service boatyard with transient slips. Located in **Whittaker Creek**, the facility is a short walk from shops and town, but if you prefer to drive, a courtesy car is available.

The approach to Whittaker Creek is simple enough. You will find the channel a little north of flashing green 4sec "1" at flashing red 2.5sec "2." The channel is clearly marked, but proceed with caution since the water shoals on either side. Align yourself with the range markers on the bank of Whittaker Creek proper, and you should make it home safely. The most common error is to stray from the range markers as you near the end of the channel and begin to inch over towards flashing green "5."

Don't cut it close. A thumb of firm bottom juts out at the base of number "5" and will snag most wayward vessels.

Located at the head of the Neuse River, **New Bern** is a wonderful town to explore by foot. Everything of historical merit is within eight blocks of the waterfront. In addition to the Tryon Palace (800) 767-1560, North Carolina's first Governor's Mansion, there's the Fireman's Museum (919) 636-4087, a Civil War museum (919) 633-2818, the Attmore-Oliver House (919) 638-8558, and many other fine 19th-century Victorian homes. You will not need to look far for places to dine in New Bern, for there are several award-winning restaurants right on the waterfront.

From Oriental to New Bern, the Neuse River offers some of the finest sailing in North Carolina, and with the expanded marine facilities found all along the river, this passage is an extremely worthwhile diversion from the Intracoastal Waterway.

Beaufort to Southport

The **Newport River** marks your approach to Beaufort and Morehead City. These two towns introduce a change in the attitude and pace of your cruise through North Carolina. While Beaufort embodies tradition and the time-honored values of classic seamanship, Morehead represents contemporary coastal living with its easy access to a wide variety of movie theaters, nightlife, and shopping malls. Though the two towns are linked by the Morehead City bridge and share the Beaufort Inlet, they are distinct in culture and heritage.

Beaufort has long been hailed as the "Gateway to the Caribbean," and in the fall, as the western edge of the Gulf Stream veers to within 40 miles of the inlet, fresh northwest winds carry eager sailors out into the Atlantic. It is a four-day sail to the Bahamas, a day more to Bermuda, and a week to ten days to the Virgin Islands. This close proximity to the great Atlantic circle was not lost on the early settlers of Beaufort. The town was developed during colonial America as a deep water port to receive and distribute European merchandise.

Today, much of the town's historic district has been restored, thus revealing this unique blend of Caribbean and European influence. Beaufort's importance as a major seaport waned with the expanded settlement and growth in the Northeast, so the town turned to whaling, crabbing, and boatbuilding for its survival. Today, much of this rich heritage is on display at the North Carolina Maritime Museum (919) 728-7317, located behind the Gulf Docks.

Beaufort is a delight to explore. It offers several fine restaurants, antique and craft shops as well as a range of specialty and marine-related boutiques. The Beaufort Historical Society (919) 728-5225 conducts daily tours of the restored homes in the area and includes a visit to the old cemetery. Wild ponies are frequently seen grazing near the anchorage along the shore of Carrot Island. A larger herd can often be seen roaming the miles of empty shoreline on Shackleford Banks. Across the inlet, Fort

Beaufort Docks

5 p. 218

Photo: Tom Doe

500 Front Street, Beaufort NC 28516

(919) 728-2503

Monitor VHF Channel 16

The closest marina to EVERYTHING —

JUST LOOK AT WHAT WE HAVE TO OFFER!

1. Dockmaster's Office Dockhouse Restaurant Lookout Lounge
2. Beaufort House Restaurant
3. Cluster of Specialty Shops
4. Beaufort Marine Discount
5. Fishtowne Alley Shops
6. Clawson's Restaurant
7. Mike Smith Memorial
8. Net House Restaurant
9. Front Street Grill
10. (not marked) Numerous old homes dating to 1700s
11. Mike's Restaurant
12. Bank
13. N.C. Museum Watercraft Center
14. Spouter Inn Restaurant and Shops
15. Restored 18th Century Village Square
16. Old Burying Ground - 18th Century Town Graveyard
17. Beaufort Grocery Restaurant
18. Sailmaker
19. Inlet Inn Hotel
20. Post Office
21. Machine Shop
22. Commercial Ice Plant
23. Diesel Fuel Delivered at each slip
24. Hardee's 24-hour Restaurant
25. Greek Pizza Restaurant
26. N.C. Maritime Museum NOAA Charts
27. Prop Shop
28. Western Union
29. Royal James Cafe
30. Liquor Store
31. Car Rentals
32. Beaufort Inn
33. Electronics Distributor

Hampton Roads Tides; High: -1:40, Low: -1:41, Mean: 3.2 feet	Fresh Water 120V, 220V, 50 Max Amps
$1.05/ foot per day 81 Berths, 250' LOA	Byrd's Grocery 919-728-4471 Pak A Sak Grocery 919-728-7262
Prop., Sail, Elec., Rigging, Gas, Diesel, Wood, Fiberglass	12 Restaurants within walking distance - various prices
BP Gas, Diesel	MasterCard, Visa, BP

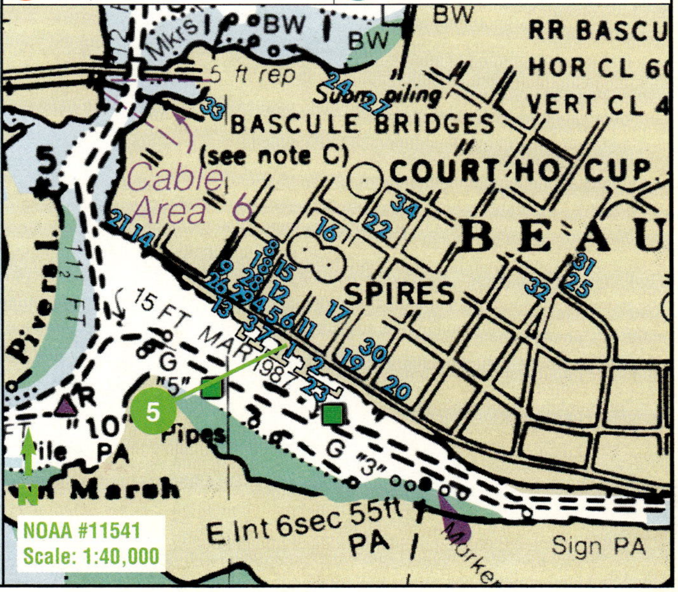

Macon warrants an afternoon of exploration. Still, it is the quiet walks along the narrow streets of Beaufort, adorned as they are with ancient oaks and unique shops, that bring you back to Beaufort.

There are two approaches to Beaufort. The shorter route takes you down Gallants Channel to Town Creek. This course departs the Newport River at quick flashing red "34," and quick flashing green "35." Follow the channel to port, giving "35" ample room, and keep an eye out for the unmarked shoal to starboard. Hug the green markers at the entrance to the channel—it has shoaled considerably. Head towards **Town Creek,** passing red "2" to starboard. At quick flashing red "4," you can fall off to starboard and await the opening of the Beaufort drawbridge, which opens on the half hour. Because of this restrictive opening, many vessels find it just as convenient to proceed down the channel of Newport River in the direction of the Morehead City bridge.

On the other approach, you will pass red "36" and flashing red 4sec "38" to starboard and will most certainly notice the dry storage silos and long building to starboard. The route to Beaufort and Morehead passes beneath the 65-foot highway bridge and the railroad bascule, leaving you in the turbulent waters of the State Port Terminal turning basin. Here you must decide whether to continue on past **Radio Island** to Beaufort, proceed out of the inlet, or turn to starboard for the Morehead City waterfront. If you are heading to Beaufort, continue out of the inlet channel until you've cleared Radio Island and its concrete landing ramps, keeping red "18" to port and flashing red 2.5sec "2" to starboard. Here you will turn back to port, passing flashing green "3" to port, red "4" to starboard and green "5" to port. As the channel bends back to starboard, you will see the Beaufort waterfront blossom before you.

Due to the intimacy of the waterfront, you will likely have an audience if you dock at the Beaufort Docks. The preferred perch is the porch of the Dock House, but any position along the boardwalk railing seems to attract onlookers. **Beaufort Docks** provides easy access to shoreside diversions and offers shower and restroom facilities. **Bock Marine** also offers slips and specializes in repairs; it can be found in **Core Creek**. For those wishing to anchor out in **Taylor Creek**, we suggest you stay clear of the main channel which runs along the waterfront. Space in the harbor may be limited, and two anchors are recommended to constrict swinging. A dinghy dock is provided in front of the post office.

Morehead City offers the benefits of a major commercial seaport, an active sportfishing fleet, and easy access to such familiar diversions as retail shopping and fine dining. Located on a narrow peninsula, its waterfront restaurants and shops are on the south side facing Bogue Sound. Across the street in the protected waters of Calico Creek is where you'll find **Morehead City Yacht Basin**. Head west in the channel just north of the State Port Terminal and high-rise bridge. As you clear green daymark "5" to port, you'll see the large docks of the Yacht Basin to port.

As you might expect, a number of seafood markets can be found along the waterfront in Morehead City, as well as fine restaurants and unique gift shops. Numerous dive shops operate out of Morehead. The branch post office, a book store, and a convenience store are within easy walking distance of the waterfront. A few of the restaurants have transient dockage, though you will probably fare better at the **Morehead City Yacht Basin**.

Aside from the Newport River and the Intracoastal Waterway route, Beaufort and Morehead City can be reached via the **Core Sound**. Long and narrow and shallow throughout, the Core is certainly the most secluded, if not the most attractive, body of water in North Carolina. If you draw more than 4 feet and are in need of facilities, look elsewhere. But if an adventurous, albeit risky, run is what you're after, this is it. Just keep a watchful eye on your chart and your soundings. **Cape Lookout** on the south end of the Sound is an ideal spot for an outing. As part of the Cape Lookout National Seashore, commercial development is prohibited. On the southwest shore is Cape Lookout Bight, which averages over 20-foot depths—perfect for anchoring.

You will not find many coastal communities more visually appealing than **Swansboro**. After the

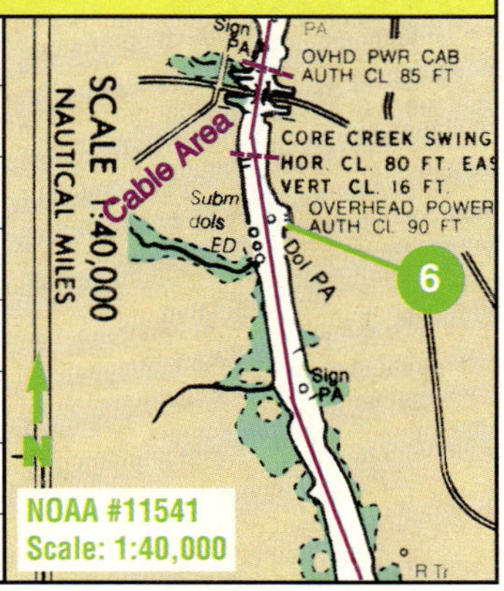

208 Arendell St., P.O. Drawer A • Morehead City, NC 28557 • VHF Channel 16, 71

(919) 726-6862

7 p. 218

MOREHEAD CITY YACHT BASIN

Located on the ICW in a protected harbor on the north side of the High Rise Bridge, just 2 miles from the Beaufort Inlet, Morehead City Yacht Basin offers many conveniences:

- Owned/operated by Paul C. Kerwin since 1977
- James A. Hester, Dockmaster since 1952
  - 12 feet on approach, 8 feet alongside
  - Short walk to any of 12 restaurants
  - 1/4 mile walk to early breakfast
  - Convenient to downtown banks and waterfront shops
  - Laundromat
  - Covington/Detroit Diesel on premises
  - Almost no current
  - Renowned for courteous, professional service
  - Multiple 240V, 50A service for larger yachts
  - Clean shower and restrooms
  - Exxon gasoline and diesel
  - Sailboats and trawlers welcome
  - Ship's store featuring marine supplies and Topsider shoes
  - Cable TV and on-dock phones
  - Courtesy car to shopping centers and major grocery stores
  - Gas grill for customer cookouts

Visit Morehead City Yacht Basin for the ultimate in quality, service and convenience!

Hampton Rds. Tides; High: -1:21, Low: -1:25, Mean: 3.1 feet	Fresh Water, TV, Telephone 110/220V, 30/50 Max Amps
$1.00/ foot per day 15 Berths, 150' LOA	Major chain groceries nearby - courtesy car
Propeller, Gas, Diesel	Numerous nearby - call marina for details
Exxon Gas, Diesel	MasterCard, Visa, American Express, Exxon

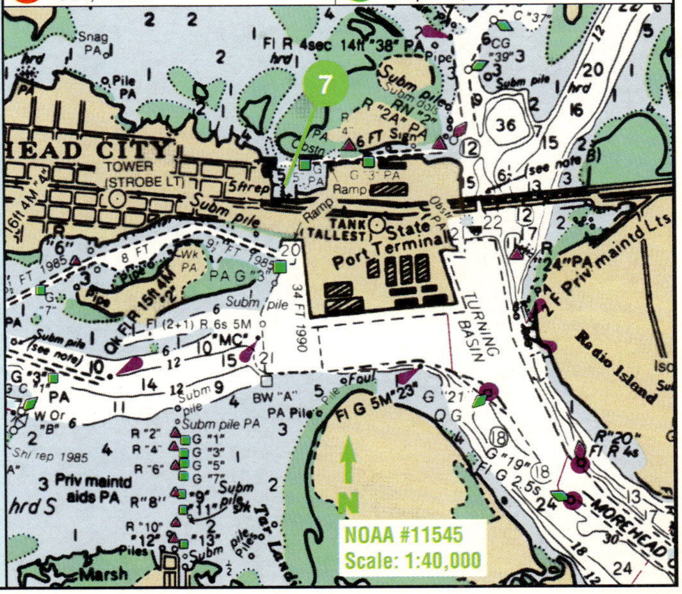

Revolutionary War, this colonial seaport was home to a thriving timber industry that depended on **Bogue Inlet** and the deep water of the **White Oak River**. Sawmills and shipbuilding were the major industries, fueled by large windmills that harnessed the prevailing onshore winds. These tripod-mounted fixtures had a wheel attached to the end of a long shaft that turned the mill-house to face the wind. As with so many cultures that have been ruined by the fickle fortunes of a shifting inlet, the windmills and their culture perished with the shoaling of Bogue Inlet. Now, Swansboro caters to the simple needs of fishermen and boatbuilders.

Piloting in this area can be tedious due to the current and constant shoaling. Keep a watchful eye astern to make sure the current does not carry you onto one of the spoils of the dredged channel. You will pick up the entrance to Swansboro at flashing red 4sec "46" just south of the entrance to Bogue Inlet. Keep green "45B" to port, and red "46A" to starboard as you approach the town. A shoal to starboard is marked by "46B." Clear this mark and red "2" before turning to starboard, since red "2" serves as the junction buoy for the river and is not a part of the waterway numbering sequence. A good anchorage can be found in the White Oak River, just off the waterfront and north of red "4."

From Swansboro to the **Cape Fear River**, the only significant body of water is **New River**, a wide stream running from the Intracoastal Waterway to **Jacksonville**. Its generous depths and charming natural landscape will undoubtedly lull you into a false sense of seclusion and safety. Don't be fooled—most of the river's shoreline is part of Camp Lejeune Marine Base and is restricted.

As you continue south, the Intracoastal Waterway cuts through an area of shoaled inlets, trickling streams, and marshland known as Middle Sound. At the south end of the "Sound," you'll find **Wrightsville Beach**, which combines the best of waterway travel, superb restaurants, nightlife, marine facilities, and grocery stores, all within walking distance of your berth. Wrightsville Beach has an annual population of about 4,000, and as you pass under the bridge, looking to port and starboard, you get the feeling every resident has a slip.

From **Snows Cut** at **Carolina Beach**, you may head south along the Cape Fear River or turn north to explore the exciting port of **Wilmington**. With its historic sights and many fine restaurants along the waterfront, this old riverboat town has much to offer.

As you enter the Cape Fear River, you will see the old Orton Plantation on the western shore. This estate was built back in the early 1700s as a rice plantation, and though currently a private residence, its lush gardens are open to the public during the spring and summer months. Also located nearby are the excavated remains of the early colonial village of Brunswick Town. This community was destroyed by fire in 1776, yet many of the original foundations have been uncovered, revealing significant historical artifacts. In order to visit either, you'll need a car, which you can rent if you continue down towards Southport.

Much of Wilmington's charm can be found in the restoration of its riverfront homes and early new-world commerce. Many of the homes are open to the public and the Chandler's Wharf (919) 763-2551, with its collection of restored sailing craft, shops and fine restaurants, preserves Wilmington's rich seaport heritage. Riverfront Park links Chandler's Wharf with The Cotton Exchange. This old warehouse has been transformed into a gallery of shops and restaurants and is worth an afternoon of browsing.

Passengers may board the ferry *Captain Moffit* and visit the battleship *USS North Carolina* across the river. Tickets to the paddleboat *Henrietta II* may also be purchased nearby and the St. John's Museum of Art (919) 763-0281 houses an outstanding collection of art in an old Masonic lodge.

In your approach from Snows Cut, you will see flashing 4sec "163" to port. Continue on a northwest heading, leaving green "1" and "3" to port and flashing red 6sec "B" to starboard. Favor the port side of the channel as you near flashing green 4sec "5," keeping red "4" to starboard. At "5," the channel turns north and widens somewhat. Green "7" and "9" should be left to port before the channel bends back west at flashing red 4sec "10." Leave "12" to starboard, then turn a little north as you enter the main channel of Cape Fear. The city of Wilmington is 12nm upriver. You are advised to make this passage with a rising tide and southerly winds to reduce the effects of the current and cross-chop.

Because dockage is limited in Wilmington, many boaters traveling down the waterway choose to lay over in Southport and visit Wilmington by car. **Southport** has the feel of an old New England fishing village, but with the temperate climate of the subtropics. Once a secure harbor for southern blockade runners, this small village is adorned with live water oaks, antique shops, and historic old homes. The Southport Maritime Museum (919) 457-0003 preserves the rich maritime history of the lower Cape Fear region and Fort Caswell on Oak Island, the site of the last Civil War naval battle. For all its slow southern charm, Southport is a spirited port of departure, since it is the last harbor north of Charleston, S.C., that can be reached by a secure, deep-water channel.

If you choose to stay at Southport or continue south down the Cape Fear River, exit Snows Cut at marker "163." On a southwest heading, you will leave "165" and "167" to port and quick flashing red "B" to starboard. Turn slightly to the south at quick flashing "A," then leave "170," "172," "174," and "176" to starboard. Flashing 4sec "177" to port marks a shoal and the entrance to Cape Fear. It is best to make the 12-mile run downriver against the tide.

After you have cleared flashing green 4sec "15" to starboard, you must remember to revert back to red-right-returning if you are continuing down the waterway. This will place red "2" to starboard before the Southport waterfront and flashing green "1" to port as you continue down the ditch. Don't attempt to leave flashing green "1" to starboard, as you will undoubtedly run aground. Flashing red 4sec "14A" and red "14" will both be seen on your port side off Battery Island, since this route continues your passage out of the Cape Fear channel via Bald Head Island.

Bald Head Island has over 2,000 acres of natural beauty, including 14 miles of wonderfully secluded beaches. Once the refuge of such pirates as Stede Bonnet, Blackbeard, and Mary Ann Blythe, Bald Head is a refreshing alternative to the glitzy resorts found along the waterway. In addition to the 2,000 acres of tastefully developed land, the marina, restaurant, golf course, and boutiques are framed by the creeks and marshes of a 10,000 acre natural habitat that makes a cruise past this island an eyeful. Bald Head Island is on the Waterway and easily accessible from Southport.

As you continue south from Southport, your passage becomes more restricted as the Waterway winds behind the Golden Isles of North Carolina. Long Beach, Ocean Isle Beach, Holden Beach, Yaupon Beach, and Sunset Beach are wonderful family beaches, but they offer little in the way of cruising amenities or marine facilities. From Southport to Charleston, the opportunity for side excursions will be limited, so enjoy the open waters of North Carolina before you continue on in your coastal cruise.

Shoreside and Emergency Services

Ambulance, Fire, Police: 911

Coast Guard:

Coinjock (919) 453-3411
Oregon Inlet (919) 987-2311
Hatteras Inlet (919) 986-2175
Ocracoke Inlet (919) 928-3711
Hobucken (919) 745-3131
Swansboro (919) 354-2462
Wrightsville Beach (919) 343-4881
Oak Island (919) 278-5945

Radio Telephone:

VHF 24; I.D. Elizabeth City Marine Operator
VHF 26; I.D. Wilmington Marine Operator
VHF 28; I.D. Morehead City Marine Operator

Tow Service:

Salty Dawg Marina, Manteo
(919) 473-3405 or VHF 16
River Forest Marina, Belhaven
(800) 346-2151 or VHF 16, 9
Robb's Boatyard & Marina, Belhaven
(919) 943-2110 or VHF 16
Deaton Yacht Service Inc., Oriental
(919) 249-1180 or VHF 16
Minnesott Yacht Basin, Minnesott Beach
(919) 249-1424 or VHF 16
Beaufort Towing, Beaufort
(919) 728-5088 or VHF 16
Marine Towing & Salvage, Beaufort
(919) 728-3033 or VHF 16
Sea Tow Cape Lookout, Atlantic Beach
(919) 726-6977 or VHF 16
Sea Tow East Coast Marine, Wrightsville Beach
(919) 452-3798 or VHF 16
Specialized Marine Services, Carolina Beach
(919) 256-5780 or VHF 16

Local Services

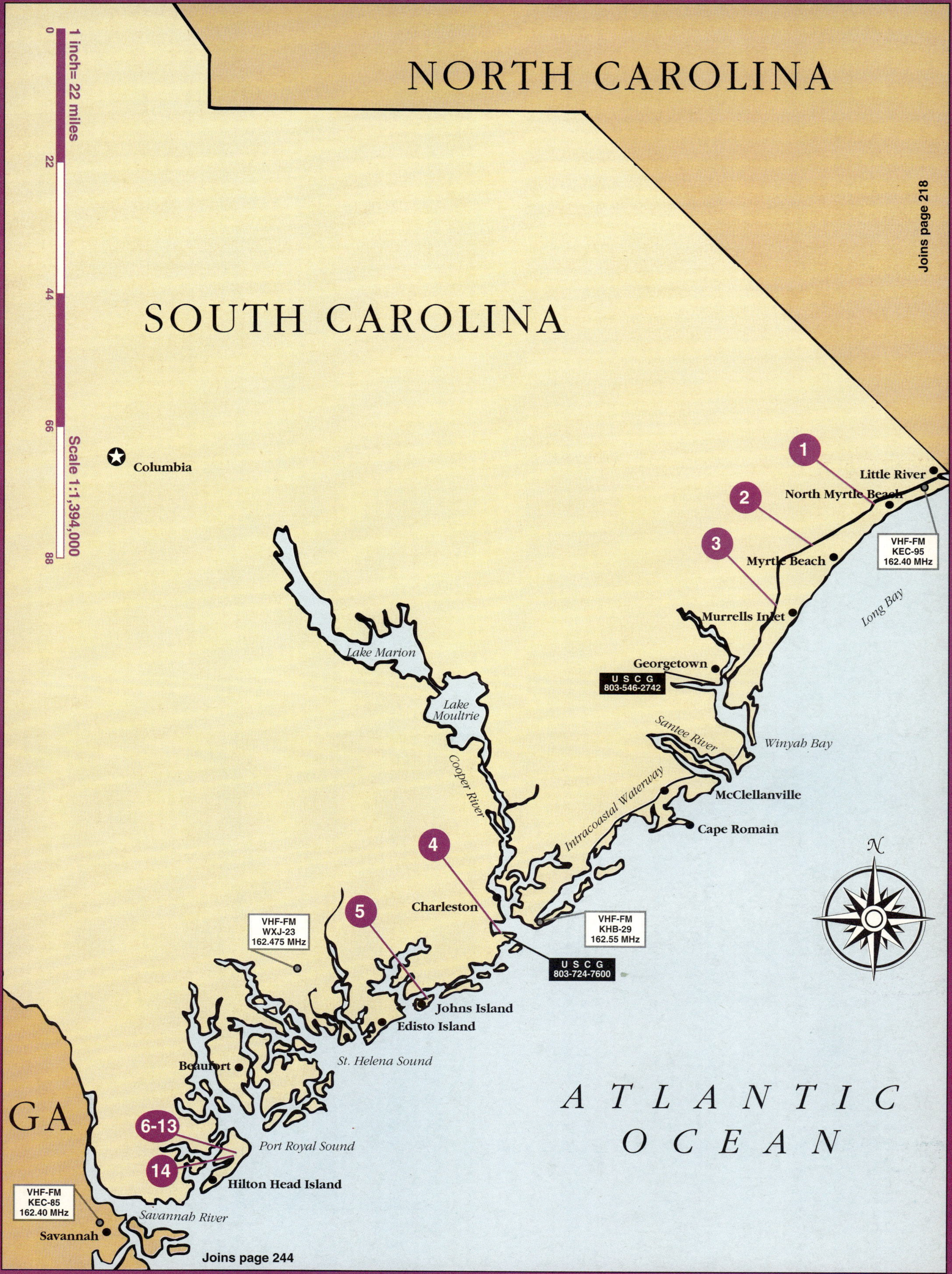
NORTH CAROLINA
SOUTH CAROLINA
GA
ATLANTIC OCEAN
1 inch= 22 miles
0
22
44
66
88
Scale 1:1,394,000
Joins page 218
Joins page 244
Columbia
Little River
North Myrtle Beach
Myrtle Beach
Murrells Inlet
Georgetown
Long Bay
Winyah Bay
Santee River
McClellanville
Cape Romain
Intracoastal Waterway
Lake Marion
Lake Moultrie
Cooper River
Charleston
Johns Island
Edisto Island
St. Helena Sound
Beaufort
Port Royal Sound
Hilton Head Island
Savannah River
Savannah
VHF-FM KEC-95 162.40 MHz
USCG 803-546-2742
VHF-FM KHB-29 162.55 MHz
USCG 803-724-7600
VHF-FM WXJ-23 162.475 MHz
VHF-FM KEC-85 162.40 MHz
N
1
2
3
4
5
6-13
14

The Palmetto State

Other towns in other states erect museums to remember a bygone era. In South Carolina, the towns are the museums. There is an elegance of the past—a way of living that has matured with progress, yet retained the values of a simpler time—which draws you to places like Charleston, Georgetown, and Beaufort. In short, it's the Southern way.

You approach this state along a coastline that expands slowly from the restricted boundaries of the Intracoastal Waterway to the sweeping delta of lush sea islands. The shores of South Carolina blossom as you cruise south, and the narrow river that held you at dawn swells beyond its banks by dusk, rushing to fill deep green tidal marshes of the coastal plain. This endless tidal flow nourishes the environment, feeding fish and fowl.

As you're traveling these waters, you'll come upon an old rice plantation or the grand home of a cotton grower in a bend of the river, and you're reminded that South Carolina is steeped in tradition and history. Enjoy the uniqueness of this state.

Summer is the busiest boating season in the Palmetto State, but the heat and humidity can make the evenings uncomfortable. Autumn is ideal for cruising, and though temperatures may dip into the thirties in January and February, there is little chance of a severe frost.

© J.H. Peterson

Old homes along the Charleston waterfront.

Navigation through South Carolina is straightforward. There are few bodies of open water, and most of the inland rivers are deep and well marked. You'll find numerous sidewaters for exploration, many with ample depths right up to shore. Keep in mind that the tidal

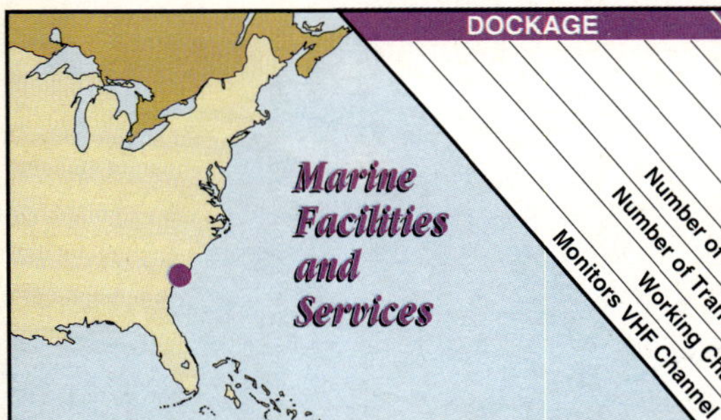

		DOCKAGE								SERVICES					SUPPLIES			AMENITIES			COAST GUARD
		Monitors VHF Channel	Working Channel	Number of Transient Berths	Number of Transient Moorings	Seasonal / Year-round	Maximum Length Over-All	Approach Depth in Feet at MLW	Dockside Depth in Feet at MLW	Hookups: Fresh Water / Phone / Cable TV	110V★ 220V▲ 3 Phase ■ Maximum Amps	Ramp / Dinghy Dock / Launch Service	Railway / Lift / Crane: Capacity (tons)	Repairs: Propeller / Sail / Electronics / Rigging / Gas / Diesel / Wood / Fiberglass	SCUBA: Air / Rentals / Equipment / Charters	Gas / Diesel Fuel	Fuel Brand	Marine / Groceries / Bait / Ice / LPG / CNG	Showers / Laundry / Pumpout / Restrooms	Hotel / Pool / Tennis / Golf	Public Phone / Restaurant / Snack Bar	MasterCard / Visa / Discover / AmEx
1 The Marina at Dock Holidays (p. 234)	803-280-6354	16	10			Y	125	10	7	FPC	50				C	GD	Ph.66	MGBI	SLR	All	All	All
2 Hague Marina (p. 235)	803-293-2141	16	9	40	6	Y	100	6	6	F	★▲■ 50		L35	All		GD	Gulf	MGIL	LSR	H	PS	MV
3 Wacca Wache Marina (p. 235)	1-800-395-6694	16	10/68	20		Y	150	15	10	FC	★▲ 100	All		PEGD		GD	TEX	All	All		All	All
4 Ashley Marina (p. 237)	803-722-1996	16	71	40		Y	150	25	25	FP	★▲ 50			All	ARE	GD	Tex	MGBI	All		P	All
5 Bohicket Marina Village (p. 239)	803-768-1280	16	68	20		Y	185	15	22	F	★▲ 50	L		All	RC	GD	Tex	MGBI	SLR	All	All	MVA
6 Windmill Harbour (p. 241)	803-681-9235	16	14	40		Y	75	8	8		50	R				GD		M	PRS	PT	PS	MV
7 Outdoor Resorts Marina (p. 241)	1-800-845-9560	16	68	5		Y	130	25	20	F	▲■ 50	RD		on call		GD	BP	MGBIL	All	PT	All	MV
8 Skull Creek Marina & Yacht Club (p. 241)	1-800-237-4096	16	68	50		Y		8	8	CP	100		L30			GD	Tex	M	All	PTG	PR	MVA
9 Schilling Boathouse (p. 241)	803-681-2628	16	69			Y	30	6	6					PGF	C	G	BP	MBI	S		R	MVA
10 Shelter Cove Harbour (p. 241)	803-842-7002	16	68	8			200	9	9	CP	50					GD	BP	MB	All	All	PR	MVA
11 Wexford Harbour (p. 241)	803-686-8813	16	14	10			75	8	8	CP	50								All	PTG	P	MVA
12 Harbourtown Yacht Basin (p. 241)	803-671-2704	16	19	25			130	6.5	8	CP	50					GD	Tex	MG	All	PTG	PR	All
13 Melrose Landing (p. 241)	803-842-9998	16	71	40		Y	100	20	20	F	50					GD	Tex	MIGL	R		SP	MVA
14 Palmetto Bay Marina (p. 243)	803-448-3875	16	68	12			100	20	20		50		L16			GD	Tex	MB	All	P	All	All

See Adjacent Chart for Coast Guard Information

Information in these listings is provided by the facilities themselves.

range south of Charleston usually runs about 8 feet, so what appears to be deep water at noon may leave you tired and listing by dusk. This large variation in tidal depths can also produce swift currents.

South Carolina boasts three good inlets to the Atlantic: at Winyah Bay near Georgetown, Charleston Harbor, and the Savannah River on the Georgia state line. But the Intracoastal Waterway is the lifeblood of coastal exploration and the artery of travel preferred by most.

Little River Inlet to Georgetown

Once you leave Southport, N.C., there are two attractive alternatives as you continue south down the Carolina coast. If the wind is out of the north and the sea is calm, the 80-mile passage from Southport to Georgetown Inlet and Winyah Bay presents an excellent opportunity to shake out the sails or power up the ponies as you run offshore across the sloping shelf of **Long Bay**. But if you choose this outside route to Georgetown, you'll miss the beach communities along the inland route.

Your second option is using **Little River Inlet** into South Carolina and following the inside passage to Georgetown. This is the "Grand Strand" of South Carolina. Nestled among this ribbon of golden isles, plush hotels, and world-class golf resorts are some of the Waterway's finest amenities. And despite the abundance of shoal inlets along this state's crescent shoreline, Little River Inlet is the only safe access to the Intracoastal Waterway north of Winyah Bay.

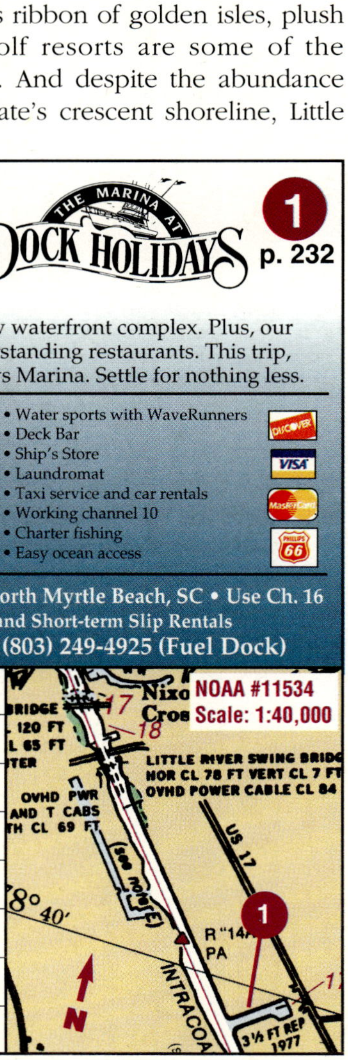

It's a 30-mile run offshore from flashing green 2.5sec "5" at the entrance of the Cape Fear channel to red and white whistle "LR" at Little River Inlet. From flashing red and white "LR," leave flashing red 4sec "2" to starboard and flashing green 4sec "1" to port at the entrance to the jetties. Once inside the breakwater, you will pick up a series of daymarkers marking the channel. Favor the port side of the channel until you reach the junction of Little River and the ICW at flashing green 4sec "119." As you head southwest again, you'll enter the Little River and the Pine Island Cut, part of the Intracoastal Waterway.

North Myrtle Beach is home to beach music, great restaurants, shoreside amusement parks, and all the golf and tennis you can stand. You'll find several marinas here, located right on the ICW. **The Marina at Dock Holidays** is a full-service facility with all the amenities, including five restaurants. North Myrtle Beach is at mile marker 350, which is just west of the Nixon Crossroads fixed bridge and the Little River swing bridge.

The soul of **Myrtle Beach** includes the communities of **Cherry Grove, Ocean Drive**, **Windy Hill**, and **Garden City**. These communities have restaurants, night clubs, and stores, but the real passion of the area is golf. You could easily spend two weeks here just playing golf. The annual Murrells Inlet Seafood Festival is held in April, there are fishing tournaments in May and June, and the South Carolina State Bluegrass Festival is held in late November. For a taste of true southern cooking, don't miss the annual Gumbo Stew Festival, also in October. For more information on these and other events, call the Myrtle Beach Chamber of Commerce (803) 626-7444.

Myrtle Beach and the Pine Island Cut offer little in the way of suitable anchorages, but there are marinas in the area. We recommend **Hague Marina,** located three miles north of the Socastee Bridge. The full-service boatyard has laundry facilities, showers, and courtesy transportation.

Bucksport, a small fishing village, is a pleasant noontime destination. Named after Captain Henry Buck, a lumber entrepreneur, Bucksport represents a prelude to South Carolina's low country cruising. From Myrtle Beach, it's a short run down the Waterway to Bucksport, located just after flashing red 4sec "36."

As you cruise south, the Waccamaw River meanders into the tidal creeks and marshlands of coastal Carolina as the rich wilderness of the Palmetto State blossoms along the banks of **Sandy Island**. Sandy Island is similar in many ways to Daufuskie Island farther south. Both reflect the pure and natural state of South Carolina's coastal tributaries, and Sandy Island has retained much native splendor. As recently as 1960, the children of the island hurried down sandy lanes each morning to catch the school boat. You will find Brookgreen Gardens (803) 237-4218 across the Waccamaw and in the shadow of the Mount Rena bluffs. This is the largest collection of original American statuary, in an outdoor setting, in the world. The sculptured gardens include several former rice plantations and have become the focal point of cultural events in the area.

Marine facilities are fairly scarce in this area but just east of Sandy Island on the ICW you'll find **Wacca Wache Marina**, a fresh-water facility in a protected cove that offers a fuel dock and an extensive ships store. There are also several anchorages worth exploring in and around Sandy Island. Two miles south of Bucksport, leave flashing green 4sec "43" to port and flashing red 4sec "44" to starboard. The ICW continues south to the right of the fork. To port is **Prince Creek,** a delightful tributary that encircles Longwood Island and rejoins the Waccamaw at Mulberry Landing. There is a minimum depth of 14 feet around the island, and because Prince Creek is a popular anchorage, you may find yourself joined by others as night falls. **Bull Creek** is another choice anchorage that offers solitude and good holding. In Bull Creek and Prince Creek you are advised to stay clear of the main channel, and maintain an anchor light since fishermen may be running the creeks.

Georgetown's past is intimately linked to the **Waccamaw**, **Great Pee Dee**, and **Sampit Rivers**. In 1732, King George of England declared Georgetown a Colonial Port of Entry and today it remains one of the oldest ports in continuous operation.

The historic significance of Georgetown shows in its ante-bellum homes and pre-Revolutionary architecture.

WACCA WACHE MARINA

3 p. 232

1950 Wachesaw Rd, Murrells Inlet, SC
1-800-395-6694

At Wachesaw Landing, Mile 383.5, six miles below Bucksport, the Wacca Wache Marina provides a fuel dock on the waterway and a deep-water channel leading into its extensive protected dockage area. This fresh water marina is understandably a favorite of ICW veterans. Conveniently, restaurants in the area provide free transportation. This "starport" facility offers Texaco products and has a large marine store stocked with the latest marine supplies, groceries, tackle and features a snack bar.

- Charleston Tides; High: + 3:11, Low: + 4:00, Mean 2.74'
- $.90 / ft per day, 20 Berths,150' Max LOA
- Propeller, Electronics, Gas, Diesel
- *Texaco* Gas, Diesel
- Fresh Water, Cable TV 240V, 100 Max Amps.
- Marine Supplies, Light Groceries 1-800-395-6694
- Several nearby: Transportation Provided, Casual, $10-14
- Mastercard, Visa, Discover, American Express, Texaco

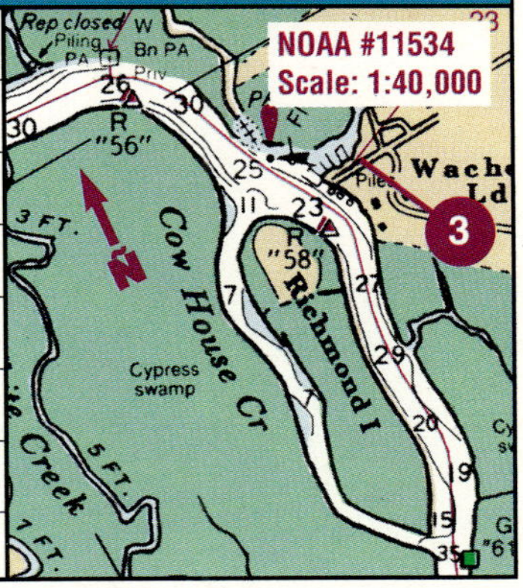

HAGUE MARINA

2 p. 232

"The Myrtle Beach Yacht Basin"

P.O. Box 835, Myrtle Beach, South Carolina 29577 • (803) 293-2141

A full-service boatyard featuring:

- **Fastest Prop & Shaft Repair**
- **Custom AWLGRIP Application**
- **35-Ton Lift — 65' x 18' Yacht**
- **Wet & Dry Storage**
- **Hardware & Paint**
- **Marine Supplies & Fittings**
- **Charts & Accessories**
- **Laundry — Showers & Tubs**
- **Courtesy Transportation**

J H Westerbeke · Cummins · CAT · Yanmar · Crusader · Cruisair · Onan · Chrysler Engines · Volvo Penta · Marine Gulf Products

VISA · MasterCard

3 Miles North of the Socastee Bridge

- Charleston Tides; High: +3:26, Low: +4:49, Mean 1.97 feet
- $0.75 / ft per day, 40 Berths, 100+ LOA
- Prop., Sail, Elec., Rigging, Gas Diesel, Wood, Fiberglass
- *Gulf* Gas, Diesel
- Fresh Water 115/220V, 30/50 Max Amps
- Food Lion, Winn Dixie nearby
- Mancuso's Casual, $13
- MasterCard, Visa

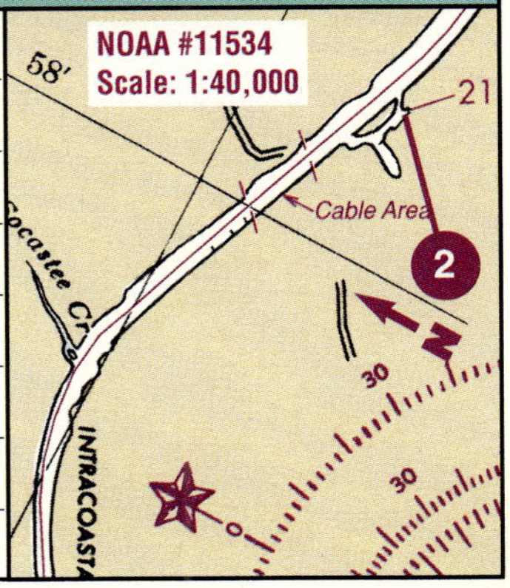

Sunrise over the Waccamaw River.

Front Street has been restored to its waterfront prominence. It's the starting point for Miss Nell's guided walking tour and a tram tour; both include more than 50 homes, buildings, and churches on the National Register of Historic Places. Call the Chamber of Commerce (803) 546-8437 for details. The Kaminski House (803) 546-7706, built in 1760, contains a collection of antiques and is one of two colonial homes open to the public.

A visit to the Rice Museum (803) 546-7423 on Front Street reflects Georgetown's stature as a rice empire. Hopsewee (803) 546-7891 and Hampton (803) 546-9361 plantations are also open to the public.

The aroma is one part of Georgetown that's not fondly remembered by most. International Paper's pulp facility emits a sulfuric odor that permeates the air even on the best days, and while offensive to visitors, it is the smell of industry and jobs to the town's people.

The approach to Georgetown from the Waccamaw may be confusing to some as the Pee Dee and Sampit Rivers merge into **Winyah Bay**. As you pass under the Lafayette Bridge, clear red "94" to starboard, then look ahead for flashing red 6sec "W." To the right you may see the red daymarkers marking the entrance to Great Pee Dee River. The most common mistake is to confuse the red daymarkers to the far right with the channel marked by daymarkers. Avoid the red daymarkers and maintain a southwest heading until you pass "W" and quick flashing red "40." This links the waterway with the entrance to the Sampit River. Leave quick flashing red "40" to starboard, then turn back to the northwest, and look for flashing red 4sec "42" to starboard. You should leave flashing green 2.5sec "41" off Sampit Point to port, then proceed upriver. You'll find several marinas on the Georgetown waterfront.

If you're coming from Southport using the offshore route, approach the entrance to **Winyah Bay Inlet** at flashing red 4sec "4." The channel is narrow, flanked by shoals on either side, and should only be run in calm seas as the swells may carry you out of the channel. Once you are abeam of **North Island**, the channel bends back to the northwest and continues up Winyah Bay with red right returning. The primary route across Winyah Bay continues north between the spoil areas and marsh islands. Green "17" joins the southern entrance of Western Channel here as the channel bears off westward. Western Channel is unmarked until you are abeam of Western Channel Island. Western Channel borders several historic plantations, and while a longer route than the main channel, its beauty is worth the effort. To get to Georgetown, follow the channel until you reach quick flashing red "40" off the entrance to the Sampit River.

Winyah Bay to Charleston

South of Georgetown, the Waterway passes through the Cape Romain National Wildlife Refuge, a long stretch of undeveloped area similar to the barrier islands of North Carolina. Here you'll find several good overnight anchorages. Jeremy Creek runs along the waterfront of **McClellanville,** holds a minimum depth of 7 feet, and is a popular stop. McClellanville offers some conveniences, but many visit just to enjoy an afternoon of fishing from the dock, or a stroll down a country lane. McClellanville is best known for its people who rebuilt after Hurricane Hugo and their festival, the Lowcountry Shrimp Festival featuring the Blessing of the Fleet, held on the first Saturday in May.

Harbor River is another good anchorage just off the Waterway that can be found a little farther south. **Hamlin Creek** north of **Capers Island** and **Dewees Creek** along the banks of **Dewees Island** also lend themselves to a quiet night away from civilization.

As you head south past the **Isle of Palms**, which is mostly residential, you emerge from the restricted borders of the Waterway and enter the sweeping panorama of **Charleston Harbor**. Passing through the swinging bridge on Sullivans Island, you will clear red "122" to starboard and green "123" to port. Be especially careful on a rising tide that you are not swept into **Jeanette Creek** by the current. Favor the port side of the channel, with an eye toward flashing green 4sec "125." Here, you will come abeam of the south end of **Sullivans Island** and Fort Moultrie, which has a history that remains an inspiration to South Carolinians.

In June of 1776, the British Navy dispatched a fleet of warships from New York intent on establishing a base of

Slip Into Something More Comfortable

4
p. 232

ASHLEY MARINA

(803) 722-1996 • VHF 16

Charleston Tides; High: +0:04, Low: +0:07, Mean: 5.34 feet	Fresh Water, Phone 110/208V, 50 Max Amps
$1.00/ foot per day 40 Berths, 150' Max LOA	Supermarket nearby
Prop., Sail, Elec., Rigging, Gas, Diesel, Wood, Fiberglass	
Texaco Gas, Diesel	MasterCard, VISA, Discover, American Express, Texaco

Located in picturesque Charleston, South Carolina, one of America's most beautiful historic cities, Ashley Marina's brand new floating docks provide a comfortable berth for your boat and the convenience of Charleston for you. You are within minutes of fresh seafood restaurants, theaters, historic homes and flowering alleyways. Your boat is carefully tended in a year-round frost-free harbor under professional, full-service supervision.

SERVICES

- ✓ All Floating Docks
- ✓ Texaco Diesel Fuel on 860' Dock
- ✓ Gas
- ✓ Ice (Crushed/Block)
- ✓ 110/208V - 30/50 Amp Electrical Power
- ✓ Touring Kit
- ✓ Captain's Lounge
- ✓ Rental Cars
- ✓ Facsimile Machine
- ✓ Copy Machine
- ✓ Notary Public
- ✓ Showers, Laundromat
- ✓ Ship's Store

With special winter rates for your boat and the charm of Charleston for you, we promise... you'll both be more comfortable.

ASHLEY MARINA
Charleston, SC

Mile Marker 470 ICW on the Ashley River
33 Lockwood Boulevard, Charleston, S.C. 29401

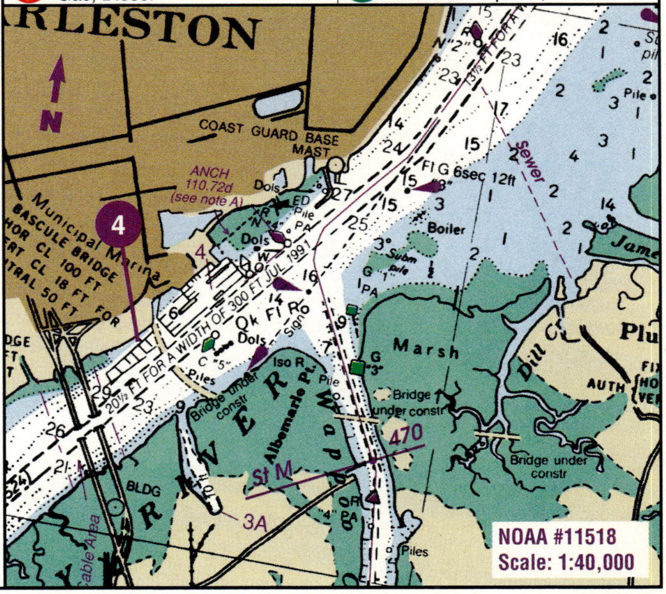

NOAA #11518
Scale: 1:40,000

military operation in Charleston. As word of this invasion spread, Colonel William Moultrie sought to check the British fleet off **Sullivans Island** and ordered his troops to construct an earthen fortress fashioned from sand and palmetto logs. Citizens came out to jeer Moultrie and his men, suggesting the palmetto planks would soon become their coffins. Moultrie and his men grew more determined, and when the British fleet weighed anchor off Charles Town bar, the fort was finished, at least on three sides. On the morning of June 16, over 300 guns opened fire on the tiny garrison. When the smoke cleared from the initial barrage, Moultrie and his men were heartened to see that the spongy palmetto logs had absorbed most of the cannonballs, leaving the fort unscathed. Then the garrison returned the fire, shattering the hulls and riggings of the British warships. The battle raged throughout the day, and by nightfall, the crippled and humiliated British fleet retreated offshore. The victory marked one of the soundest British defeats of the war. In honor of the brave men at Fort Moultrie, South Carolina adopted the crescent-moon banner that had flown over the fort and the palmetto tree as part of the state flag. Much of Fort Moultrie was modified during the Civil War, but you can still visit this historic monument.

Most enjoy **Charleston**'s grace and hospitality by bike or horse-drawn carriage. The historic district is large, with the many pubs and restaurants scattered throughout the downtown quarter. It's a stout 30-minute walk to the Battery and Bay Street from the anchorage off the municipal marina. The Battery marks the extreme tip of the Charleston peninsula and in the early years was little more than a salt marsh. White Point Gardens was created in later years, and it was from this venue that the people of Charleston viewed the bombardment of Fort Sumter, a battle between Union and Confederate forces that marked the beginning of the Civil War.

Fort Sumter is the most prominent feature in Charleston Harbor. In early 1860, Major Robert Anderson of the United States Army transferred a small garrison of men from Fort Moultrie to Fort Sumter in anticipation of southern opposition to President Lincoln's inauguration. In early April, 1861, General P.T. Beauregard ordered Anderson to surrender the fort to his South Carolina troops. When Anderson refused, Beauregard's troops fired upon Anderson's Union forces, thus igniting the Civil War. Anderson surrendered the next day and Confederate troops occupied Fort Sumter until 1865. Boat tours of Fort Sumter originate at Patriots Point and Charleston Municipal Marina. The Maritime Museum at Patriots Point (803) 884-2727 features the aircraft carrier *Yorktown* as well.

Also on the Charleston waterfront, you'll see multistory mansions from the past. Two of these homes, the Calhoun Mansion and the Edmondston-Alston House (803) 722-7171, are open to the public. As you gaze out across the harbor from the second floor balcony of the Edmondston-Alston home, palms braced against the wood railing, you are cooled by the ocean breeze and gain a deep appreciation for Charleston.

The abundance of steeples piercing the Charleston skyline gave rise to the town's reputation as the "Holy City." The religious roots of the city can be seen in the number of stately churches listed on the walking tours. St. Philip's Episcopal Church was built in the early 1700s, burned, and was rebuilt again in 1835. According to tradition, only native Charlestonians may be buried in its adjoining cemetery. Near the corner of East Bay and Elliot Street, you can see a row of houses known as "Rainbow Row." These brick shops, with their varied hues and colors, were built by merchants who worked downstairs and maintained living quarters on the upper floors. It was in this area that DuBose Heyward was inspired by the cheerful demeanor of the black residents living along Cabbage Row, along Church St., and composed his novel "Porgy." George Gershwin later adapted the novel into the opera, "Porgy and Bess." For a smorgasbord of handicrafts, dolls, and straw baskets woven from native sweet grass stroll down to the straw market on Market Street.

As you turn northwest away from Fort Moultrie, you will pass between red nun "126" and flashing green 4sec "127" to enter Charleston Harbor. If you intend to proceed up the **Ashley River** along the ICW, pass west of flashing red 2.5sec bell buoy "26." Swing to port, leaving red "26" to starboard, and this will place you in the main channel of the Ashley River, taking you past Battery Park and the historic district of Charleston. **Ashley Marina** is a full-service facility that is close to the theaters, restaurants, and historic homes of Charleston. The marina has full dockage and repairs.

As you come abeam of flashing red 4sec "32" off Potts Shoal, you should see the South Channel Range. The range is marked by a quick flashing green buoy in the foreground and a green flashing 6sec buoy in the background. Veer to starboard as you approach the quick flashing green mark, and head upriver until you see red nun "2" to starboard that indicates the entrance to the Ashley River. Ahead on your starboard side you will find the municipal dock.

An alternative to the Ashley River is to leave the ICW and head to the outside via the **Charleston Inlet**. After clearing Sullivans Island, head to the southeast and pick up the harbor entrance channel at flashing green 4sec bell "25", leaving it to port, and proceed to sea.

Charleston to Hilton Head

South of Charleston, the tenor of your cruise through South Carolina slows, inviting a more refined appreciation for the tidal flats and lush habitat of the "low country." Do not expect a swift passage through these marshlands, for the gnarled route of the waterway seems to turn back on itself every few miles. It is possible to exhaust an entire morning pushing south through the creeks and rivers and never escape the shadow of a high-rise bridge or radio tower. Your passage through the "low country" is best savored with an emphasis on the *quality* of miles made good, not *quantity*. Relish the simple lifestyle of a boy fishing from a finger pier, or ponder the graceful flight of an egret above an emerald carpet of sea grass. These sea islands offer a rare opportunity for gunkholing and some of the finest cruising on the waterway. The interior of the islands is heavily wooded, though the shoreline is marsh and sea grass.

As you leave Charleston and the Ashley River, leave green "1" to port off Albemarle Point. Proceed down **Wappoo Creek** until it converges with Elliott Cut just north of Edgewater Park. Here you will enter **Stono River** as it turns west, dividing James and John Island. **Wadmalaw River** and North Edisto River offer a myriad of quiet anchorages from which to select. While some entrances may be shoaled in places and require careful navigation, many creeks have plenty of water up to their banks. Continue down the waterway until you leave flashing green 4sec "63" to port at Church Flats. Here, **New Cut Creek** splits off to the south, just abeam of red "64." Do not attempt to enter New Cut. Though charts may suggest sufficient depths, the channel is not marked and is shoaled.

If you wish to access New Cut and **Church Creek**, continue south until you leave flashing red 6sec "A" and quick flashing red "A" to starboard. Proceed until flashing green "77" is to port, then turn back to the east, and follow the channel up Church Creek. This shelter offers one of the better overnight stops along the waterway, since the entrance is broad and deep, carrying a minimum depth of 10 feet well upstream. The high ground off New Cut Landing offers good protection and affords a scenic view. **Oyster House Creek**, a little farther south, is riddled with shoals at its entrance and accessible only to vessels drawing 4 feet or less.

Toogoodoo Creek is a better choice for anchoring. The channel's entrance carries at least 7 feet, although you are advised to stay east of the center of the channel as you proceed upriver. The best anchorage is just past the first big bend to the west where the northern shore is lined with tall pines and hardwoods. The upper reaches carry good depths, and here you will find the ruins of an old car ferry. As with all anchorages around the sea islands, the tidal currents run strong, so make sure your anchor is set securely before retiring for the evening.

You'll find the **North Edisto River** to be one of the more delightful cruising grounds in South Carolina. The river divides **Wadmalaw Island** from Edisto Island and offers several overnight sidewaters for exploring. North Edisto River is your access to **Bohicket Creek** and the resorts of **Kiawah Island** and **Seabrook Island.** Here you may wish to sample the plush accommodations and fine dining, or play a round of golf on one of Kiawah's championship courses.

The mouth of Bohicket Creek is located just south of the 16 foot quick flashing white light that is part of a set of range markers locating the inlet channel at Deveaux Bank. **Bohicket Marina Village**, located in a sheltered cove on the creek, has all the amenities of a full-service marina, plus shops and boutiques, fine restaurants, and a complete ships store.

The planters of **Edisto Island** made their fortunes from cotton, and today much of that heritage can be viewed in stately old homes. The first settlers tried to grow rice and failed. Next they tried indigo and cotton and grew rich for a time, but the economy soon suffered from war and pests. Most of the white population abandoned the island during the Civil War, and the Federal troops who occupied the island freed the slaves and resided for a time in the plantation homes. For this reason, few were destroyed. Oak Island Plantation and Cassina Point Plantation (now a bed and breakfast) can both be seen on **Westbank Creek**, at the mouth of the North Edisto River.

The mouth of Westbank Creek is approximately 2.5nm south of flashing red 4sec "110" at White Point. Westbank Creek is a convenient anchorage with better than 8 feet of water and good protection in the shelter of high ground.

Steamboat Creek is another anchorage of historical significance. Up the creek you will find two attractions worth exploring. An octagonal brick building with high

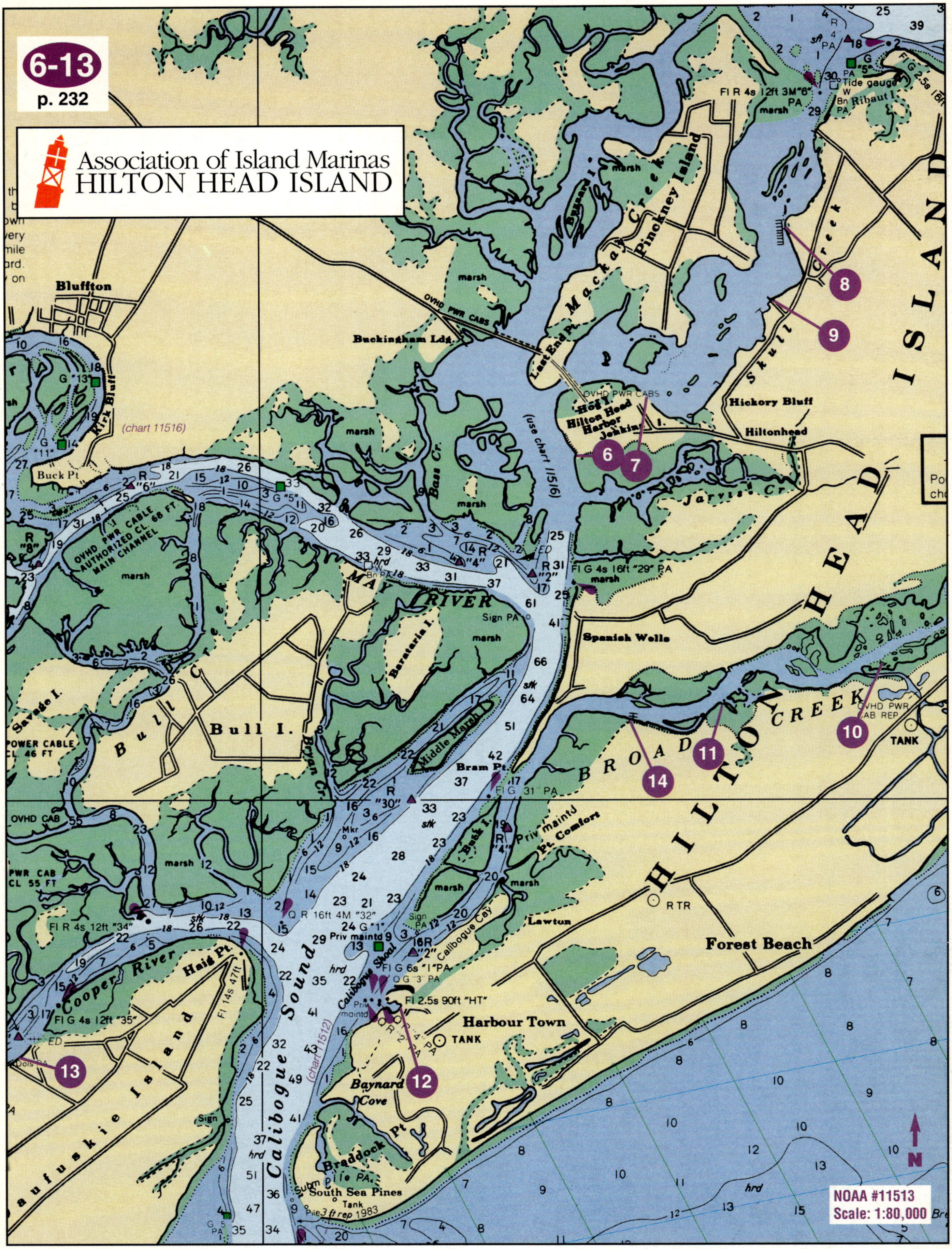
6-13
p. 232
Association of Island Marinas
HILTON HEAD ISLAND
Bluffton
Buckingham Ldg.
Pinckney Island
Mackay Creek
Skull Creek
Hickory Bluff
Hiltonhead
Jarvis Cr.
May River
Spanish Wells
Bull I.
Bull Cr.
Savage I.
Buck Pt.
Kirk Bluff
Bram Pt.
Middle Marsh
Broad Creek
Hilton Head Island
Pt. Comfort
Lawton
Forest Beach
Cooper River
Haig Pt.
Daufuskie Island
Calibogue Sound
Calibogue Shoal
Harbour Town
Baynard Cove
Braddock Pt.
South Sea Pines
NOAA #11513
Scale: 1:80,000

© Mary Jane Hayes

A mansion in Beaufort decorated for Halloween.

windows was once used by locals as they waited for steamboats to sail up the Edisto. Behind this structure is the Dodge Plantation, built in 1810 and a testament to the fortunes cotton brought to the fortunate of this land. The entrance to Steamboat Creek is indicated by red daymarker "2." Leave this mark to starboard when entering the creek.

Continuing down the Waterway along the upper reaches of North Edisto, you will pass flashing red 4sec "110." Use caution as you enter **Dawho River**, since the entrance is shoaled in several places, and the channel makes several turns before narrowing near North Creek. Leave "110" to starboard off White Point, and proceed southwest favoring the east side of the channel as you clear green "111." To starboard just abeam of green "113," you will see red "112" marking the southern edge of a shoal. Favor the south side of the channel until you're abeam of flashing green 4sec "115." Bear off to the northwest, passing between red "118" and flashing green 4sec "119," and your anxiety should ease as you round the bend at flashing green 4sec "121."

As you exit Watts Cut at the entrance to **South Edisto River**, flashing green 4sec "143" will be seen to port. Here, you will be aided by three sets of range markers that will guide you safely to Fenwick Cut. The first set becomes visible as you pass green "147." As you approach red "152" to starboard, the second set comes into view just south of **Alligator Creek**. The last set is seen on the east bank of **Fenwick Island** as you come abeam of red "156." The passage from Fenwick Cut to Ashepoo Coosaw Cutoff takes you through **Hutchinson Island** and up **Rock Creek**, where navigation is straightforward.

Through **St. Helena Sound** until just north of Beaufort, the "low country" offers an assortment of day excursions and overnight anchorages. The many creeks and rivers provide nourishment for the tidal flats, helping to support a large fleet of shrimp boats. On a quiet night, anchored in the elbow of **Back Creek** on **Morgan Island**, you can hear the steady drone of the working diesels, straining to lift the trawls filled with the shrimp of **St. Helena Sound**. On St. Helena Island, live oaks envelop the sandy lanes, and the ocean breeze rustles the sea grass along the marsh. In the past, this delta concealed pirates and rum-runners, ex-patriots, and shrewd Indians. Today, the mansions and natural beauty of the islands of St. Helena Sound stand as a tribute to a simpler time.

Access to Back Creek is marked by flashing green 4sec "13" on the east side of Morgan Island. There are no markers in the creek, so keep a close eye on your depth sounder.

Beaufort might well have been the hub of South Carolina's coastal commerce had it not been for the Cusabo Indians. In 1670, they suggested to the English colonization party that the lands to the north were more fertile and the rivers better suited for the white man's large ships. With this knowledge, they marched north to the headlands of the Ashley and **Cooper Rivers** and founded Charleston.

While Charleston may be the heart of South Carolina cruising, Beaufort is the soul. Despite a turbulent past, Beaufort has remained dignified. Lucas Vasquez de Ayllon claimed her for Spain, while Jean Ribaut settled her for the French. William Hilton explored her for the British, calling the air "clear and sweet and the country very pleasant." A band of Scotsmen settled **Stuart Town** near **Port Royal**, and in 1715, the Yamassees Indians, angered by the encroachment of the white man, slaughtered every settler in the region and reclaimed their sovereignty.

The nostalgia in Beaufort is addictive. You don't walk the streets of Beaufort, you stroll. The character of the town is reflected in such street names as Port Republic, King, Prince, Duke, and Washington. The town's heritage envelops you like the Spanish Moss clinging to huge oaks on the waterfront. This is a town of old money, really old money. This was the home of Edmund Rhett, the United States senator who lead the call for South Carolina's secession from the Union. The Castle was the home of Dr. Joseph Johnson and takes up an entire city block. The structure boasts 79 windows, four chimneys, and a porch so massive that six pillars are required for its support. The Tidalholm, a private residence, was the house used in the film "The Big Chill." Also on the walking tour of Beaufort is the John Mark Verdier House Museum (803) 524-6334, the George Elliott House Museum (803) 524-8450, the Riverview, and the St.

Helena Episcopal Church (803) 522-1712. For more information on these and other house museums, call the Chamber of Commerce (803) 524-3163.

As with all low country towns, the pace is slow but the tide runs fast, so caution is urged whenever you secure dockage. Berths are available along the Waterside Park and there is a fair anchorage west of Downtown Marina. A better choice is up **Factory Creek**, east of Ladies Island Bridge. It is protected from the wake of boat traffic, and the bluffs offer some protection from the wind. Due to the current, a Bahamian moor is recommended. Factory Creek is reached by making a sharp turn to port just before the Ladies Island Bridge, leaving green "237" and green "1" to port. If you opt for the bulkhead at Waterside Park, or a marina slip, proceed through the swing bridge and look off to starboard for both.

Visitors to **Hilton Head** come for the opulent shops, thrilling golf, and elegant accommodations. They ascend Harbor Town's famed red and white candy striped lighthouse and gaze in awe at the panoramic view. Hilton Head is an island at play, a refuge for the young at heart and stout of wallet. Even the dolphins frolicking in **Calibogue Sound** elevate their play, performing tricks in exchange for fish or dinner scraps. The Yamassee Indians also came to Hilton Head for game, but of a different sort. In the 1600s the Yamassee lived off the abundant fowl, shellfish, and wild game found on the island. While the original tenants of this pristine landscape, the Yamassee became caught in a territorial conflict between the French, English, and Spanish, and by 1663, Great Britain claimed the island as its own, renaming it in honor of Captain William Hilton. Indigo, rice, tobacco, and sugar cane flourished during the colonial period, but it was cotton that brought true wealth to the planters of Hilton Head. With the onslaught of the boll weevil and destruction of the southern economy after the Civil War, Hilton Head faltered until its development as a resort community began in the mid-1950s.

As the Beaufort River empties into **Port Royal Sound**, you will leave flashing green 4sec "27" to starboard and red "26" to port, and bear west on a heading that will take you off Dolphin Head, the point of land marking the northwest corner of Hilton Head Island. Leave red "2" to starboard and pick up flashing green 2.5sec "3" to port off the island. Red "4" to starboard marks a large shoal that runs off **Pinckney Island**. Leave green "5" to port and flashing red 4sec "6" to starboard, and begin your majestic passage along the shores of Hilton Head Island.

Marinas are abundant on the island. The **Association of Island Marinas** operates nine marinas around the island offering every convenience imaginable. Approximately three miles up **Broad Creek**, you'll find **Palmetto Bay Marina** offering transient dockage, restaurants, and a full services. As you leave South Carolina by way of **Daufuskie Island**, you are deep in Southern coastal country and, no doubt, not minding it at all.

Shoreside and Emergency Services

Ambulance, Fire, Police: 911

Coast Guard:

Georgetown (803) 546-2742 or VHF 16 or 21
Charleston (803) 724-7600 or VHF 16 or 21

Radio Telephone:

VHF 26; I.D. Georgetown Marine Operator
VHF 26; I.D. Charleston Marine Operator

Tow Service:

Carolina Sea Service, Charleston
(803) 577-7343 or VHF 16
Sea Tow Savannah, Savannah
(912) 897-4304 or VHF 16

✦

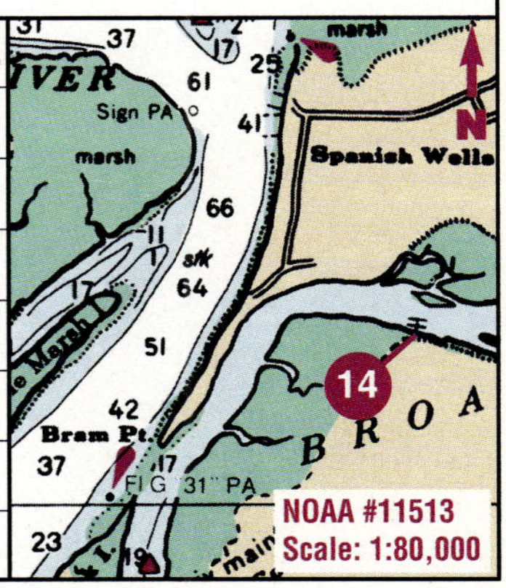

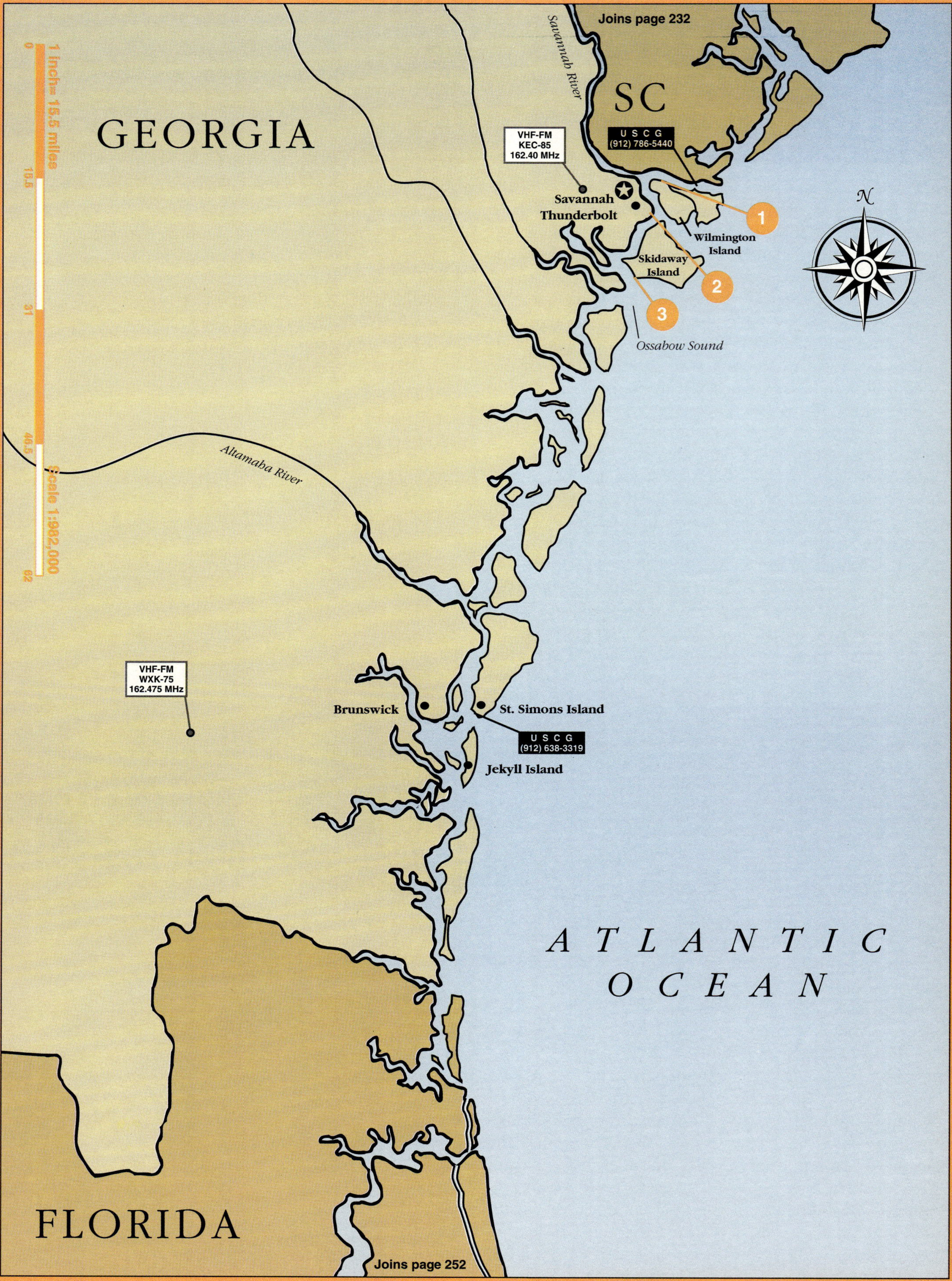
Joins page 232
Savannah River
SC
GEORGIA
U S C G
(912) 786-5440
VHF-FM
KEC-85
162.40 MHz
Savannah
Thunderbolt
Wilmington
Island
Skidaway
Island
1
2
3
Ossabow Sound
N
1 inch= 15.5 miles
0
15.5
31
46.5
62
Scale 1:982,000
Altamaha River
VHF-FM
WXK-75
162.475 MHz
Brunswick
St. Simons Island
U S C G
(912) 638-3319
Jekyll Island
ATLANTIC
OCEAN
FLORIDA
Joins page 252

The Peach State

Coastal Georgia has largely escaped the developer's bulldozer. You won't find miles of high-rise apartments or overcommercialized areas. What you will find is the quiet, awesome beauty of nature—miles of coastline fringed by marshes, serpentine rivers emptying into vast sounds, and uninhabited barrier islands.

While many people do live on or near the coast, residential communities are concentrated in two major areas: Savannah at the northern border and the Brunswick/Golden Isles area near the southern border. Tybee, Jekyll, St. Simons, and Sea Island—the only barrier islands that are developed—sport active communities. Most of the remaining barrier islands are private reserves or have been set aside by the state or federal government as sanctuaries. So unlike other parts of Georgia—such as Atlanta which has experienced remarkable growth in recent years—much of the coast remains the same as it has been for centuries.

When entering the Intracoastal Waterway in Georgia, expect some of the highest tides in the United States, sometimes yielding 10-foot tidal differences. A strong wind out of the northeast will push more water in through the inlets and make the tides even higher. High tides can be deceptive, hiding mud banks and luring you away from the marked channel to a supposed shortcut. Low tides are similarly affected by winds. A strong west wind will push out more water and cause the tide to fall even lower than normal. The highest and lowest tides normally occur in October and November.

When on the Intracoastal, plan your route ahead of time, and cross off each marker on your chart as you pass it. During the frequent rainstorms of July and August, it's easy to become disoriented, and strong winds can make positioning difficult. If you've marked your chart, you won't have any trouble knowing where you are.

© J.H. Peterson

Savannah's historic district displays architecture unique to the South.

If you're following an offshore route, keep in mind that many of the inlets along Georgia's coast are shallow and difficult to navigate. Bars and breakers often reach far out into the Atlantic, and even in the best conditions, you're likely to encounter quite a chop. Opposing winds and tides can make your passage downright dangerous. In short, be prepared if you're approaching The Peach State from the ocean.

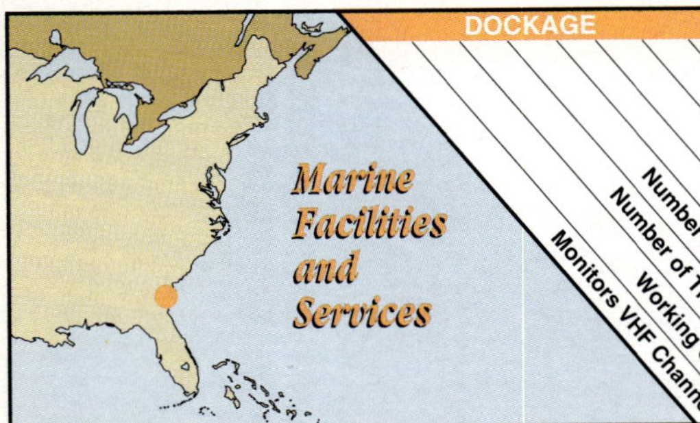

Marine Facilities and Services

Marina	Phone	DOCKAGE: Monitors VHF Channel	Working Channel	Number of Transient Berths	Number of Transient Moorings	Seasonal / Year-round	Maximum Length Over-All	Approach Depth in Feet at MLW	Dockside Depth in Feet at MLW	SERVICES: Hookups: Fresh Water / Phone / Cable TV	110V★ 220V▲ 3 Phase ■ Maximum Amps	Ramp / Dinghy Dock / Launch Service	Railway / Lift / Crane: Capacity (tons)	Repairs: Propeller / Sail / Electronics / Rigging / Gas / Diesel / Wood / Fiberglass	SCUBA: Air / Rentals / Equipment / Charters	SUPPLIES: Gas / Diesel Fuel	Fuel Brand	Marine / Groceries / Bait / Ice / LPG / CNG	AMENITIES: Showers / Laundry / Pumpout / Restrooms	Hotel / Pool / Tennis / Golf	Public Phone / Restaurant / Snack Bar	MasterCard / Visa / Discover / AmEx
1 Bull River Marina (p. 246)	912-897-7300	16	7	10		Y	150	25	18	F	★▲ 100	D		FEGD		GD	Tex	MGBIL	SPR		PR	All
2 Palmer Johnson Marina (p. 247)	912-352-4956	16	9	30		Y	100	16	16	FC	★50	All	L150	PSRGDF		GD		MI	All		PR	All
3 Delegal Creek Marina (p. 249)	912-598-0023	16	68	30		Y	100	7	18	F	★▲■ 50			PGDF		GD	Chev	MGBIL	All		PR	MVA

COAST GUARD: See Adjacent Chart for Coast Guard Information

Information in these listings is provided by the facilities themselves.

Savannah River to Ossabaw Island

In 1633, James Oglethorpe, the founder of Georgia, sailed up the **Savannah River** and established the city of **Savannah** on Yamacraw Bluff. His vision for the city was a series of symmetrical squares with a common green in the center. This careful planning by Oglethorpe and his successors is obvious in the city today. The historic district is the nation's largest, encompassing more than 2,350 architecturally and historically significant buildings in its 2.5 square miles.

Savannah is very much a walking city. Excellent antique shops are scattered throughout the downtown area. Your tour should include the Colonial Cemetery dating from the city's earliest days, Old Christ Church, and the Cathedral of St. John the Baptist (912) 233-4709. Also, essential to your exploration of the historic district is a trek to the Savannah Visitor's Center (912) 944-0456 on Martin Luther King Boulevard situated in the Old Central of Georgia Railroad Station. The Savannah History Museum (912) 238-1779 in another part of the station features artifacts, displays, and theater presentations about the city's evolution.

While you're in town, don't miss Johnson Square in the heart of the city. Vendors hawking everything from lemonade to local artwork ring the area. At lunch time, there is usually some type of live entertainment, often a jazz band. Several blocks west of Johnson Square is City Market, four blocks of restored buildings built between 1830 and 1850 which once housed produce markets, feed stores, blacksmiths, and barbershops. Today, the lower level contains restaurants and shops. Upstairs is an art gallery and studios, while the center promenade is the scene of numerous festivals and musical events, such as the Savannah Maritime Festival and frequent jazz concerts.

Scattered throughout the city's historic district are numerous museums and restored buildings. The Davenport House (912) 236-8097, an exceptional example of Federal architecture, was built in the early 19th century. Located in the basement is the Historic Savannah Foundation's museum shop which boasts one of the best selections of publications about Savannah. Ships of the Sea Museum (912) 232-1511 offers a large collection of shipping memorabilia including a collection of over 75 ships-in-bottles. Telfair Academy of Arts and Science (912) 232-1177 on Barnard Street is the oldest public art museum in the South with a superb collection of American and European paintings and sculptures.

Just outside the city proper is Fort Jackson (912) 232-3945, the oldest standing fort in Georgia. Originally an earthen battery constructed during the Revolutionary War, the fort has been rebuilt many times. During the Civil War it served as the headquarters for the Confederate defenses in this area. Its vantage point is clear: Every ship that enters or leaves the port passes within shouting distance.

Enter Georgia by way of Fields Cut. As you exit the cut to the Savannah River, leave flashing red 4sec "48" to starboard. Currents are swift at the mouth of Cut, and the river is busy with commercial traffic, especially at high tide. The ICW continues across the river, but if you're headed to downtown Savannah's riverfront, turn west when you exit Fields Cut, and follow the river channel markers. There is dockage available in the center of downtown, but they charge by the hour, and navigating the river may be difficult with strong currents and heavy commercial traffic. We suggest docking in Thunderbolt and visiting Savannah by car.

Thunderbolt is a quaint shrimping village on a high bluff dotted with modern condos and apartments. But it still maintains its small-town aura. Along the sidewalk that edges the bluff, you'll be welcomed by the friendly locals. You'll also find modern marinas, marine repair facilities, and marine stores. You can eat at informal restaurants or dine elegantly overlooking the Wilmington River.

From the junction of Fields Cut and the Savannah River, head for the flashing white 2.5sec marker at the mouth of Elba Island Cut, a straight passage leading to the Wilmington River. Be especially careful between red 4sec "4" and red 4sec "6" as they mark the passage across South Channel, a spoil area quite shallow in places. If you're in need of services, turn southeast just after red marker "8" into St. Augustine Creek, which turns into Bull River. Nestled in the beautiful salt marshes at the north end of Wilmington Island, **Bull River Marina** has all the amenities and is a 10-minute drive from downtown Savannah.

Continuing down the ICW, follow the channel leaving red marker "12" to starboard. Green "27" is just north of the Causton Bluff Bascule Bridge, a double span with a vertical clearance of 26 feet. Time your approach to the bridge as it closes to water traffic Monday through Friday, year round between 6:30 and 9:00 a.m. and 4:30 and 6:00 p.m. The bridge will open for waiting traffic at 7 a.m., 8 a.m., and 5:30 p.m.

From the Causton Bluff bridge, follow the channel—it's clearly marked and fairly straightforward. On the west shore below quick red "30" lies Bonaventure Cemetery, one of the most picturesque sights on the coast. Stay close to flashing green 4sec "31," leaving it to port, and head for red "32." Hug the right shore here to avoid the submerged pilings just before green "33." Before the Memorial Bridge (65-foot vertical clearance) in Thunderbolt, you'll find docking facilities. Just south of the bridge is where you'll find **Palmer Johnson Marina**, a full-service facility with all the dockside amenities, as well as restaurants, transportation to and from Savannah, and a loaded repair yard.

From Thunderbolt, you might also want to take a trip to **Tybee Island**. Because of limited accessibility by boat, we suggest getting a car at one of the marinas in Thunderbolt. The island lies at the mouth of the Savannah River and is the site of one of the oldest lighthouses—the third built in America. At 154 feet, it offers a panoramic view of the entire area. Fort Screven (912) 786-5444, dating from the Spanish American War, is on the north end of the island and is also open to the public. Like most of Georgia, Tybee is a town to tour on foot or by bike. Enjoy the beach or the tree-shaded streets. Come early for a magnificent sunrise. Eat a seafood dish in one of the local restaurants, and talk to the locals. They can fill your ear and your afternoon with tales of the hurricanes of 1893 and 1896 that destroyed much of the island.

On the way to Tybee, you'll pass Fort Pulaski National Monument (912) 786-5787, a pre-Civil War installation on Cockspur Island, which was used as a Confederate defense barracks during the Civil War. The fort and the island are good areas for picnics or exploration.

Ante-bellum homes grace the bluff at **Isle of Hope**. Take a stroll to examine this unique, pre-Civil War architecture. Pay a visit to Wormsloe (912) 353-3023, a colonial estate constructed by Noble Jones, one of Georgia's first settlers. Some of the earliest tabby ruins can be found here, too. Tabby is a building material native to coastal Georgia. Its principal ingredient is crushed oyster shells and is an example of how settlers learned to use what was available to produce sturdy, inexpensive building materials. Your stay on the island would not be complete without a visit to the nearby Byrd Cookie Shanty (912) 355-1716, maker of the famous Benne Wafer.

From Thunderbolt, follow the channel down the Wilmington River. Stay close to flashing green 4sec "37," leaving it to port. The opposite side of the river is shallow. But after red "38," the channel widens and presents few problems. Continue on the Intracoastal Waterway, giving a wide berth to flashing red 4sec "40" as you enter the Skidaway River. In the Isle of Hope area, there's a well-protected anchorage below the marina on the western shore, just past red "46A."

The next stop on the waterway is **Skidaway Island**, site of several large, carefully planned golf communities. Most are private clubs, but they do allow non-members to play—for a nominal fee, of course. The best way to tour the island is by bicycle. Visit Skidaway Island State Park (912) 598-2300 where you can follow nature trails through the salt marsh. In the fall, the marsh blooms and tall waving heads of grasses and numerous tiny flowers appear. If marine life is what you're after, tour the Marine Science Aquarium (912) 598-3474 featuring a smattering of marine life from Georgia waters. If you're passing through in April, pay attention to the meadows along the road to the aquarium. If the thistles are blooming, you're likely to see swarms of swallowtail butterflies.

The passage south from Isle of Hope heading to Skidaway Island is narrow and twisty. Shoaling occurs on the starboard side between red "50" and red "52," and a no-wake speed is recommended, especially at low tide in the Skidaway Narrows section. The bridge here is a bascule bridge with a clearance of 22 feet which opens on demand. Just south of the bridge is a popular public boat ramp and manmade beach with a roped off section for swimming. Spring to fall, this area is a favorite for boaters and water skiers. **Delegal Creek Marina** is located at the southern tip of Skidaway Island and has fuel, showers, laundry facilities, repairs, groceries, and towing services.

At green "65" follow the wide curve of the river. Flashing red 4sec "74" at the tip of Marsh Island marks the entrance to Moon River, immortalized by songwriter and Savannah native Johnny Mercer. Although he described it as "wider than a mile," it isn't. It is navigable in a small boat, but we don't recommend it in a cruising vessel. Opposite green "77" is Vernon View, another picturesque suburb. Stay close to the starboard shore past flashing green 4sec "79," and then move toward center channel.

You'll want to stock up on fuel and supplies at Skidaway before heading south. Except for Kilkenny, you won't find facilities until you reach the Golden Isles area.

Ossabaw Island is one of the largest barrier islands in Georgia and is now a Georgia Heritage Wildlife Reserve managed by the Georgia Department of Natural Resources (912) 598-0279. Whenever you pass this or any of the other barrier islands, you're likely to catch a glimpse of the only natives: whitetail deer and wild hogs. They can frequently be seen feeding in the marshes. Egrets and herons pick their way cautiously along mud flats in search of minnows. Rails, commonly called marsh hens, scurry along the banks. And don't be surprised if your passing vessel sends a whole flock of grackles winging through the air.

The waterway leaves the wide and deep Vernon River at Hell Gate. Leave flashing green 4sec "87" close to port. Continue almost to flashing green 4sec "91" at the end of the cut before making a hard turn to starboard to enter the Ogeechee River. Beware of Middle Marsh Island on the south side of the river—it's covered at high tide.

The current in the Ogeechee, one of Georgia's largest rivers, is quite swift, especially on the outgoing tide. Leave the Ogeechee at flashing red 4sec "98," and follow the waterway through Florida Passage. At the southern entrance to this passage, give red nun "102" a wide berth, approaching the flashing marker on the south shore before making your hard turn to starboard into Bear River. Ossabaw Island is to port.

If you'd like some time enjoying isolation, head west into Kilkenny Creek, just past flashing red 4sec "106" and before green "107." There is a marina 1.5nm off the waterway. There are no markers in the creek, but the dockmaster advises that it's easily navigable if you stay slightly north of center.

St. Catherines Sound to Altamaha Sound

Cruising the mid-coast region of Georgia will provide you with a close-up view of nature. Because of the warm climate in this area of the South, the marsh grass grows all year round, and the bird populations flourish. You're likely to see birds of prey circling overhead—any variety of hawk and even eagle—ready to swoop down on an unsuspecting victim. The key to enjoying this undisturbed natural habitat? Relax in the quiet serenity that surrounds you, and remember to leave it the way you found it.

Below Kilkenny and just before the Bear River empties into St. Catherines Sound, you need to be wary of shoaling between flashing red 4sec "110" and flashing red 4sec "112." There is a bar that sticks out between the markers, and you can easily run aground if you're cutting the channel too close.

St. Catherines Sound is generally exposed and fairly wide. This combination can make passage difficult for sail boats if a stiff wind is blowing. We suggest you time your passage with a moderate tide. During shrimping season, be careful of the shrimp boats—their major concern is their nets, not passing vessels. **St. Catherines Island** in the southeast corner of the sound is privately owned with limited visitation. It serves as a research site for the American Museum and the New York Zoological Society. Walburg Creek is off to port and provides safe anchorage.

Accessing St. Catherines Sound from the outside, stay between flashing 2.5sec red "2" and green can "3." Be careful around red "2"—the area north of the marker is shallow. From here, pick up the ICW at flashing red 4sec "114A."

South of Mile Marker 620, leave St. Catherines and enter the North Newport River. At green "121" stay close to the eastern shore, and then head central channel between red "122" and flashing green 4sec "123." Johnson Creek at flashing green 4sec "125" is a narrow but deep passage which empties into South Newport River at green "133." Watch for shoaling on the starboard side between flashing red 4sec "132" and green "133." Center channel has plenty of water except for the shoal on the starboard side between red "136" and quick flashing red "138."

Across the sound is the northern end of **Blackbeard Island**, a pirate haven of years past which is now a federal wildlife preserve. Named for Edward Teach, the pirate known as "Blackbeard," it's the supposed site of buried treasure. But the creek is shallow, so don't plan on doing any treasure hunting in your cruising boat.

The South Newport River leads you to **Sapelo Sound**, another wide body of water. Go beyond quick flashing red "138" before making a turn to starboard. Leave flashing red 4sec "140" and flashing red 4sec "142" well off to starboard. North of these two markers is shallow, as well as the area west of red "150." Beware of the submerged pilings on the western shore before red "152." Here, as in all of the sounds along the Georgia coast, look for bottle-nosed dolphin, which can be seen year-round. Sapelo Sound is also the next feasible inlet if you're traveling the outside route. At the red and white spherical buoy "S," turn to starboard, and follow the buoys. There's a shallow area north of red nun "6" known as Experiment Shoal, so leave the buoy to starboard. Then pick up the ICW at red "138."

The next section of the waterway is narrow, so slow your speed. Leave red "156" to starboard and flashing green 4sec "157" to port as you exit Creighton Narrows. **Sapelo Island**, site of several research facilities including the University of Georgia Marine Institute (912) 485-2221, is off to port. Follow the markers carefully through Old Tea Kettle Creek. The channel hugs the southeast shore at the entrance and moves to the other side at the next turn. Beyond that, the creek widens and empties into **Doboy Sound**. Except for the shallow area to port around green "167," you'll have easy cruising.

In Doboy Sound, be sure to give quick flashing red "178" a wide berth to avoid the bar which juts out from the western shore. Follow range markers carefully through Rockdedundy River Range. This area and Little Mud River are both narrow and shallow. Pay close attention between flashing green 4sec "185" and red "188"—there's only five feet of water at low tide. The inlet at Doboy Sound is narrow and shoals, and it requires a considerable amount of travel to pick up the ICW. So if you're traveling the outside route, we suggest using the inlet at St. Simons Sound.

At green "195," the Intracoastal enters the **Altamaha Sound**. Numerous bars and shallow spots dot the sound. The dredged area between red "194" and green "195" has strong currents and some shoals. From green "195," steer a straight course directly to quick flashing red "198." Keep

Savannah River Entr. Tides; High: +0:04, Low: +0:10, Mean 7.2 feet

$.95 / ft per day
72 Berths, 100' Max LOA

Propeller, Gas, Diesel, Fiberglass

Chevron
Gas, Diesel

Fresh Water
220V, 50 Max Amps

Liquor Store, Grocery Store and Bank Nearby

Upper Crust Restaurant
Casual $ Moderate

MasterCard, Visa, American Express

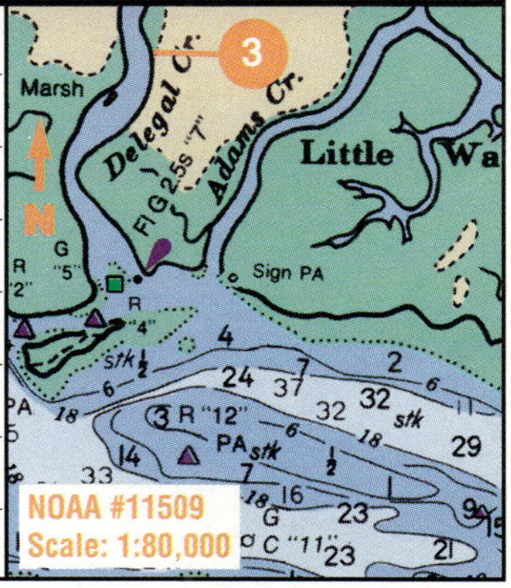

© J.H. Peterson

The Museum of Coastal History on St. Simons Island.

well away from the buoy; if you try to cut it, you will surely run aground. The course is then straight to red "208" and heads through Buttermilk Sound.

Golden Isles to Florida

Next stop on the ICW is the Brunswick/Golden Isles area. According to legend, once the sands of **The Golden Isles** are in your shoes, you'll always want to come back. It's like moving back into history—to a less hurried, more peaceful time—yet still having all the conveniences of modern life.

Between the islands and the mainland of Brunswick lie the *Marshes of Glynn.* From his vantage point under what is now known as Lanier Oak, poet Sidney Lanier described them as stretching "leisurely off, in a pleasant plain, to the eternal blue of the main." Today the marshes are protected by law and retain their same pristine beauty. Their role in coastal ecology is crucial: They serve as a buffer between ocean and mainland, prevent erosion, and provide a spawning ground for shrimp and other marine life.

St. Simons, the largest and northernmost of The Golden Isles, has maintained its captivating character. The island was first occupied by the Spanish in the 16th century. It was later colonized by James Oglethorpe who established Georgia's first military outpost at Fort Frederica (912) 638-3639. The fort is now a national monument open to visitors year round.

During the Revolutionary War, the island's highly prized, hard oak timbers were milled for use in constructing warships. The U.S. Frigate *Constitution* ("Old Ironsides" from the War of 1812) was built with this wood. During the 19th century, dozens of thriving plantations grew high-quality cotton. Hampton Plantation, one of the most famous of these, is now the site of Epworth-by-the-Sea, Methodist conference center and religious retreat (912) 638-8688. Christ Church at Frederica (912) 638-8683, the second oldest Episcopal Church in Georgia, dates its congregation from the time of John and Charles Wesley who established Methodism in the mid-17th century. The church's present sanctuary was built in 1886 by Rev. Anson G. P. Dodge, Jr., as a memorial to his first wife Ellen. Its stained glass windows are magnificent.

The St. Simons Lighthouse, originally built in 1808, is one of the nation's oldest continuously working lighthouses. The current structure was completed in 1872. Visitors can climb 129 steps to the top for a breathtaking view of the ocean, islands, marshes, and mainland. The Museum of Coastal History (912) 638-4666 is in the keeper's house. St. Simons is also the site of the Battle of Bloody Marsh, fought in 1742 and so named because the marsh was said to have run red with Spanish blood. This victory by the small English force over the Spaniards coming up from Florida ended the threat of Spanish invasion.

If you're traveling the outside route and plan to spend the night in the Brunswick/Golden Isles area, the inlet at St. Simons Sound is the place to re-enter. The red and white "STS" whistle marks the entrance. There's a spoil area just south of the buoy, so continue out to flashing red 4sec whistle "2B" before turning and following the markers into the Sound.

The first marine facilities in this area are on the northern tip of St. Simons. If you take the first river after red "222," entering at green "27," a marina lies about three miles upriver between the Intracoastal and the ocean. There is a strong current so exercise care when docking. Another tight spot is past green "227." The passage is shallow on either side at green "229." Entrance to the Frederica River is to the east of green "229." This passage takes you by Fort Frederica National Monument where landing is prohibited. The river provides safe anchorage and reconnects with the waterway in Mackay River at flashing green 4sec "241." On the route through the Mackay River, once you're past green "235A," move toward the western shore, and leave flashing green 4sec "237" to port. There is a shoal to the west of red "238."

Sea Island, another of The Golden Isles, is reached by causeway from St. Simons. Its surrounding waters are shallow, so we suggest getting a car at St. Simons and driving to the island. It's the site of The Cloister (912) 638-3611, an internationally acclaimed resort. The island is almost entirely residential and is renowned for its elegance and isolation.

South of St. Simons and six miles off the mainland lies **Jekyll Island**. Here, even more than on St. Simons, time slows down and visitors enjoy every minute. Jekyll is the essence of daydreams—a golden beach, swaying palms, moss-draped oaks, and echoes of the bygone era when the fabulously rich kept it all to themselves. Turn-of-the-century American millionaires like Pulitzer, Rockefeller, Morgan, McCormick, and Gould bought the island, formed the elite Jekyll Island Club in 1886, and virtually closed it to the outside world. For 56 years, club members and their families vacationed here. Most stayed in the Victorian Club House (912) 265-3175, but some built mansions which they ever-so-modestly called "cottages."

Today the 240-acre National Historic Landmark District holds a treasure of knowledge about America's early industrialists and their lifestyles. Many of the 33 structures are restored or are being restored. Several of the cottages, including the Goodyear Cottage and the DuBignon Cottage with its unique wrap-around porch, can be seen on tours. Faith Chapel, with stained glass windows by Louis Comfort Tiffany and D. Maitland Armstrong, is also open to the public.

Also on Jekyll you'll find the Horton Brewery site. These are tabby ruins of Georgia's first brewery which supplied ale to troops and settlers on Fort Frederica. The Horton House, a two-story ruin dating from 1742, is one of the oldest standing tabby structures in the state.

Brunswick on the mainland was named for the German ancestral home of King George II, grantor of Georgia's original charter. The streets and squares of this quiet port city were laid out before the American Revolution. The names of the streets—Newcastle, Norwich, and Gloucester—give Brunswick a decidedly English flavor that you won't find in other Georgia towns. Homes in the area are noted for their fine Victorian architecture. Brunswick is touted as one of the shrimp capitals of the world, and the picturesque shrimp fleet can be seen along the waterfront. Other than shrimp, trees are, oddly enough, an important part of Brunswick culture. In addition to Lanier Oak, there is Lovers' Oak, a massive 900 year old tree said to be the spot where Indian braves and their maidens met. Again, we suggest visiting by car.

At the northern end of **Lanier Island**, which is mostly marshland, the Intracoastal Waterway now runs to the west around the island. Even if you've traveled here before, pay attention to the markers. There's access to a marina on Jekyll Island beyond the 65-foot fixed bridge over the Mackay River. Give green "249," the turning buoy, a wide berth—it's located on top of a shoal. Leave flashing red 4sec "2" to starboard, and enter the channel.

St. Simons Sound and the **Brunswick River**, like the Savannah River, are busy with shipping traffic. Stay near the markers in the sound, but be careful crossing—the sound can be rough. The entrance to Jekyll Creek is narrow. You'll find a marina just off the channel.

The Jekyll Island Lift Bridge is past flashing red 4sec "20." The vertical clearance is 9 feet, and it opens on demand. The currents run strong here, so proceed with care. The amenities of **Jekyll Harbor** are located on the east side of the waterway at marker "23." At flashing green 4sec "25," the creek empties into **Jekyll Sound.** Beginning at flashing green 4sec "25," give the markers a wide berth past Jekyll Point, the southern tip of the island. Shoaling occurs here, especially on the east side between flashing green 2.5sec "29" and can "29A." Once past the can, you're in **St. Andrews Sound**. As Horseshoe Shoal is in the middle of this, resist the temptation to cross the sound—follow the markers, making a hard turn to starboard around red "32." Passage through this area can be aggravated by a brisk wind and strong tides. Slack water is the best time for crossing the sound into Cumberland River.

On your port side is **Cumberland Island**, largest of the barrier islands. The magnificent treasures of the island's wilderness have been protected from development since 1972 when it was established as a National Seashore. Sixteen miles long, Cumberland is the southernmost barrier island in Georgia's chain. It teems with wildlife: female loggerhead turtles come ashore here to lay eggs; herds of horses graze openly in the marsh and forest areas of the island; and wild turkeys and other birds, alligators, deer, mink, and otter also inhabit the island. If you're lucky, you'll witness this untamed atmosphere as you pass the island, but don't disrupt the wildlife by stopping.

Cumberland River is generally deep, and navigation is easy until you reach red "40." Leave it to starboard to avoid a bar that juts out from the mainland. Enter Cumberland Sound between red "78" and green "79." Numerous reports of manatee sightings have been made from time to time, so keep a lookout for these gentle creatures.

Starting at flashing green 2.5sec "51," leave the green markers to starboard and the red markers to port because now you are heading toward sea. After passing between red "30" and green "31," make a sharp turn south to continue down the Intracoastal Waterway by way of the Amelia River into Florida.

Shoreside and Emergency Services

Ambulance, Police, Fire: 911
Coast Guard:
Tybee Station (912) 786-5440 or VHF 16 or 21
St. Simons Island Station (912) 638-3319 or VHF 16
Radio Telephone:
VHF 28; I.D. Savannah Marine Operator
Tow Service:
Crescent Towing and Salvage, Savannah, GA (912) 236-2571
Savannah Marine Service, Savannah, GA (912) 232-3943
Sea Tow, Savannah, GA (912) 897-4304
Turecarno of Savannah, Savannah, GA (912) 232-8103

GEORGIA
Joins page 244
VHF-FM WXK-75 162.475 MHz
Tallahassee
VHF-FM KHB-39 162.55 MHz
USCG (904) 247-7311
Jacksonville
ATLANTIC OCEAN
St. Johns River
St. Augustine
Daytona Beach
VHF-FM KIH-26 162.40 MHz
USCG (904) 428-9085
N
VHF-FM KIH-63 162.475 MHz
Titusville
Cape Canaveral
USCG (407) 853-7601
VHF-FM WXJ-70 162.55 MHz
Melbourne
FLORIDA
Vero Beach
USCG (407) 464-6100
Ft. Pierce
Stuart
VHF-FM KEC-50 162.475 MHz
Lake Okeechobee
Palm Beach
GULF OF MEXICO
VHF-FM WXM-58 162.400 MHz
USCG (407) 844-5030
Ft. Lauderdale
USCG (305) 927-1611
Miami
VHF-FM KHB-34 162.55 MHz
USCG (305) 536-5611
1 inch= 45 miles
Scale 1:2,851,000
0
45
90
135
180
Key Largo
VHF-FM WXJ-95 162.40 MHz
USCG (305) 292-8856
USCG (305) 743-6388
USCG (305) 664-4404
Key West

The Sunshine State

Swinging in the Florida palms.

© Monroe County Tourism

While cruising down the Intracoastal Waterway in Florida, there's a recurring thought about this country's history: We've come so far in so little time. So much about Florida is new. It is the oldest part of the New World, yet it has only been a part of the United States slightly more than 170 years. In a geological perspective—a perspective that puts most of Florida under the sea several times over the last 100 million years—170 years is but a short, quick breath.

It is therefore fitting that Florida's Cape Canaveral is home to the National Air and Space Administration (NASA) which rockets us forward in time and continues to explore new worlds. This same pioneering spirit enabled Florida's first settlers to envision a modern tropical paradise in the swamps they found. Perhaps while cruising along Florida's coast, you too will be consumed by this spirit and inspired to do some exploring of your own.

Cruising conditions along the Intracoastal Waterway (ICW) are generally sufficient for average cruising vessels, but just like the landscape of Florida, these conditions are subject to change at any time. Project depths are 12 feet all the way to Fort Pierce, and 10 feet south to Miami. Be aware of current conditions and the most recently dredged depths as you travel—these are ideal depths and may not be the same from season to season. Tides in the ICW can range from 1 foot to 7 feet; currents normally run at about 2 knots, increasing to 4 knots at inlets.

Bridges are always a potential navigational problem for Florida cruisers. Between Fernandina and Coconut Grove, there are 81 bridges to contend with, and each one of them requires communication with the bridge tender. Every vessel is required to signal for the opening of a bridge (if indeed, the bridge needs to be opened), and this is done by either radio contact (VHF 13), visual signal, or horn signal. A visual signal is either a white flag or light raised and lowered vertically. The horn signal consists of one prolonged blast of four to six seconds, followed by one short blast.

Cruising Florida's east coast gives you the opportunity to see change as it happens, to experience events that will undoubtedly shape our country's future. The modern architecture and ever-expanding skyline of Florida's coastal cities makes us feel far less removed from our

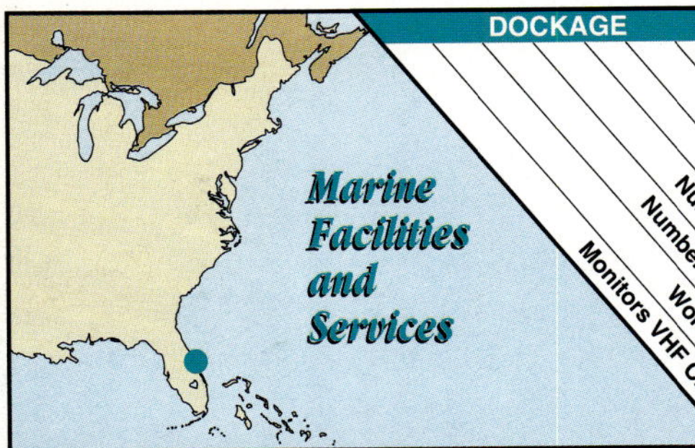

Marina	Phone	DOCKAGE: Monitors VHF Channel	Working Channel	Number of Transient Berths	Number of Transient Moorings	Seasonal / Year-round	Maximum Length Over-All	Approach Depth in Feet at MLW	Dockside Depth in Feet at MLW	SERVICES: Hookups: Fresh Water / Phone / Cable TV	110V★ 220V▲ 3 Phase ■ Maximum Amps	Ramp / Dinghy Dock / Launch Service	Railway / Lift / Crane: Capacity (tons)	Repairs: Propeller / Sail / Electronics / Rigging / Gas / Diesel / Wood / Fiberglass	SCUBA: Air / Rentals / Equipment / Charters	SUPPLIES: Gas / Diesel Fuel	Fuel Brand	Marine / Groceries / Bait / Ice / LPG / CNG	AMENITIES: Showers / Laundry / Pumpout / Restrooms	Hotel / Pool / Tennis / Golf	Public Phone / Restaurant / Snack Bar	MasterCard / Visa / Discover / AmEx
1 Halifax Harbor Marina (p. 257)	904-253-0575	16	71	50		Y	120	7	7	FC	50/208	R		Nearby		GD	TEX	I	All		RS	All
2 Harbortown Marina Boatyard (p. 259)	407-466-7300	16	68/9	40		Y	145	9	9	All	★▲■		60/150	All	RC	GD	Fina	MGBI	All	P	All	MVD
3 Northside Marina (p. 258)	407-692-4000	9/16	69	96		Y	120	7	7	All	★▲■ 50			All		GD	76	MBI	All	P	PR	MV
4 Stuart Yacht Builders, Inc. (p. 258)	407-283-1947	9	68			Y	80	6	6		★▲50		L40	All				M	SR		P	MV
5 Cannonsport Marina (p. 261)	407-848-7469	16	9/68	5		Y	150	12	20	All	★▲50					GD	Exx	MGBI	SLR	HP	P	All
6 Waterways Marina (p. 264)	305-935-4295	16		12		Y	100	21	10	All	★■							G	SPR		PR	MV
7 Turnberry Isle Resort & Club (p. 263)	305-933-6934	16	68	25		Y	200	10	15	All	★▲■ 100				C			MI	All	All	PR	All
8 Biscayne Bay Marriott Hotel & Marina (p. 264)	305-536-6335	16	68	50		Y	120	7	16	All	★▲50			On Call	C	GD	TEX	MGBI	All	HP	All	All

COAST GUARD: See Adjacent Chart for Coast Guard Information

Information in these listings is provided by the facilities themselves.

history than we usually consider ourselves. At the same time, some of the Spanish-influenced styles found in the cities, and Native American artifacts at museums and historical societies remind us that our history is shared with the histories of other cultures and that indeed we were not the first to discover the splendor of Florida.

The First Coast: St. Mary's River to Tomoka Basin

Perhaps nowhere in the country has there been more dispute over a parcel of land than in **Fernandina**. Before the Spanish ceded Florida to the United States in 1821, at least eight different flags had flown over Fernandina, typifying the frequent land struggles that characterize Florida's past. You can visit the Amelia Island Museum of History (904) 261-7378 for further insights, as well as a display on the Timucuan tribe of Native Americans who populated Florida from Jekyll Island down to Daytona. Or you can spend the day playing a round of golf or enjoying one of the many water activities in the area. Fort Clinch State Park (904) 277-7274, located in the northeast corner of Amelia Island, offers swimming off pristine beaches, fishing from shore or pier, and hiking through nearly 1,000 acres of preserved land filled with birds and butterflies.

From the shipping channel in **Cumberland Sound**, head south after passing green 4sec flasher "31," and pick up the **Amelia River** channel between red 2.5sec flasher "2" and green 4sec flasher "1." The naturally deep, well-marked channel leads past **Fernandina Beach** and through marshland before cutting sharply west, then south again.

Heading down past **Amelia City** and out toward **Nassau Sound** is a serene stretch of the ICW that meanders through unspoiled marshes and hammocks. Amelia City is actually more of a small town than a city. Not even the sight of a marina will distract you from your enjoyment of this desolate cruise—because there aren't any! In fact, just about the only site to visit in this area is the Kingsley Plantation State Historic Site (904) 251-3537, which was once the home of one of the most progressive and liberal slave owners in Florida. You'll find the site on Fort George Island, just southeast of **Sawpit Creek**. You can anchor on Fort George Inlet in front of the site.

If you choose to visit the Kingsley Plantation, you'll have to backtrack in order to pick up the ICW where Sawpit Creek branches off to the west at red 4sec flasher "72." From here, the narrow channel in **Sisters Creek** leads south to the **St. Johns River**. The **St. Johns Inlet** is regarded as the best inlet in Florida, particularly due to its protective breakwaters that extend well into the Atlantic. One of the few rivers in the country that flows northward, the St. Johns is the access route for the towns of Mayport and Jacksonville.

Mayport is not exactly a tourist trap for the visiting yachtsman, but it does offer a convenient stopover between destinations. This small town is characterized by fishing boats and a naval station (904) 270-5011 where tours are available. Otherwise, Mayport offers very little in the way of attractions and activity, aside from hiking and picnicking in various state parks in the area.

The city of **Jacksonville** is the first representation of what most of Florida's coast is like—progressive. Ever since a period of hardship during the Civil War, Jacksonville has persistently opposed stagnation. As a result, it has become the most important commercial port in northern Florida. The refreshing twist is that this industry has not wreaked havoc on the city's appearance. Jacksonville thrives as a favorite destination for cruisers. Shopping and public transportation are both easy to find. Visit the Jacksonville Art Museum (904) 398-8336 and the Jacksonville Historical Center & Maritime Museum (904) 355-9011, and if you happen to be in Jacksonville during the fall, the city hosts a jazz festival in October.

Of course, marine facilities are abundant in and around Jacksonville. The St. Johns River is easily navigable, with generous depths throughout. Just south of Jacksonville, navigational aids become less frequent, then resume after the **Ortega River** and continue to **Palatka**. This stretch is rather wide and can get rough in heavy winds.

In order to get back to the ICW, backtrack to where Sisters Creek meets the St. Johns River, and head south through **Pablo Creek**. This section of ditch is narrow but manageable, cutting past **Jacksonville Beach** and Palm Valley Landing before meeting up with the headwaters of the Tolomato River. Marinas are plentiful in this area, all the way to St. Augustine. The scenery spans the gamut, from the business-oriented shores of Jacksonville Beach to the preserved natural state of the Tolomato. Markers are infrequent until you reach the green 4sec flasher "3."

Founded in 1565 by the Spanish, **St. Augustine** stands alone in more ways than one. Not only is it the oldest city in America, but it is also the only city of its kind that you'll encounter along the Florida coast. Like all Spanish cities, St. Augustine is built around a plaza and has a bridge connecting it to **Anastasia Island**. As such, St. Augustine is its own attraction. The architecture and the history it reveals should not be missed.

Navigating the area is fairly straightforward, aside from a few shoaling areas around the **St. Augustine Inlet**. As you head south past the Vilano Bridge, keep quick green "57A" to port. Leave flashing red 2.5sec "58" and red nun "60" to port and red nun "2" to starboard. The river breaks west, then turns south under the Bridge of Lions before passing the mouth of the **San Sebastian River**.

The stretch from St. Augustine to Daytona is similar to Fernandina. The mid-16th century brought land struggles and bloodshed to the area around the **Matanzas River**, and most of the blood was French. Both Spanish and French troops arrived on Florida's East Coast in late August 1565. A month later, after several efforts by both sides to gain position, the Spanish prevailed, capturing and eventually slaughtering nearly 250 French troops. And so Matanzas, meaning "place of slaughter," received its name. The irony in all of this is that today, this last stretch of the First Coast remains largely unsettled.

It's difficult to find a more beautiful stretch of the ICW, and equally difficult to find facilities in this area. So if you missed the marinas at St. Augustine, you won't find any until you reach Daytona Beach. Most choose to visit this area by car, stopping at the Visitors Center on the barrier island side of the ICW and taking a ferry across to Fort Matanzas on **Rattlesnake Island**. Farther south in Palm Coast, Marineland (904) 471-1111 attracts coastal cruisers with its dolphin shows and other marine life exhibits. Still farther south, Flagler Beach State Recreation Area is the only attraction.

The Space Coast: Ormond Beach to Melbourne

If the Space Coast wasn't named for the space exploration program at home in Cape Canaveral, it may have been named for the vast waterways that define it—**Tomoka Basin**, the **Halifax River**, **Mosquito Lagoon** and the **Indian River**—which offer so much space they feel like vast bays.

Daytona Beach is the most popular vacation town on the central East Coast of Florida, attracting university students by the thousands every spring. The traditional Spring Break festivities turn the streets and beaches along this part of the coast into a flood of excitement, pouring millions of dollars into the local economy. If the thought of Spring Break doesn't thrill you, don't worry—most of the activity is limited to the beachside and only lasts three weeks.

Students weren't the first to turn this strip into a fast lane. Since the turn of the century, people have been testing their "metal" on the hard sand beaches of Daytona, pushing race cars to their limits in search of the land speed record. Racing has since moved to the Daytona International Speedway (904) 254-2700 where the famous Daytona 500 is held every February. Daytona certainly has a propensity for speed!

Being the tourist-oriented town that it is, Daytona does its best to accommodate visitors, offering several marine facilities. **Halifax Harbor Marina** is the newest, biggest, and best and is located in Daytona Beach on the west shore of the Halifax River. Along with all the amenities, the marina has a restaurant, a waterfront park, and is a short walk to downtown.

There are five bridges to contend with as you travel along the ICW here, beginning with the Ormond Beach fixed bridge, which has a vertical clearance of 65 feet. The average vertical clearance of the next four bascule bridges is only about 21 feet. Aside from that, navigating through Daytona Beach is straightforward. Just watch your chart, and don't stray from the channel—the waters to either side can be shallow.

After passing through Daytona Beach, you'll be surprised when you see how easily the ICW lazes back into the more serene, less populated area of **New Smyrna**. Sprinkled with residential outcroppings, this area is certainly a change of pace from Daytona. The beach is the focal point here. Protected by a bulwark of rock ledges extending up to 40 miles into the Atlantic, you won't feel the pull of the ocean currents at New Smyrna Beach. Otherwise, a bike ride through town or a visit to the Ponce de Léon Inlet Lighthouse can make for a pleasant afternoon. A visit to the barrier island side of town (across the Coronado Bridge) will fill a lazy afternoon stroll with beach shops and restaurants.

At red daymarker "2," the ICW continues southeast, but a route exists heading east to the **Ponce de Léon Inlet**. This channel is not maintained by the Coast Guard, though, so proceed carefully. South of the inlet in the ICW, keep the red daymarkers to starboard until you swing hard to the west at flashing red "30." The channel doglegs south again at flashing red "38," and it's straight ditch south of here for several miles.

The next large body of water you'll encounter as you head south is **Mosquito Lagoon**, located just south of New Smyrna. This area is host to a legend of sorts, a story not universally accepted by the Florida community. Although known as a haven for manatee and porpoises, Mosquito Lagoon is purportedly the home of another creature, a lagoon monster described as a gray-brown serpent larger than a manatee. Perhaps while you're cruising along, relaxing in this peaceful bit of primordial Florida, you'll stumble upon the elusive beast. Just don't expect anyone to believe you!

Whether you see the mysterious creature or not, you'll remember Mosquito Lagoon as the most pristine water in all of Florida. While this is certainly an enjoyable trip, be prepared to travel about 35 miles without a stop for fuel—this area is part of the Merritt Island National Wildlife Refuge and doesn't have any facilities.

From the flashing green 4sec "5," it's straight ditch until you start your approach to the **Haulover Canal** at red 4sec flasher "36." From here, the channel turns slightly east, then east again, then cuts sharply west into the canal. Then, at the flashing 4sec 16-foot "12," the channel swings sharply south and meanders to Titusville.

Titusville is a city whose identity is difficult to define: most of its 50,000 residents are engineers for NASA, but at the same time, its proximity to the Space Center and Disney World has made it an accommodating stopover for transients. As such, Titusville hovers on the line between a tourist trap and a residential haven.

Golf is quite popular in the area, and a scenic drive in the Merritt Island National Wildlife Refuge comes highly recommended. (You might even be witness to a space shuttle launch.) If you're lucky enough to visit during the last week in April, Titusville sponsors the Indian River Festival, a waterborne parade, arts and crafts show, and food enough for the 50,000 people who attend.

Not since Daytona Beach has there been better reason for you to leave your boat and your worries on the ICW and head to the open Atlantic. **Cocoa Beach** has long been touted as having the best waves and weather on the

East Coast and has been called the Small-Wave Surfing Capital of the World. Getting there, however, is not simple. If you dock in **Cocoa** to the west or **Port Canaveral** to the north, you'll need a cab to get to the beach. But it's worth it. In stark contrast to the carnival atmosphere of Cocoa Beach is Olde Cocoa Village where antiquing and shopping replace sunbathing and surfing. The area south of Titusville is where the Indian River opens to its widest point, extending like a lake to the east. From here to Cocoa, the river remains vast, narrowing as you head south. Don't be tempted by the open waters here to stray from the channel—depths to either side are suspect.

It's straight ditch all the way to 12-foot green 4sec flasher "67" where the entrance to the **Canaveral Barge Canal** opens up to the east. The canal cuts through **Merritt Island** and across **Banana River** to Port Canaveral. If you remain in the ICW, Cocoa lies just south, recognizable by the large water tower adorned with an enormous American flag. The town is connected to Merritt Island by twin fixed bridges with a vertical clearance of 65 feet.

Spaceport USA at the Kennedy Space Center (407) 452-2121 on **Cape Canaveral** is perhaps the most popular tourist attraction on the eastern seaboard. After all, not even the best rides at Disney World (407) 824-4321 are equipped to send *their* passengers into space. And although admission into Spaceport USA doesn't include rides, the IMAX movies will bring you as close to actually being in space as you will ever experience.

But not all of the activity on Cape Canaveral centers around space exploration. It's also one of the largest commercial shipping ports in Florida with one of the largest fleets of cruise ships in the state. Disney, located just 40 miles west in **Orlando**, even has its own cruise ship in the port, playfully named *Big Red Boat.*

Accessing Cape Canaveral via the Canaveral Barge Canal is suggested. Start from the red 4sec flasher "12" in the Indian River. The canal offers generous depths throughout, 10 to 12 feet, but try not to stray too far from the center of the channel. Green flasher "5" marks a crossroads at which you can either head northward or southward in the Banana River, or you may opt to continue east to the Atlantic. This area is prime viewing grounds for catching a glimpse of a spectacular space shuttle launch.

Now, if you'll kindly turn your attention from the sky to the water, **Melbourne** is a self-made watersports capital that will surely keep you busy. Home of the U.S. Boardsailing Team, as well as host to the annual Aquafest (407) 724-5400 held every third weekend in June, Melbourne certainly has its share of activity. And if watersports don't float your boat, so to speak, there are seven golf courses, four parks, and two museums to choose from in the greater Melbourne area. For more information, call the Melbourne Chamber of Commerce (407) 728-2211. Marinas are plentiful in this area, but they tend to be spread out along each shore of the Indian River.

There are two feasible approaches to Melbourne. Heading south from Port Canaveral in the Banana River, you'll find adequate depths, but little, sometimes nothing, in the way of navigational assistance. We suggest sticking to the ICW, which is deeper and certainly better marked. The channel is straight and surrounded by deep water, but it is best to stay within the confines of the daymarkers until you pass the Palm Shores bridge. Here the waters are deep from shore to shore, making this a popular spot for cruisers, as well as locals who frequently tack back and forth. The only obstacle along this route is a spoil area extending from the Palm Shores bridge to about flashing red 4sec "100."

The Treasure Coast: Sebastian to Jupiter

The early 18th century saw fleets of Spanish galleons returning from their conquests, their ships filled with everything from silks to chests of gold. As one fleet traveled along the East Coast of Florida in 1715, a hurricane struck, wrecking 10 ships between Sebastian and Fort Pierce and sending precious cargo to the ocean floor. To this day bits and pieces of this treasure are being discovered. But along with sunken goods, this section of Florida's coast is replete with another kind of treasure—small islands lining the ICW. Although not worth anything to the scavenger in search of treasure, these islands will provide you with hours of exploration and discovery.

Fortunately for the town of **Sebastian**, money-hungry divers and beachcombers do not descend on the area like executives on Wall Street. Instead, it remains a tranquil stopover for transients in search of an anchorage. The main activity here? Relaxation. Along with the majestic scenery, you're likely to see manatees and dolphins whose docile behavior is testament to this peaceful beginning of the Treasure Coast. Marine facilities suitable for transients are somewhat scarce in this area, so anchoring may be your best choice.

The ICW to Sebastian is straight and deep from **Palm Bay** south. There's a dogleg west at flashing green "21," then another east at flashing red "28." It's straight ditch from here to red daymarker "32" and zig-zags from there to Sebastian, approaching the western shore more and more along the way. East of green flasher "63" lies the **Sebastian Inlet**, considered by Coast Guard and cruiser alike to be the most dangerous inlet on Florida. Currents can reach up to 10 knots, and the bridge that spans the opening has a vertical clearance of only 37 feet. Take our advice—steer clear of it.

There is nothing pretentious about this next town. After all, with a name like **Vero Beach**—"vero" meaning "truth" in Latin and Italian—pretense would not be appropriate. One of the first settlers of this area built a house here out of driftwood as a seaside retreat for his family and friends. That ramshackle house is now a very popular inn. Nothing pretentious about that.

Your Home Port or Your Port Away From Home

1 p. 252

Halifax Harbor Marina

at Daytona Beach

This season, discover Columbia Plaza at Halifax Harbor. Delight in Spanish Cuisine at world famous Columbia Restaurant and go native with Caribbean food and fun at Cha Cha Coconuts. Then, meander through our waterfront park or stroll to other restaurants and shops in downtown Daytona Beach.

Reservations Recommended: **1-800-343-2899** or **904-253-0575**

TEXACO STARPORT: All Major Credit Cards

450 Basin Street, Daytona Beach, Florida 32114

Fernandina Beach; High: -0:53, Low: -0:39, Mean: 3.95 feet	Fresh Water, Cable TV 50/30V, 208/110 Max Amps
$.85/ foot per day 520 Berths, 120' Max LOA	Downtown Daytona Beach 904-250-8972
	Columbia Restaurant 904-248-1200 Spanish Atmosphere
Texaco Gas, Diesel	MasterCard, American Express, Discover

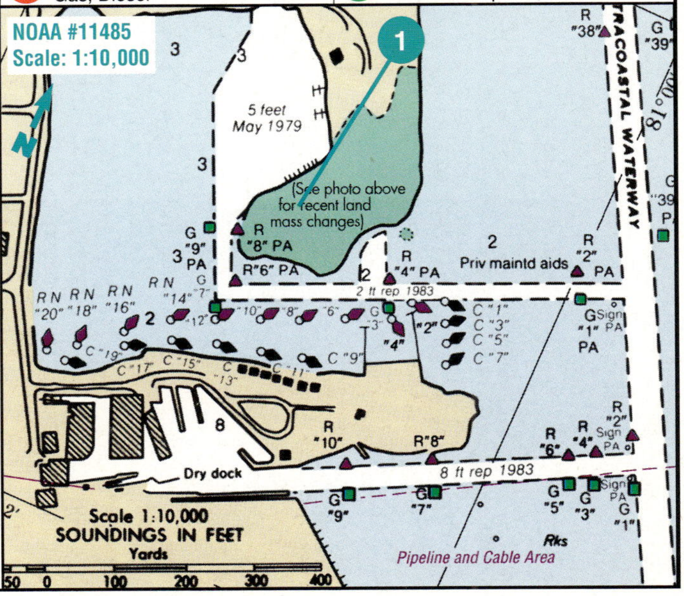

Despite the conspicuous power plant, developments and bridges that arise from its shores, Vero Beach has all the appeal of Smalltown, USA. Hence, tourist attractions are not easily found in this area. Baseball fans will want to head out to Dodgertown (407) 569-4900 in March when the L.A. Dodgers begin their spring training. The number of marinas in Vero Beach indicates one thing—this is the perfect place for a stopover between destinations.

From Sebastian, it's more or less straight until you reach flashing red "76" where the channel doglegs east along Pelican Island National Wildlife Refuge. At flashing red "82," following a hard turn east, the ICW winds its way along the shore of the barrier island and through **Barker Island**. There is little room to play with through here, so don't deviate from the channel. Vero Beach sits on the western shore, but the marinas are in a cove to the east. The channel to the cove is marked by flashing green 4sec "137" and green daymarker "139."

Fort Pierce is not exactly pretentious either, but it is a bit deceptive. At first glance, Fort Pierce seems to be just a quaint old town, free from the industry of the urban areas found farther south. But a closer look reveals a town on the verge of major growth, evidenced by the blue power plant between the two bridges and the twin nuclear reactors south of town. And indeed there are long-range plans to enlarge the port and the incoming waterways, which would allow for serious commercial traffic. While in town you can visit Fort Pierce Jai-Alai (407) 464-7500, as well as **North Jetty Beach** at Fort Pierce State Park (407) 468-3985.

After Vero Beach, the ICW zig-zags slightly until it reaches green 4sec flasher "167." From there, you've got straight ditch for several miles, and at green daymarker "181," it turns ever-so-slightly east. From here, you'll see the North Fort Pierce Bascule Bridge, with a vertical clearance of 26 feet. Once past it, you'll begin to feel the effects of the currents coming in from the Fort Pierce Inlet. And although it can get rough through here, this is the inlet of choice for boaters from miles around.

This area of the ICW is lined with marinas willing and able to accommodate transients. We recommend the fine facilities at **Harbortown Marina Boatyard**, located along the west shore in a protected, deep-water inlet. Their staff is equipped to handle all repairs, from mechanical trouble to extensive carpentry work, and is always available.

South of Fort Pierce, the Indian River remains quite wide, running along the extensive Hutchinson Island all the way to the town of **Jensen Beach**. In a historical sense, this can be considered the pineapple capital of America. When Captain Thomas Richardson arrived here, he planted the fragrant fruit throughout Jensen Beach, and it wasn't long before he was shipping out more than a million boxes each year.

So, in keeping with tradition, the Jensen Beach Chamber of Commerce (407) 334-3444 sponsors the annual Pineapple Festival, held every second week in November. Also of interest to visitors is **Sea Turtle Beach**, which owes its name to the more than 6,000 sea turtles that nest here each summer. The Chamber also hosts Turtle Watches during June and July so visitors can catch a glimpse of the nesting process.

The town of **Stuart**, just northwest of the ICW on the **St. Lucie River**, is perhaps the turning point from the vast open waters of the Indian River to the ports farther south that are constantly on the move. The boat traffic along this stretch is dense and diverse. Considered to be the "Sailfish Capital of the World," Stuart has fishing charters leaving from various marinas during prime sailfish season (December to May). Stuart Beach is also a popular center of activity, known locally as the best beach around.

Perhaps the strongest evidence that Stuart is, indeed, a turning point is the sheer number of marinas. According to our figures, there are nearly 30 crammed into the St. Lucie River and its tributaries. We recommend visiting **Northside Marina**, conveniently located on the St. Lucie River. A full-service facility, Northside offers all the amenities including a restaurant and swimming pool. If you're in need of repairs, stop in at **Stuart Yacht**, located on the south fork of the St. Lucie River and renowned for their designs as well as repairs.

Flashing red 4sec "4" marks a crossroads, where the ICW, the St. Lucie River, and the **St. Lucie Inlet** converge (we do not recommend that inexperienced boaters use the inlet). Swing west at this buoy to access the St. Lucie River, and keep in mind that not every red daymarker has a corresponding green. Keep the red markers to starboard and you shouldn't have any trouble, but be careful to note red "2" which marks a different channel into Manatee Pocket.

After flashing green "17," the first stretch of the St. Lucie River heads almost due north, then turns sharply west after the Evans Crary Bridge. After flashing green 4sec "23" near Britt Point, the river splits into North Fork and South Fork, the latter of which leads to the **Okeechobee Waterway**.

You'll come across the waters of **Hobe Sound** as you continue down the coast. Known for Jonathan Dickinson State Park (407) 546-2771 which lines its shores, Hobe Sound offers calm waters on a wild background. The 10,200-acre park is one of the largest and best natural preserves in Florida, and since "natural" is the operative word, Hobe Sound offers very little in the way of man-made entertainment. We think that's just fine—the beautiful surroundings are entertainment enough. Since a natural preserve wouldn't be all that natural with buildings and docks lining its shores, marine facilities are somewhat scarce in Hobe Sound.

Across Hobe Sound to the east is **Jupiter**, a perfect example of how Florida's formative years are not so far removed. Up until the 1950s, we were told by the historical society, Jupiter's telephone directory was no more than half a page! Since then, however, Jupiter's popularity has skyrocketed (no pun intended) with the help of Perry Como and Burt Reynolds, the latter of which built a dinner theater (407) 746-5566 (now known as the Jupiter Theater), that still attracts celebrities.

At the southern tip of Jupiter Island is the **Jupiter Inlet** and the entrance to the **Loxahatchee River**. This area marks the end of the Treasure Coast. Boaters who are unfamiliar with the area should not attempt to use Jupiter Inlet—shoaling almost always makes the buoys obsolete, and only immediate local knowledge can remedy the situation. Also difficult in this area is the entrance to the Loxahatchee River, sometimes confused with the ICW. After heading west away from the inlet, the ICW turns sharply south at the mouth of the river. Look for flashing red 4sec "8."

The Gold Coast: Palm Beach to Coconut Grove

It is said that the Gold Coast got its name the same way as the Treasure Coast—from the recovered goods of Spanish galleons that sank offshore. But as you travel along this stretch of the coast, you will be consumed by the splendor and richness of the area, and you'll be convinced, as we were, that from Palm Beach south, Florida is a veritable gold mine of sights and activity.

The construction of Henry Flagler's Royal Poinciana Hotel ultimately marked the birth of **Palm Beach** as a resort community. The hotel gained immediate fame as having the finest food and service, and soon America's elite were traveling from distant reaches of the country to visit and stay in Palm Beach. Business was so good that Flagler built the Palm Beach Inn (now called The Breakers) (800) 833-3141, which is still in operation today. Whitehall (407) 655-2833, another of Flagler's structural accomplishments, has been converted into an ornate and glamorous museum with telling exhibits.

Historical architecture is not the only attraction in Palm Beach. As the name implies, the beaches in the area are abundant and are some of the best in Florida. But across the ICW in **West Palm Beach**, the tide shifts from big fun to big business. Acknowledged as the financial center of the area, West Palm Beach is also the service center catering to Palm Beach—exactly as Flagler intended. Of interest to tourists is the Norton Gallery of Art (407) 832-5194, featuring works of Renoir and Picasso, the Dreher Park Zoo (407) 533-0887, featuring a planetarium and science museum, and five golf courses. Of course, the Palm Beach Seafood Festival (407) 832-6397 every February is a favorite among visiting boaters.

And visiting boaters are plentiful. The many marinas in and around the Palm Beaches recognize this and will bend over backwards to provide you with what you need. **Singer Island** is where you'll find **Cannonsport Marina**, a small, private marina providing first-class accommodations for guests.

From the northern end of Lake Worth, the ICW is fairly straightforward. The waters here are wide, and you may be tempted to stray from the channel. We advise against it. You may see locals flying around outside the channel, but stick to the ICW. At flashing green 4sec "39," the channel doglegs southwest alongside **Peanut Island**, then resumes a southerly direction at red 4sec flasher "42." From here, the easy passage of the **Lake Worth Inlet** can be reached by swinging east around Peanut Island and following the series of red and green flashers. If you choose to continue south, the ICW is straight until you pass West Palm Beach.

It's ironic that the towns of **Lake Worth** and **Lantana** offer little more than antique shopping as their tourist attractions. Considering neither was established until the early 1920s, there's really nothing antique about them. Lantana is also home to some of the best (or worst, depending on how you look at it) gossip in the U.S., being home to the offices of the *National Enquirer*. Lake Worth is home to the Hallpatee Seminole Village (800) 624-2730 which features authentic displays of native Seminole Indian culture. And as always in Florida, the beach is never too far—Lake Worth Municipal Beach is the site of a fishing pier, wind and board surfing, and of course, swimming.

The Waterway is narrow, but well-marked and straight, and you will likely only be using it to reach another destination. Marine facilities are few and far between, so we suggest continuing on to points south. But don't hurry—keep your eyes open for manatee and signs posting manatee season, at which time a no-wake speed limit is in effect.

Before you immerse yourself in the carnival atmosphere of Fort Lauderdale and Miami, you will cherish time spent in **Boynton** and **Delray Beaches**. These peaceful reprieves offer little in the way of activity, save the beaches and golf courses. But you will draw on memories here with adoration as you reach the busier ports on your itinerary. So relax and spend a day on the beach. If the beach doesn't appeal to you, the Loxahatchee National Wildlife Refuge is located in Boynton Beach and the Delray Beach Tennis Center (407) 243-7360 opened in March. You will not have trouble finding dockage unless it's high season. A few marinas are scattered along the west shore of the Waterway, and the Waterway is just about all the water you get here.

The namesake of the next town is a mystery. Meaning "mouth of the rat" in Spanish, **Boca Raton** has an illustrious history of ups and downs, successes and failures. Addison Mizner envisioned Boca Raton in the image of Palm Beach, and he set out to establish the town as a resort community. He rounded up investors, purchased 16,000 acres of land, and unveiled the Cloister Inn in

1926, replete with ornate fountains, chandeliers, private patios, and trellises.

But fate dealt a blow to Mizner as well as Boca Raton—the Great Hurricane of '26 and the Great Depression beginning in '29 shattered Mizner's dream like fine crystal. But gradually the crystal coalesced around a reincarnation of the Cloister Inn, the Boca Raton Resort and Club (407) 395-3000, and Boca Raton regained its popularity.

Golf and beach activities are the most prominent source of entertainment in this town. Most of the golf courses are private, but there are a few that cater to the public. Residents in the area claim that Spanish River Park (407) 393-7840 and Red Reef Park (407) 391-5014 are the best beaches, so pay them a visit. Farther south in **Pompano Beach** you'll find Pompano Park (305) 972-2000, the only harness track in Florida, the International Sportfishing Museum, and the Pompano Beach Seafood Festival and Fishing Rodeo, which is held the last week in April.

Cruising the ICW in this area is like walking through a haunted house—periods of calm interrupted by sudden fear. Conditions are generally fair, but certain areas get suddenly quite rough, such as the **Boca Raton Inlet** and the waters beneath the Rte. 798 bascule bridge. But approaching Boca Raton from the north is the only way—the inlet is extremely dangerous, and we recommend against using it. Marinas are plentiful along this route and increase as you head south.

And now you've arrived. The self-proclaimed pleasure-boating capital of the universe, **Fort Lauderdale**. Known around the country as the "Venice of America," this town has it all. If the sheer number of vessels doesn't impress you, the size and splendor of the luxury yachts surely will. Built around its canals, you won't be disappointed with the "feel" of Fort Lauderdale. Watersports are the most popular activity, rivaled only by golf and tennis. Ocean World (305) 525-6611, the Broward Center for the Performing Arts (305) 522-5334, the Fort Lauderdale Museum of Art (305) 763-6464, and the Museum of Discovery and Science (305) 467-6637 are just a few of the places to visit. The Fort Lauderdale International Boat Show (305) 764-7642 is held in late October and the Fort Lauderdale Seafood Festival (305) 463-4431 every first week in April.

With so many canals and waterways, it's no wonder Lauderdale has an abundance of marinas. In fact, some of them are so lavish that they're considered attractions themselves. **Turnberry Isle Resort & Club** is just such a place. Located between Miami and Fort Lauderdale on the ICW, this facility has all the amenities, plus golf courses, tennis courts, a spa, and several restaurants.

Heading into Fort Lauderdale, the channel remains fairly straight until you reach the area just above the **Middle River**. After a couple of easy doglegs, the ICW takes a wide but decisive turn to the west for a short stretch, leading directly to the mouth of the **New River**. At the red and green junction marker "A," the ICW splits, with one branch swinging south again toward **Port Everglades** and the other heading into the New River.

Navigating the New River, the buoys continue with red-right-returning, but keep an eye on the chart just to be sure. There are plans in the works to dredge the river to 10 feet to allow for shipping vessels, but the current depths should be adequate for most cruisers.

If you're continuing toward Port Everglades along the ICW, follow the markers until you see the 17th Street Bridge which divides the ICW from the wider water of Port Everglades. The bridge has a vertical clearance of 25 feet and opens every 15 minutes to allow for cars. A large digital clock on both sides of the bridge shows waiting boats how long they must wait for the next opening. Below Port Everglades, you will soon see the entrance to the **Dania Cut-off Canal** to the west at quick-flashing green "35."

At first glance, speed seems to be the attracting force in the town of **Dania**, evidenced by the jai-alai fronton and the Florida Derby, the precursor to the Kentucky Derby. But Dania is also the self-proclaimed antique center of southern Florida, and speed does not exactly go hand-in-hand with antique shopping. Amusement parks such as Grand Prix Race-O-Rama (305) 921-1411 and Blockbuster Golf & Games (305) 846-7650 will provide the kids (as well as the kid in every adult) with a thrilling day. For anglers, the last weekends of April and October are the dates for the semi-annual Billfish Tournament (305) 563-0385.

Marine facilities are plentiful in the Dania Cut-off Canal, marked by quick flashing green "35," which carries deep water all the way to the fixed bridge. The only navigational aid in the canal is red nun "2," but the ample depths should not give you any trouble. South of the canal, the ICW is fairly straight all the way down to flashing green 4sec "49," with a slight dogleg at flashing red 4sec "36." Beyond green flasher "49," the channel swings west, and at red daymarker "54" you can either swing back south or turn to starboard toward **Maule Lake**.

It's difficult to imagine **Miami Beach** as it was a hundred years ago—all dunes and swampland waiting to be discovered by someone with a dream. That someone was Carl Fisher. He imagined Miami Beach as a playground for the rich, surpassing even Palm Beach. And what he imagined came true. This burgeoning offshore paradise became a summer haven for the wealthy. A building boom peaked in the late 1930s and early 40s with the Art Deco craze, a style characterized by symmetrical designs and pastel colors. Today the now-restored Art Deco District is a very popular place to visit.

Miami Beach is divided into three parts: north, central, and south. The north is predominantly a residential area. Central Miami Beach is mostly apartments and hotels, coupled with upscale condos and hotels to the north, the most lavish of which is the Fontainbleau Hilton (305) 538-2000. This part of Miami is also home of the Miami International Boat Show (305) 531-8410 in February. South Miami Beach is where the nightlife can be found, as well as the Art Deco district. The hundreds of restored buildings and the beginnings of Miami Beach at South Pointe Park attract visitors to this end of town. Call the Greater Miami Convention & Visitors Bureau (305) 539-3000 for a list of events throughout the year.

Not surprisingly, Miami Beach has many marinas. Still, if you plan to be there during February, be sure to make reservations. The boat show has a tendency to overflow into several of the marinas. Just north of Miami Beach in Aventura is where you'll find **The Waterways Marina**, a full-service facility that offers transient slips and all the amenities of a resort minutes from the night life in Miami.

The ICW is a series of straight shots through Miami Beach and the upper parts of **Biscayne Bay**. From flashing green 4sec "5," make a beeline in a southwesterly direction. Keep in mind that red "6A" and green "7" mark an area of shifting shoals pushed along by the incoming currents of **Bakers Haulover Inlet**. The inlet can get rough, and the fixed bridge has a vertical clearance of only 32 feet, so most sailboats and some fishing vessels with tuna towers won't want to use it.

From flashing red 4sec "26," the ICW channel then cuts south under the 79th St. bascule bridge (vertical clearance 25 feet) and along Miami's north shore to flashing green 4sec "33." Another stretch heading southwest and passing under the Interstate 195 fixed bridge (vertical clearance 56 feet) will bring you to flashing green 4sec "45," just north of **Biscayne Island**. The channel then turns slightly to the west before heading due south under a pair of bascule bridges. From here, the Port of Miami lies ahead, edging the main channel from **Government Cut** which runs southeast. The waters along the shoreline are mined with submerged pilings, so keep a watch out.

Just like Miami Beach, the city of **Miami** got its start roughly a hundred years ago. And just like several other parts of southern Florida, Henry Flagler's railroad put it on the map. Julia Tuttle, who arrived here in 1891, was convinced that Miami had the potential to be a tropical resort, but she had to convince Flagler. When a freeze destroyed crops throughout Florida in 1894-95, Tuttle sent Flagler a bouquet of fresh orange blossoms wrapped in moist cotton, and he was convinced. By 1896, the railroad reached Miami, and the rest is history.

You will need a car to get to most of Miami's attractions, although the Metro-Dade bus system is usually sufficient. You'll want to visit Bayside, Miami's answer to Faneuil Hall Marketplace in Boston. In addition to the shopping and the nightlife associated with a metropolitan area, Miami offers a number of sporting events. Jai-alai is a way of life down here, as well as football. The Orange Bowl Classic (305) 643-7100 kicks off the year. And of course, the Miami Dolphins games at Joe Robbie Stadium (305) 623-6100 put Miami on the sporting world's map.

TURNBERRY ISLE RESORT & CLUB

7
p. 252

I deally located midway between Miami and Fort Lauderdale on the Intracoastal Waterway, Turnberry Isle Resort & Club offers vessels from 30 to 200 feet a beautiful setting and all dockage services including cable TV, switchboard-monitored and private phone lines, mail service, ice, Ship's Store, Crew's Lounge, and 24-hour security, as well as all guest privileges at our exclusive resort and club.

Our Marina guests enjoy two Robert Trent Jones golf courses, 24 tennis courts, spa, Ocean Club and 11 restaurants and lounges. This Mediterranean-style Resort with 340 grand deluxe rooms and suites was recently rated one of America's top resorts by the new U.S. Zagat Hotel, Resort and Spa survey.

Turnberry Isle Resort & Club

Between Miami and Fort Lauderdale (mile marker 1076)
19999 West Country Club Drive, Aventura, Florida 33180

1-800-327-7028, 1-305-933-6934

Miami Harbor Entr. Tides; High: +1:34, Low: +2:05, Mean: 2.1 feet	110/220V, 100 Max Amps
$1.75/foot per day, 40' minimum 117 Berths, 200' Max LOA	Ship's Store/Crew Lounge Convenient Shuttle Service
	A variety of restaurants from casual to elegant
	MasterCard, Visa, Discover, American Express

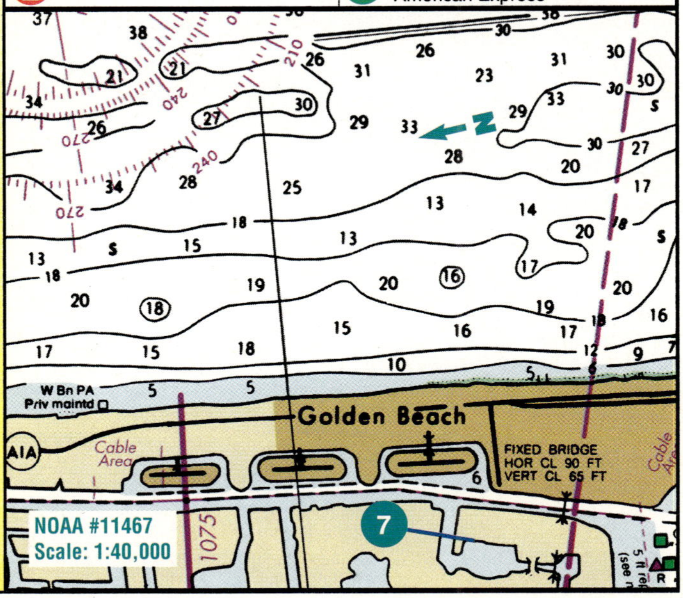

Heading south along the ICW past Biscayne Island to the northwest end of **Dodge Island**, head southwest at red flasher "50" along the shores of Bay Front Park. The entrance to the **Miami River** lies between the park and **Claughton Island** and is used mostly by commercial traffic. Marinas are as plentiful here as anywhere in Florida. We recommend using **Biscayne Bay Marriott Hotel & Marina**, a first-class resort and marine facility with all the amenities. Located just west of Biscayne Island on the mainland, the marina is also home to the in-water powerboat section of the Miami Boat Show.

Although known as **Virginia Key** and **Key Biscayne**, these islands are usually associated more with Miami than with the Keys. Yet the busy and hurried atmosphere of the mainland metropolis will rarely invade these peaceful sites, even though they remain connected to the mainland by the Rickenbacker Causeway. The beaches and woodland of both islands offer activities as well as refuge from the bustle of the city. The sails of windsurfers quietly slice the waters of Biscayne Bay, and scuba divers constantly explore the reefs. On land, the Miami Seaquarium (305) 361-5705 on Virginia Key is usually crowded with visitors eager to see the dolphin and killer whale shows.

Coming from Miami, navigating the ICW down to the Keys can be tricky, as this is Miami's busiest boat traffic area. Be sure to heed the markers as you head south to Rickenbacker Causeway. You will find a few marinas

A favorite stopover is fast becoming a significant refit and service yard

Call the Dockmaster on CH 16 VHF or phone 912 352 4956 and you'll get a warm invitation to refuel and stopover at our very pleasant marina.

Located south of the Thunderbolt fixed Bridge, close to ICW Mile 583 and between ICW markers G "35" and R "36".(With all floating docking.) Entry and exit are very easy, we can accommodate craft of 200 ft. and draft of 15 ft.

Facilities include a bright laundry and showers, a gift shop that offers an exclusive range of PJ branded clothing as well as charts and general boat supplies.

All docks have 110/220 electricity and water, there's optional cable/HBO hook-up if desired.

Rivers End Restaurant on the complex comes highly recommended, and there are two fish markets close to the marina.

We are only twenty minutes to Savannah International Airport and ten minutes to down town, Savannah's Historic District.

Transportation is available.

If you do stopover we'll deliver fresh doughnuts and a newspaper to you in the morning. How's that for Southern hospitality!

Our service and refit facilities are now competing internationally, bringing to Savannah some of the worlds most famous luxury yachts. But no yacht is too small and we welcome all owners.

If you are thinking of having a paint job done or a major refit, talk to manager Eric Haberli who will be delighted to discuss the work you want done and show you our facilities. These include a 1150 ton Syncrolift, a 150 ton wide beam Travelift,modern workshops and a most amazing service building that will take yachts over 200ft (60m).

Our reputation as America's premier yacht builder assures you the finest craftsmen will be working on your yacht. All this at an attractive hourly rate.

While you are in the area, do not miss the opportunity to visit Historic Savannah, like Palmer Johnson service, it's rather special.

PALMER JOHNSON
Savannah

P.O.Box 5860, Savannah, Georgia 3124 USA.Tel: 912 352 4956 Fax: 912 352 0593.

© Embassy

The neon lights of Miami Beach.

beyond the causeway along the southwest shore of Virginia Key, but most cruisers choose to anchor off of the Keys—this area of Biscayne Bay offers great protection in any weather.

There is a shallow area off the southern tip of Virginia Key, so head for quick flashing green "1," and giving it a wide berth to port, turn northeast. Split the green and red markers, and you should have no trouble reaching **Bear Cut** between the two islands. The cove on the north end of Key Biscayne is home to a few marinas, and is also a great place to drop your hook. Farther south on Key Biscayne, **Hurricane Harbor** is a favorite anchorage of anyone familiar with these waters. Head back through the green and red markers south of Virginia Key, and head south at flashing green 6sec "73," leaving the buoy to port. Then pick up the red 4sec flasher "2" just northwest of the harbor.

Appropriately the last destination along the Gold Coast, **Coconut Grove** is a microcosm of the world around it, a melting pot of different cultures. The streets are always lined with groups of people, leisurely making their way to restaurants and cafés to hear live music. Several different languages can be heard echoing through town, testament to the diversity found here. And this diversity can be experienced through theater and art at the Coconut Grove Playhouse (305) 442-4000 and the Coconut Grove Art Festival (305) 447-0401 in February.

Several marinas can be found along the shores of Coconut Grove, just south of Miami. Once you've passed the Rickenbacker Causeway, look for **Dinner Key Channel** marked by flashing green 2.5sec "1." The channel will bring you into the downtown area, where you'll find the center of Coconut Grove's activity.

The Florida Keys: Inside-and-Out to Key West

There are many characteristics of the **Florida Keys** that make them essential log entries. The fishing here is perhaps the best in the world (with more than 500 varieties to catch). The cruising conditions along the inside route are nearly perfect, protected in the lee of the islands most of the time, and the natural surroundings are nothing short of breathtaking.

But perhaps the most striking characteristic of the Keys is that they are so vastly different from mainland Florida. The post offices here have two mailboxes—one that says "KEYS" and another that says "WORLD." People here know these islands are different. Very different. Down in Key West at Mallory Square, people gather on a daily basis for the "Sunset Celebration." As the sun goes down, everyone watches and applauds as an unusual hodgepodge of street acts are performed at the square. Where else in the world do people take the time to celebrate a sunset? Perhaps the reason these islands are called Keys is because they open your mind to what's around you, the keys to appreciating Florida.

There are two routes you can take as you cruise along the Keys. **Hawk Channel** on the outside has plenty of water, but during foul weather, the winds can bring a good chop. The advantage of this route is that the offshore reef extending along the outside of the Keys acts as a natural breakwater, so the wave activity during an average day of sailing is moderate. Water depths range from 9 to 34 feet, and there is usually quite a distance between markers. So unless otherwise noted, navigating through Hawk Channel is straightforward—keep an eye on your chart, and you should be fine. Also keep in mind that many of the buoys are set in hard coral and should be given a wide berth.

The ICW continues from Coconut Grove down the inside of the Keys and can be run in almost any kind of weather. The depths are a bit scant along this inside route so if you draw more than 5 feet, you're better off sticking to the Hawk Channel. The tides in **Florida Bay** are difficult to predict. Wind blowing into the bay or out of the bay toward the **Gulf of Mexico** can cause the change in water level to be drastic. At the same time, when the tide goes out on the outside of the Keys, the water on the inside can't pour out between the islands fast enough. So before a significant tidal difference can be seen on the inside, the tide on the outside is on its way back in.

The inside route through Biscayne Bay begins at quick flashing green "71" just south of the Rickenbacker Causeway. Keep in mind, as you continue along this inside route, that there are several spots in the Keys offering safe anchorage. As is always the case, straying from the ICW involves a certain amount of risk. So if you choose to pull into a cove and intend to anchor for the night, be aware of surrounding depths and the condition of the bottom—you don't want to run aground on a coral reef.

From green "71," the ICW heads southwest through Biscayne National Park, in which swimming and scuba diving are prohibited. At flashing yellow 4sec "B," you can turn toward 2.5sec green flasher "21" which marks the first of several passages between the outside and inside routes. **Biscayne Channel** carries about 13 feet and is lined with red daymarkers to port.

If you choose to stick to the inside, **Featherbed Bank** is the first obstruction you'll encounter. The narrow channel is marked by red "4" to starboard and flashing green 4sec "3" to port. Your run south of here is 7nm along **Elliott Key** to the end of Biscayne Bay and the beginning of **Cutter Bank**. You may feel a cross current in this area from **Broad Creek** and **Angelfish Creek** which open up to the Hawk Channel.

At the end of the channel through **Cutter Bank**, the ICW doglegs west through **Card Sound** toward a narrow cut into **Little Card Sound**. The dredged channel turns eastward at flashing red 2.5sec "24," just beyond a 65-foot fixed bridge connecting **Key Largo** to the mainland. Green "27" marks another dogleg back west through **Barnes Sound** to **Jewfish Creek**, separating **Cross Key** from Key Largo. Be careful navigating through Jewfish Creek, especially when the winds are out of the north and the currents increase dramatically.

Blackwater Sound is next, and you should keep in mind that the purple course line is 400 yards east of red "40." But as long as you don't draw more than 5 feet, don't think twice about it. You'll pick up **Dusenberry Creek** which has all red markers leading around to starboard and heading due west before cutting southwest again through **Grouper Creek**. You may see a trail of silt following your stern as you pass through **Buttonwood Sound**, but don't be alarmed—the bottom is 6 feet down and soft.

Flashing green 4sec "55" marks the start of a channel between **Pelican Key** and the **Swash Keys** known as **Baker Cut**. Stick to the marked channel until you pass **Pigeon Key**, then swing hard south around quick green "61." After another bend to starboard around all red markers, the ICW leads through **Cowpens Cut** and **Cotton Key Basin**, north of Cowpens Anchorage. Again, the anchorage should be approached carefully.

The stretch of the ICW west of red "80" can get pretty thin. Winds out of the east or northeast blow the water toward the Gulf of Mexico, dropping water levels by a foot. But once through **Steamboat Channel** above **Shell Key**, depths shoot up to 8 feet all the way to **Bowlegs Cut**. As you approach the cut, continue on course until the channel markers line up, then swing to port into the channel and follow red right returning.

Beyond flashing red 2.5sec "98," you have a choice of continuing along the ICW, or turning slightly southwest toward **Channel Five**. There's a 65-foot fixed bridge separating Florida Bay from the Hawk Channel, so if you choose to switch over to the outside route, this is a good place to do it. You'll pick up the Hawk Channel at flashing red 6sec "44" and continue southwest.

Back in the ICW, follow the purple course line down to flashing green 4sec "1," down through **Channel Key Pass**, and out to flashing green 4sec "15A." From here, turn due south, leaving the buoy wide to port, and you'll approach the town of **Marathon** on **Vaca Key**.

Several marinas can also be found on Marathon's south shore, but you'll need to be in the Hawk Channel to access them. If you're not already on the ocean side of

© J.H. Perterson

Racing yachts spar off Key West.

the island, head down along the ICW past **Pigeon Key** to the fixed bridge at **Moser Channel**. Once past the bridge, stay between green "5" and red "6." You'll want to leave red "4" to port before you swing northwest toward the channel between Marathon and **Boot Key**. Here is where you'll find several facilities.

The rest of your trip through the Keys is better left to the Hawk Channel. Your only option in Florida Bay is the **Big Spanish Channel** which heads northwest from flashing red "20" and cuts through the Great White Heron National Wildlife Refuge and the National Key Deer Refuge. This passage can be pretty tricky, so unless you draw less than 5 feet or have good local knowledge, we advise you to stay ocean-side.

And Hawk Channel couldn't be more direct. From flashing green 6sec "49A," it's a straight shot for almost 18nm, all the way to Ninefoot Shoal. Green "57" is another 11.7nm, and quick flashing red "12" on **Key West Channel** is only another 3nm or so. Follow the channel buoys to the ranges until you reach red "24." Off to starboard in **Key West Bight** is where you'll find several marine facilities.

Ah, **Key West**. The western-most point on the Atlantic seaboard, replete with all that makes a cruise through coastal waters an unforgettable trip. In the same way the tide in Florida Bay never quite catches up with the tide on the outside, your mind will surely lag behind as you

© J.H. Peterson

The southernmost house in Key West.

depart this glorious island. Nowhere will this be more apparent than in **Mallory Square** during the daily "Sunset Celebration." As the awe-struck gatherers cheer at the last bit of sunlight, take a deep breath and hold it, for that breath will surely be a peaceful reprieve from your worries back home.

Shoreside and Emergency Services

Ambulance, Fire, & Police: 911

Coast Guard:

Mayport Station (904) 247-7311 or VHF 16
New Smyrna Beach Station (904) 428-9085
Emergency (904) 428-9094 or VHF 16
Port Canaveral Station (407) 853-7601 or VHF 16
Fort Pierce Station (407) 464-6100 or VHF 16
Lake Worth Inlet Station (407) 844-5030 or VHF 16
Fort Lauderdale Station (305) 927-1611 or VHF 16
Miami Regional Station (305) 536-5611 or VHF 16
Islamorada Station (305) 664-4404 or VHF 16
Marathon Station (305) 743-6388 or VHF 16
Key West Station (305) 292-8856 or VHF 16

Radio Telephone:

VHF 26; I.D. Jacksonville Marine Operator
VHF 28; I.D. Daytona Beach Marine Operator
VHF 26; I.D. Cape Canaveral Marine Operator
VHF 26; I.D. KTR-945, Cocoa Beach
VHF 27; I.D. KVY-628, Vero Beach
VHF 26; I.D. St. Lucie Marine Operator
VHF 28; I.D. Palm Beach Marine Operator
VHF 26 or 28; I.D. Port Everglades Marine Operator
VHF 24 or 25; I.D. Miami Marine Operator
VHF 28; I.D. KLU-791, Homestead
VHF 27; I.D. KSK-210, Marathon
VHF 26 or 84; I.D. KQU-411, Key West

Tow Services:

Sea Tow Jacksonville, Jacksonville Beach
(904) 249-0309 or VHF 16
Sea Tow St. Augustine, St. Augustine
(904) 824-8978 or VHF 16
Sea Tow Treasure Coast, Port Salerno
(407) 288-2900 or VHF 16
Sea Tow Palm Beach, West Palm Beach
(407) 844-8056 or VHF 16
Sea Tow Pompano, Pompano Beach
(305) 781-0063 or VHF 16
Sea Tow Fort Lauderdale, Fort Lauderdale
(305) 525-0037 or VHF 16
Sea Tow Miami, North Bay Village
(305) 864-9481 or VHF 16
Sea Tow Key Largo, Key Largo
(305) 451-4910 or VHF 16
Sea Tow Islamorada, Islamorada
(305) 664-4493 or VHF 16
Sea Tow Lower Florida Keys, Big Pine Key
(305) 872-2752 or VHF 16

Embassy's Product & Service Directory

AIRLINES & AIR CHARTER

New England Airlines (p. 131)
Buisness, casino, & vacation charters, scheduled Block I. flights
Airport Rd., Westerly, RI----------------------------------401-596-2460

BOAT & YACHT BUILDERS

Freedom Yachts (p. 4)
Exterior state of the art construction, exquisite interiors
305 Oliphant Lane, Middletown, RI--------------------401-848-2900

Migrator Yachts (p. 23)
Cruise the Atlantic coast in a boat built just the way you want it.
72 Sandwich Road, Wareham, MA---------------------508-295-8000

Stuart Yacht Builders, Inc. (p. 253, 258)
Full service yacht repair facility. Agents - Yanmar, Onan, Kohler
250 SW Salerno Rd., Stuart, FL------------------------407-283-1947

BOOKS & CHARTS

Bluewater Books & Charts (p. 9, 269)
Over 25,000 Nautical Books & Charts. . . Same Day Shipping!
1481 SE 17th St., Ft. Lauderdale, FL------------------800-942-2583

Chase Leavitt & Co. (p. 97)
NOAA, DNA, CHS, supplies, engines, electronics and more!
10 Dana St , Portland, ME 207-772-3751------------800-638-8906

Embassy Marine Publishing (p. 12, 32)
The finest cruising guides, waterproof & NOAA charts
142 Ferry Road, Old Saybrook, CT-----------------1-800-999-1075

JT'S Ship Chandlery (p. 131)
All Charts & Piloting References (Shipping World-Wide)
364 Thames St., Newport, RI (FAX-847-4215)------401-846-7256

Fred L. Woods Nautical Supplies (p. 20)
Lots of cruising guides and a wide selection of navigation tools
76 Washington Street, Marblehead, MA -------------617-631-0221

Mystic Seaport Stores (p. 156)
America's largest maritime bookstore
47 Greenmanville Ave., Mystic, CT 06355---------800-331-BOOK

Scuttlebutt Nautical Books (p.231)
Charts, Chartbooks, Cruising Guides, Gifts and Gear
126 Turner St., Beaufort, NC----------------------------919-728-7765

CANVAS GOODS

Clark Sailmakers, Inc. (p. 156)
Highest quality sails, canvas & rigging for your boat
37 Pratt St., Essex, CT ----------------------------------203-767-8278

Fit 'n' Stitch, Inc.
A tradition of craftsmanship since 1959
650 Ten Rod Rd., N. Kingstown, RI -------------------401-294-3492

CATAMARANS

McGrew's Boats Inc.
Hobie - Auth. Dealer. Complete Parts & Acc. Mail Order Serv.
5211 Old Post Rd, Charlestown, RI -------------------401-322-1150

CHARTER AGENTS

Greenwich Cove Marina
Charters, Sail/Power bareboat/Captain New England, Carribean
45 Water St., PO Box 1651, East Greenwich, RI ---401-885-6611

CHARTS & PUBLICATIONS

see "BOOKS & CHARTS"

COMPASSES

Connecticut Compass Service (p. 157)
Complete magnetic compass repair
3 Anchorage Lane, Old Saybrook,CT-----------------203-388-2019

CREW PLACEMENT

Crew Unlimited
Professional crew placement and yacht deliveries
Fort Lauderdale, FL----------- 305-462-4624 Fax: 305-523-6712

DAVITS, CRANES, & LIFTS

Forespar (p. 31)
Motormate- lits anything to 100 lbs. Cargomate- 300 lb. capacity
22322 Gilberto Rancho, Santa Margarita, CA -------714-858-8820

DIESEL ENGINES

see "ENGINE PARTS & SERVICE'

DINGHIES & INFLATABLES

Inflatable Services (p. 269)
Sales and Repair • Recertification • All Major Brands
990 West State Rd 84, Ft. Lauderdale, Fl ------------305-779-7000

Landrigan Corporation (p. 63)
Sales, Rentals and Service for Inflatable Rafts and Boats
2-12 Jeffries Street, Boston, MA -----------------------617-567-2182

Sound Inflatables (p. 67, 157)
Boats-Engines-Life Rafts-Accessories-Repair-Warranty Center
304 Boston Post Road, Old Saybrook, CT -----------203-388-6634

ELECTRONIC EQUIPMENT

Electronautical Equipment, Inc. (p. 156)
Complete marine electronics, sales and service
67 Boston Post Rd., Waterford, CT--------------------203-442-5522

ENGINE PARTS & SERVICE

Atlantic Detroit Diesel (p. 157)
Full Service, Detroit Diesel, Perkins, EMD MerCruiser, Twin Disc
Noank Shipyard, Noank, CT-----------------------------203-536-6726

Caterpillar (p. 11)
Info on Cat parts/service/nearest Cat service location, call...
Engine Div., PO Box 610, Mossville, IL----------------800-447-4986

River Landing Marina (p. 133, 145)
Full-service yard - Mercruiser, Detroit Diesel Service
142 Ferry Road, Old Saybrook, CT--------------------203-388-1431

RPM Diesel Engine Company, Inc. (p. 65)
Complete fuel injection and turbo charger repair facility
2555 State Road 84, Fort Lauderdale, FL ------------305-587-1620

Sunpower Diesel (p. 269)
Engine & Generator Specialists on New River Since 1949
413 SW 3rd Ave, Ft. Lauderdale, FL------------------305-522-4775

FINANCE & LEASING

Total Dollar Management Effort. Ltd. (p. 9)
The boat insurance specialists
325 Lexington Avenue, New Yrok, NY-----------------212-689-4477

FLAGS, BANNERS & SIGNS

Bluewater Books & Charts (p. 9, 269)
Over 25,000 Nautical Books & Charts. . . Same Day Shipping!
1481 SE 17th St., Ft. Lauderdale, FL 33316 ------1-800-942-2583

Surfset (p. 157)
The little flag store "on the corner," but not just flags
2 Essex Square, Essex, CT -----------------------------203-767-FLAG

FOUL-WEATHER GEAR

JT'S Ship Chandlery (p. 131)
Exclusive Helly-Hansen SEAGEAR Showroom
364 Thames St., Newport, RI (FAX-847-4215)------401-846-7256

FUEL, OIL & SUPPLIES

Norm's Fuel Co., LTD. (p. 123, 130)
Marine Diesel Dockside Delivery 7 Days a Week Ch. 80 VHF
316 Great Island Rd., Narragansett, RI 02882 ------401-783-2084

INNS, HOTELS & SPAS

Danfords Inn & Conference Center (p. 159, 164)
85 Luxury Rooms & Suites on water in historic village
25 E. Broadway, Port Jefferson, LI, NY 11777 ------516-928-5200

Greenwich Harbor Inn (p. 135, 154)
Fine dining and a gracious inn serving discerning yachters
500 Steamboat Rd., Greenwich, CT 06830 ----------203-861-0000

Harborside Inn, Marina & Resort (p. 105, 118)
89 Rms, whirlpool, sauna, pool, grill, rec room, in town, dockage
3 South Water St., Edgrtn, Martha's Vineyard, MA 508-627-4321

Seaport Inn & Marina (p. 105, 116)
The Best & The Most Protected Dockage In The East
110 Middle St., Fairhaven, MA 02719 -----------------508-997-1281

INSURANCE

Essex Insurance (p. 2)
Save on your boat insurance costs
P.O. Box 364, 40 Main Street, Essex, CT ------------203-767-1424

LIFE RAFTS

see "DINGHIES & INFLATABLES"

MARINAS & BOAT YARDS

Allen Harbor Marine Service (p. 105, 120)
Tiara yachts, Nautical gifts, OMC Funport, and Texaco fuels
335 Lower Country Rd., Harwichport, MA 02646---508-432-0353

Annapolis Landing Marina (p. 197, 201)
Transients welcome at Annapolis' finest marina - all amenities!
PO Box 3350, 980 Awald Dr., Annaplois, MD -------410-263-0090

Ashley Marina (p. 233, 237)
Full Dockage, Winter Rates, Car Rental, Service & Parts Depart.
33 Lockwood Boulevard, Charleston, SC-------------803-722-1996

Beaufort Docks (p. 219, 227)
The closest marina to EVERYTHING!
500 Front St, Beaufort, NC -------------------------------919-728-2503

Biscayne Bay Marriott Hotel & Marina (p. 253, 264)
Exceptional year 'round dockage
1633 N. Bayshore Drive, Miami, FL --------------------305-536-6335

Bock Marine (p. 219, 228)
Serving the boater's needs with confidence and courtesy
Route 1, Box 633, Beaufort, NC 28516----------------919-728-6855

Bohicket Marina Village (p. 233, 239)
The marina for Kiawah and Seabrook Resorts
1880 Andell Bluff Boulevard, Johns Island, SC -----803-768-1280

Boston Boatyard and Marina (p. 105, 109)
Our Central Harbor location is your one-stop port of call
256 Marginal Street, East Boston, MA-----------------617-561-1400

Boston Waterboat Marina (p. 105, 110)
Closest to Quincy Market & downtown Boston
66 Long Wharf, Boston MA-------------------------------617-523-1027

Boston Yacht Haven (p. 105, 110)
Boston's premier marina
87 Commercial Wharf, Boston, MA--------------------617-523-7352

Brewer Yacht Yards (inside front cover)
Your Passport to New York and New England
Post Road, Mamaroneck, NY ----------------------------914-498-0295

Britannia Yacht & Racquet Club (p. 159, 165)
Full Service Marina • Clubhouse • Health Club • Tennis Courts
81A Fort Salonga Rd, Northport, NY-------------------516-261-5600

Bull River Marina (p. 245, 246)
Your Marina of Choice for the Upcoming Olympic Events
8005 Old Tybee Rd, Hwy 80 East, Savannah, GA -912-897-7300

Burr's Yacht Haven (p. 133, 140)
Transients welcome - Slips and moorings for boats 23' to 100'
244 Pequot Ave., New London, CT---203-443-8457

Cannonsport Marina (p. 253, 261)
Your friendly homeport for Bahama cruising & deep-sea fishing
178 Lake Dr., Palm Beach Shores, FL---800-627-8328

Cedar Island Marina, Inc. (p. 133, 147)
"The Family Boating Resort"—Recently dredged access
P.O. Box 181, Riverside Dr, Clinton, CT---203-669-8681

Channel Crossing Marina (p. 87, 96)
Observe the breathtaking view of Portland's skyline
231 Front Street, South Portland, ME---207-767-4729

Constitution Marina (p. 105, 111)
Boston's friendliest marina
28 Constitution Road, Charlestown, MA---617-241-9640

Cracker Boy Boat Works, Inc. (p. 269)
Do it yourself or full service, 70 ton hoist, Complete marine store
1602 N. 2nd St., Fort Pierce, FL---407-465-7031

Crocker's Boatyard, Inc. (p. 133, 140)
OMC Service, Gas & Diesel Repairs, Hull Repairs, Lift to 75T
56 Howard St., New London, Conn.---203-443-6304

Crystal Point Yacht Club (p. 179, 183)
"We'll pamper you and your Boat"
4000 River Rd, Pt. Pleasant, NJ 08742---908-892-2300

Deaton Yacht Service (p. 219, 226)
A Full Service Boatyard with Unlimited Service
PO Box 441, Oriental, NC---919-249-1180

Delegal Creek Marina (p. 245, 249)
Savannah's finest overnight dockage
1 Marina Road, Savannah, GA---912-598-0023

Dozier Marine Center (p. 207, 210)
Come Visit us for Friendly, Professional Service
Broad Creek & Rt 33 PO Box 388, Deltaville, VA---804-776-6711

Dowry Creek Marina (p. 219, 222)
Quiet, Sheltered Cove, Transients Always Welcome
P.O. Box 341, Belhaven, NC 27810---919-943-2728

Dutch Harbor Boat Yard (p. 123, 128)
Tranquil setting, Quality work, Competitive prices
252 Narragansett Ave., Jamestown, RI 02835---401-423-0630

Fiddler's Cove Marina (p. 105, 120)
Cape Cod's Yachting Center
42 Fiddler's Cove Rd., N. Falmouth, MA 02556---508-564-6327

Fred J. Dion Yacht Yard (p. 105, 107)
Full service yacht repair since 1914
23 Glendale St., Salem, MA---508-744-0844

Fort Rachel Marine Service, Inc. (p. 133, 138)
We cordially invite you to stop by and inspect our facilities
Foot of Water St., Mystic, CT---203-536-6647

Goat Island Marina (p. 123, 124)
A Complete Boating Community
5 Marina Plaza, Goat Island, Newpotr, RI 02840---401-849-5655

Hague Marina (p. 233, 235)
The Myrtle Beach Yacht Basin
P.O. Box 835, Myrtle Beach, SC---803-293-2141

Halifax Harbor Marina (p. 253, 257)
Located at beautiful Daytona Beach
450 Basin Street, Daytona Beach, FL---904-253-0575

Harbour Square Marina (p. 135, 152)
Ready to Accommodate you with Personalized Service
860 Canal Street, Stamford, CT 06902---203-324-3331

Harbortown Marina & Boatyard (p. 253, 259)
"Top Ten" at Ocean Inlet, Full destination Marina/Boatyard
1936 Harbortown Dr., Ft. Pierce, FL 34946---407-466 7300

Harbourtown Yacht Basin (p. 233, 241)
Everything Under the Sun
149 Lighthouse Rd, Hilton Head Island, SC---803-671-2704

Harrah's Casino Hotel and Marina (p. 179, 186)
Convenience of casino-side docking
1725 Brigantine Blvd., Atlantic City, NJ---609-441-5315

Hawthorne Cove Marina (p. 105, 106)
Salem Harbor's Newest Marina
10 White Street, Salem, MA 01970---508-740-9890

Hinckley Company, The (p. 87, 90)
The finest in yacht construction, service, marina accomodations
Shore Road, Southwest Harbor, ME---207-244-5531

Jonesport Shipyard (p. 97)
Winter Storage- Begin your cruise Downeast next summer!
P.O. Box 214, Jonesport, ME---207-497-2701

Kingman Marine (p. 105, 114)
Cape Cod's largest full-service marine center
PO Box 408, 1 Shipyard Lane, Cataumet, MA---508-563-7136

Lippencott Marine (p. 197, 203)
Quiet, Secluded Family Marina Convenient to Prime Ports
Rt. 2, Kent Narrows, Grasonville, Maryland---410-827-9300

Little Harbor Marine (p. 123, 127)
New England's Full Service Repair Facility & Marina
One Little Harbor Landing, Portsmouth, RI---401-683-7100

Manhasset Bay Marina (p. 159, 167)
Home of Manhasset Bay Gold Cup Races
10 Matinecock Ave., Port Washington, NY---516-883-8411

Marina at American Wharf (p. 133, 141)
World Class Marina, Dock, Dine, and Gamble!
One American Wharf, Norwich, CT 06360---203-886-6363

Marina at the Dock (p. 135, 150)
Enjoy the scenic sheltered Housatonic from our new marina
955 Ferry Blvd., Stratford, CT---203-378-9300

Marina at Dock Holidays, The (p. 233, 234)
Now your boat has more options
1525 13th Ave North, North Myrtle Beach, SC---803-280-6354

Marina at Rowes Wharf, The (p. 105, 109)
All Size Slips, Transients Welcome!
50 Rowes Wharf, Boston, MA 02110---617-439-3131

Marina Bay (p. 105, 112)
One-stop docking • Full-service marina community
333 Victory Road, North Quincy, MA---617-847-1800

Mason's Island Marina (p. 133, 138)
Transients welcome - Marine supplies - Emergency Repairs
PO Box 185 Old North Rd., Mystic, CT---203-536-2608

Melrose Landing (p. 233, 241)
On the Cooper River at Calibogue Sound Inlet
Hilton Head Island, SC---803-842-9998

Milford Boat Works (p. 135, 148)
A Total Boating Complex For Almost 50 Years!
1 High St., Milford, CT 06460---203-877-1475

Moby-Dick Wharf Restaurant (p. 105, 115)
On the Water Overnight Berthing up to 55 feet
One Bridge Road, Westport Pt, MA---508-636-6500

Montauk Yacht Club (p. 159, 163)
The Finest Resort Marina on the East Coast
P.O. Box 5048, Montauk, NY --------------------------516-668-3100

Morehead City Yacht Basin (p. 218, 229)
World famous for professional, courteous service
208 Arendell St, PO Drawr A, Morehead City, NC 919-726-6862

Morrison's Marina (p. 179, 184)
Full Serv. Marina, Ships Store, Cummins/Crusader/Onan Dealer
On the Bay, Second Street, Beach Haven, NJ ------609-492-2150

Mystic Harbor Marine (p. 133, 139)
A full-service marina located in the heart of historic Mystic
40 Water St., Mystic, CT --------------------------------203-536-1210

Nantucket Boat Basin (p. 105, 119)
A wicked good place to bring your boat and family
Box 1139, Nantucket, MA ----------------------------800-NAN-BOAT

Newport Yachting Center (p. 123, 125)
In the Heart of Downtown Newport
PO Box 550, Newport, RI (Ext. 3)----------------------800-653-3625

Noank Village Boatyard (p. 133, 136)
Personalized, full & friendly service at a fair price
Bayside Ave., Noank, CT --------------------------------203-536-1770

Northside Marina (p. 253, 258)
Slip into a very special place
400 NW Alice Ave., Stuart, FL --------------------------407-692-4000

Norton's Shipyard & Marina, Inc. (p. 123, 129)
Almost 50 years of Professional Services
P.O. Box 106 E. Greenwich, RI-------------------------401-884-8828

Norwalk Cove Marina (p. 135, 151)
Metropolitan New York's premier marina facility....in CT!
2 Beach Rd., Norwalk, CT 06855 ----------------------203-838-5894

Oakdale Yacht Service (p. 159, 161, 175)
Full service marina • Transients welcome
520 Shore Drive, Oakdale, NY--------------------------516-589-3015

Old Lyme Dock Co. (p. 133, 144)
$35 is our maximum transient dockage charge
323 Ferry Rd., Old Lyme, CT --------------------------203-434-2267

Outdoor Resorts Marina (p. 233, 241)
Volume Discounted Fuel
PO Box 1585, Hilton Head, SC-------------------------803-681-3256

Oyster Bay Marine Center (p. 159, 161, 166)
Spend your summer on Long Island's most scenic harbor
5 Bay Ave., Oyster Bay, NY -----------------------------516-922-6331

Palmer Johnson - Savannah (p. 245, 247, 265)
America's Premier Yacht Builder
3122 River Drive, Savannah, GA ----------------------912-352-4956

Palmetto Bay Marina (p. 233, 241, 243)
Full-service marina and boatyard • Since 1959
164D Palmetto Bay Rd., Hilton Head Is., SC --------800-448-3875

Pickering Wharf Marina (p. 105, 107)
Shopping, museums, attractions, a restaurant for every taste
23 Congress St., Salem, MA ----------------------------508-744-2727

Port Imperial Marina (p. 159, 171, 179)
Full Service Facility, Golf Center, Ferry Service to Manhattan
Pershing Circle, Weehawken, NJ ----------------------201-902-8787

Port Milford (p. 135, 148)
Full-service yard located on beautiful Milford Harbor
164 Rogers Ave., Milford, CT ---------------------------203-877-7802

Port Niantic (p. 133, 142)
Full Service Marina, Transient Dockage, Emergency Repairs
17 Smith Ave., Niantic, CT-------------------------------203-739-2155

River Forest Marina & Shipyard (p. 219, 223)
Full-service marina, shipyard, travelift, restaurant & hotel
1st inside breakwtr, 600 E.Main St., Belhaven, NC 919-943-2151

River Landing Marina (p. 133, 145)
Full-service yard - Slip rentals - Rack storage
142 Ferry Road, Old Saybrook, CT--------------------203-388-1431

Robb's Boatyard & Marina (p. 219, 225)
On-premise prop shop, machine & welding shop, 24-hr towing.
316 E. Front St., Belhaven, NC-------------------------919-943-2110

Sassafras Harbor Marina (p. 197, 199)
New Modern Facilities with Eloquent Dining Nearby
P.O. Box 68, Georgetown MD, -------------------------410-275-9966

Saybrook Point Inn & Marina (p. 133, 143, 144)
Luxury marina at mouth of CT. River & LI Sound
2 Bridge Street, Old Saybrook, CT --------------------203-395-3080

Schilling Boathouse (p. 233, 241)
Drystack Boat Storage, Boat & Motor Sales, Marine Ser. & Parts
145 Squire Pope Rd, Hilton Head, SC----------------803-681-2628

Seaport Marine Inc. (p. 133, 139)
One-stop yachting needs • 34th year • Short walk to town
1 Washington St., Mystic, CT --------------------------203-536-9681

Shelter Cove Harbour (p. 233, 241)
Where you can live your dreams
1 Shelter Cove Lane, Hilton Head Island, SC -------803-842-7002

Shipyard Quarters Marina (p. 105, 111)
Boston's friendliest pier group
1 Pier Eight, Charlestown, MA-------------------------617-242-2020

Skull Creek Marina & Yacht Club (p. 233, 241)
Luxurious Marina Overlooking Pinkey Wildlife Refuge
100 Skull Creek Dr. Hilton Head Island, SC---------800-237-4096

Spicer's Noank Marina (p. 133, 137)
The place to come when you need something fixed
93 Marsh Rd., Noank, CT--------------------------------203-536-4978

Spring Point Marina (p. 87, 96)
Welcome to Casco Bay! We provide 100% of your boating needs
1 Spring Point Drive, South Portland, ME------------207-767-3254

Stratford Marina, Inc. (p. 135, 149)
Transient boaters always welcome!
Foot of Broad Street, Stratford, CT -------------------203-377-4477

Summit North Marina (p. 191, 193)
The Gateway to the Bays on the C&D Canal
3000 Summit Harbour Pl., Bear, DE ----302-836-1800 Fax: 3647

Tides Inn, The (p. 207, 211)
"One of the World's Finest Small Resorts"
King Carter Drive, Irvington, VA -----------------------804-438-5000

Tides Lodge & Marina, The (p. 207, 211)
The Bay's Only 5 Bell Marina by Atlantic Cruising Club!
PO Box 309 Carter's Creek, Irvington, VA ----804-438-6400/6000

Tidewater Yacht Agency (p. 207, 215)
The best of what a marina should be • Full service • 325 slips
10 Crawford Parkway, Portsmouth, VA---------------804-393-2525

Troy Town Dock and Marina, Inc. (p. 161, 174)
Gateway to the New York State Cannals
427 River Street, Troy, NY------------------------------518-272-5341

Turnberry Isle Resort & Club (p. 253, 263)
Slip into great golf, tennis, spa, dining and total enjoyment
19755 NE 36 Court, Aventura, FL ---------------------305-933-6934

Two Mile Marina & Restaurant (p. 179, 187)
Exceptional accommodations - convenient location
Ocean Drive, Wildwood Crest, NJ ---------------------609-522-1341

Wacca Wache Marina (p. 233, 235)
The most convenient fresh water stop on the ICW! Stop in.
1950 Wachesaw Road, Murrells Inlet, SC ---------1-800-395-6694

Waterside Marina (p. 207, 214)
Fun is our Main Service
33 Waterside Drive, Norfolk, VA----------------------- 804-625-2000

Waterways Marina (p. 253, 264)
Calm & secure-the perfect stop between Ft. Lauderdale & Miami
21169 Yacht Club Dr., Aventura, FL------------------ 305-935-4295

Wentworth by the Sea Marina (p. 99, 103)
170 Slip full service marina • Bulk fuel • Transient dockage
P.O. Box 246, New Castle, NH -------------------------603-433-5050

West Shore Marina (p. 159, 165)
LI's Newest Resort Marina with Something for the Entire Family
100 West Shore Road, Huntington, NY --------------516-427-3444

Wexford Harbour (p. 233, 241)
One of South Carolina's Finest Harbors
Hilton Head Island, SC------------------------------------803-686-8813

Windmill Harbour (p. 233, 241)
A Tide-Free 10ck Protected Marina on Beautiful Hilton Head
161 Harbour Passage, Hilton Head, SC---------------803-681-9235

Windmill Point Resort (p. 207, 209)
Fun, friendly resort atmosphere; Pools, golf, tennis, great food
Where the Rappahannock meets the Chspk Bay---804-435-1166

Wright Island Marina (p. 159, 169)
A Modern, Full Service, Friendly Marina in a Protected Marina
290 Drake Ave., New Rochelle, NY ------------------914-235-8013

Yacht Haven Marine Center (p. 135, 153)
"The Best Bet for the Discriminating Yachtsman"
Ft of Washington Blvd., PO Box 1435, Stmfrd CT--203-359-4500

York River Yacht Haven (p. 207, 213)
Full Service Marina Convenient to Willamsburg and Yorktown
Rt 1210 & Sarah Creek, Glouscester Point, VA ------804-642-215

MARINE SUPPLIES

Chandlery at Essex, The (p. 133, 146)
Marine Hardware, Lumber & Electrical Specialists
Novelty Lane Essex, CT---------------------------------203-767-8267

Freeport Marine Supply (p. 177)
Electronics, Engine Parts, Charts, Navigational Equipment
47 West Merrick Rd., Freeport, NY --------------------516-379-2610

Ship Chandlers, Inc. (p. 189)
NOAA Charts, Marine Hardware, Gear, Gifts
601 Bay Avenue, Point Pleasant, NJ----------------- 908-899-0474

Wilcox Marine Supply (p. 157)
Complete Marine Store, Rigging Specialists, Emergency Repairs
19 Old Stonington Road, Mystic, CT ------------------203-536-4206

METALWORKING

Northeast Metal Works (p. 157)
Designing & building your ideas
41 Commerce Street, Clinton, CT----------------------800-235-2628

NAUTICAL INSTRUMENTS

Robert White Instruments Inc.
Traditional navigation & weather instruments of all kinds
34 Commercial Wharf, Boston, MA 800-992-3045, 617-742-3045

Swift Instruments (p. 41)
Technical excellence in optics since 1926
952 Porchester Avenue, Boston, MA -----------------617-436-2960

PROPELLERS & WEED CUTTERS

Spur's Line & Weed Cutter (Back Cover)
A guaranteed solution for entangled propellers
285 SW 33rd Street, Fort Lauderdale, FL------------800-824-5372

REAL ESTATE

Norris & Company Real Estate (p. 269)
Prime properties offering the best in fishing, boating, and golf
4725 North A1A, Vero Beach, FL ---------------------407-231-1270

RESTAURANTS

Americus on the Wharf (p. 133, 141)
A unique café on the waterfront-Authentic Italian cuisine
1 American Wharf, Norwich, CT------------------------203-887-8555

Danfords Inn/Marina/Restaurant (p. 159, 164)
Waterfront setting • Open all year • Spectacular dining and views
25 East Broadway, Port Jefferson, LI, NY -----------516-928-5200

Saybrook Fish House (p. 155)
Voted best seafood statewide by the readers of CT magazine
56 South Broad Street, Milford, CT--------------------203-878-2428
163 Ferry Road, Old Saybrook, CT--------------------203-388-5864

Saybrook Point Marina & Inn (p. 133, 143, 144)
Luxury marina at mouth of CT. River & LI Sound
2 Bridge Street, Old Saybrook, CT --------------------203-395-2000

RIGGING REPAIRS

Clark Sailmakers, Inc. (p. 156)
Highest quality sails, canvas & rigging for your boat
37 Pratt St., Essex, CT ---------------------------------203-767-8278

Wilcox Marine Supply (p. 157)
Complete marine store • Rigging specialists • Emergency repairs
19 Old Stonington Road, Mystic, CT ------------------203-536-4206

SAILMAKERS

Clark Sailmakers, Inc. (p. 156)
Highest quality sails, canvas & rigging for your boat
37 Pratt St., Essex, CT ---------------------------------203-767-8278

Haarstick Sailmakers/Carolina (p. 231)
Quick turnaround for transients • Call for pickup & delivery
705 Broad Street, P.O. Box 878, Oriental, NC------ 919-249-3001

Mack Sails (p. 21)
Suppliers of fine sails, rigging and furling systems
3129 SE Dominica Terrace, Stuart, FL---------------407-283-2306

Omar Sailmakers
Sails, overnight repairs, rigging, canvas
126 Hwy. 101, Beaufort, NC---------------------------919-728-5598

SPORTING GOODS

Diver's Cove / Sea Wolf II (p. 35)
Scuba retail sales, service, rental and instruction NAUI
7 Essex Plaza, Essex, CT -------------------------------203-767-1960

Marine Concepts (p. 3)
We ship quality worldwide
41 Commerce St., Clinton, CT -----------------------800-235-BOAT

North Cove Outfitters, Inc. (p. 156)
A store to explore-Camping-Canoeing-Fishing-Archery-Firearms
75 Main St., Old Saybrook, CT ------------------------203-388-6585

Patrick's Country Store (p. 156)
Casual country clothing for Men & Women, Penny Candy
288 Main Street, Old Saybrook, CT ------------------203-388-2384

TOWING & SALVAGE

Cape Ann Towing & Salvage (p. 19, 269)
When Only the Best Will Do
1520 SE 15th ST, Ft. Lauderdale, FL----------------305-463-2527

Marine Recovery Services (p. 19, 205)
When Only the Best Will Do
7074 Bembe Beach Rd, Annapolis, MD--------------410-263-1260

Marine Rescue Service (p. 19, 131)
When only the best will do . . . Call the best!
60 Reynolds Street, Wickford, RI ---------------------401-295-8711

Ocean Tow Boat Towing & Salvage (p. 19, 189)
When Only the Best Will Do
PO Box 455, Atlantic Highlands, NJ ----------------908-495-44190

Saybrook Towing & Salvage (p. 19, 156)
Assistance Servicing CT & LI • 24 hrs.
Old Saybrook, CT ---203-388-4065

SeaTow Lower Chesapeake Bay (p. 217)
Your Road Service at Sea - 24hr Emergency Towing Service
PO Box 634, Gloucester Pt, VA -----------------------804-642-6800

SeaTow New Jersey (p. 189)
Your Road Service at Sea
Brick, NJ---908-295-4466
Brigantine, NJ --609-266-1984
Highlands, NJ ---908-872-1835
Sea Isle City, NJ--609-263-9999

SeaTow Services International, Inc. (p. 13)
Marine towing, salvage, membership program, vessel recovery
PO Box 817, Southold, NY -----------------------------516-765-3660

Sound Towing Company (p. 19, 177)
When Only the Best Will Do
PO Box 58, City Island, NY-----------------------------718-885-3420

Southeastern Marine, Inc. (p. 19, 156)
Marine Salvage & Towing • 24 hour service
P.O. Box 9551, Noank, CT -----------------------------203-445-8381

UNIFORMS

QUALITY UNIFORMS FOR THE BOATING PROFESSIONALS

COMMODORE
UNIFORM & NAUTICAL SUPPLIES
335 Lower County Road
Harwich Port, MA 02646
(508) 430-7877 • FAX (508) 432-0487
A Division of Allen Harbor Nautical Supplies, Inc.

WATERMAKERS

Offshore Marine Labratories (p. 15)
Reverse Osmosis Water Filtration Equipment
23042 Alcalde "C," Laguna Hills, CA-----------------714-458-6500

WELDING & METALWORK

Marine Fabricators (p. 156)
Quality Yacht Metalwork since 1975
Pier 66 Marina, 55 Goodsell Pt. Rd., Branford, CT-203-488-7093

WINCHES & WINDLASSES

Ideal Windlass Company (p. 29)
The world's finest anchor windlasses
5810 Post Rd., PO Box 430, E.Greenwich RI -------401-884-2550

Maxwell Winches (p. 27)
The premier manufacturer of Innovative anchoring systems
1610 Babcock St., Costa Mesa, CA-------------------714-631-2634

WOOD, LUMBER & TOOLS

Woodworkers Store at South Norwalk (p. 157)
Boat, Lumber, Hardware, Fasteners, Tools & Mill Shop
85 Water Street, S. Norwalk, CT -----------------------800-422-2783

MAXWELL®

Brewer Chandlery East
Essex, CT 203-767-8267

Kellogg Marine
Old Saybrook, CT 203-388-4477

M & E Marine
Glasgow, DE......................................302-836-3300

Ross Yacht Service
Clearwater, FL 813-448-8191

Sailorman
Ft. Lauderdale, FL 305-522-6716

Sailors Supply Co.
Panama City, FL................................904-769-5007

Fairhaven Shipyard
Fairhaven, MA 508-996-8591

Falmouth Marine
Falmouth, MA 508-548-4600

Marine Exchange Corp.
Marblehead, MA 617-639-0797

Fawcett Boat Supplies, Inc.
Annapolis, MD 410-267-8681

Ocean Outfitters
Annapolis, MD 410-267-0021

Crockett Bros. Boatyard
Oxford, MD 301-226-5113

Wayfarer Marine
Camden, ME.......................................207-236-4376

Handy Boat Service, Inc.
Falmouth, ME 207-781-5110

Chase Leavitt Co.
Portland, ME.......................................207-772-3751

Hamiltons Marine Service
Searsport, ME.....................................207-548-2985

Jamestown Boat Yard
Jamestown, RI 401-423-0600

River Services
Mt. Pleasant, SC................................803-884-6099

M & E Marine Supply
Camden, NJ..609-858-1010

Freeport Marine
Freeport, NY 516-379-2610

Defender Industries
New Rochelle, NY..............................914-632-3001

Index

Embassy's

MARINE Marketplace

NEW & USED BOATS ◆ FISHING SUPPLIES ACCESSORIES ◆ SERVICES

SUBSCRIBE NOW!

YES! **Send me the next 11 issues of the *Marine ◆ Marketplace* for only $15.00. I'll save over 30% off the newsstand price and get New England's #1 Boating and Fishing Shopper delivered to my door!**

❏ MasterCard **Signature** ______________________

❏ VISA **Card#** ______________ **Exp. Date** ________

❏ AMERICAN EXPRESS **Name** ______________________

❏ Payment Enclosed **Address** ______________________

City ______________ **State** ______ **Zip** ______

Phone# ______________________

Embassy Marine Publishing • 142 Ferry Road • Old Saybrook, CT 06475 • 1-800-999-1075

Need A Boat? Find it HERE!